PHYSICS OF COGNITIVE PROCESSES

PHYSICS OF COGNITIVE PROCESSES

Amalfi 1986

Edited by
E. R. Caianiello

The present meeting was organized by the joint effort of the Institute for Advanced Scientific Studies of Naples, Italy, of the Istituto Italiano per gli Studi Filosofici, of the Centro Internacional de Fisica of Bogotá and of the Universities of Nijmegen and Salerno.

World Scientific
Singapore • New Jersey • Hong Kong

Published by

World Scientific Publishing Co. Pte. Ltd.
P.O. Box 128, Farrer Road, Singapore 9128

U.S.A. office: World Scientific Publishing Co., Inc.
687 Hartwell Street, Teaneck NJ 07666, USA

Library of Congress Cataloging-in-Publication data is available.

PHYSICS OF COGNITIVE PROCESSES

Copyright © 1987 by World Scientific Publishing Co Pte Ltd.

All rights reserved. This book, or parts thereof, may not be reproduced in any form or by any means, electronic or mechanical, including photocopying, recording or any information storage and retrieval system now known or to be invented, without written permission from the Publisher.

ISBN 9971-50-255-0

Printed in Singapore by Kim Hup Lee Printing Co. Pte. Ltd.

INTRODUCTION

The very association of the words "physics" and "cognitive processes" would have been unthinkable a few years ago. Progress made in many different areas of science, both "pure" by self-definition or "interdisciplinary", has led, though, to unforeseen consequences. In physics, for example, the study of phase transitions in magnetic models, in particular spin glasses, has brought to a variety of techniques which brain modelists and computer scientists think of possible relevance to the understanding of the behaviour of neural tissues or, at least, associative memories.

This is just one example out of a host: scientific journals dedicated to physics have more and more articles dealing with neural problems and computer sciences; the converse is also true. Norbert Wiener's dream of Cybernetics as an integrated approach to the study of complex dynamical systems is being revived; great effort goes into the study of "neural computers". We are thus entering a most exciting period, because many other sciences claim equal rights: linguistics, structure and organization of hierarchical systems, semantics and what is behind it, the meaning and forms of "cognition", may come, at least in their essence, under one and the same paradigm. What the anthropologist finds in his fields is now explained in terms of discrete sets of elements, not unlike the phenomes and graphemes of a natural language, or the quantized structural levels of a bureaucracy of an army: nobody will deny that phenemes or graphemes have to do with the brain!

Mathematical laws are being discovered that account equally well for the organization of parts of speech, of neural tissue or of a social "indicator", thus providing quantitative evidence that there exists basic common organizational principles which is our task to understand. They appear to be not specific, say, of classical or quantum physics, or of standard statistics; rather to govern the anatomy and physiology of all complex systems, from brain to language, societies and, some maintain, hadron jets and galaxies.

This situation makes it both fascinating to collect contributions from as many diverse fields falling into this perspective as possible, and difficult for any single individual to discriminate between solid work and repetitious or unsubstantiated elucubrations.

The success of an attempt, such as is made at this Meeting, to offer not only authoritative "last words" but also pregnant "first words" in so composite an area, rests by necessity upon the excellence of the participants.

Deep are therefore my appreciation and gratitude towards the eminent scientists who have promptly and friendly accepted to participate. The synergy that immediately arose not only on the scientific but, even more important, on the human side, has created in this getting together of old and new friends, an atmosphere whose memory will last, I hope, in all of us. For which also a perfect weather and a full moon on the Amalfi Riviera are accountable. May I thank also my daughter Orietta, who came from London to give us a piano concert in a XIV Century Cathedral in historical Ravello.

CONTENTS

Introduction   v

**COHERENT ACTIVITY IN NEURONAL POPULATIONS:
ANALYSIS AND INTERPRETATION** (*A. Aertsen, T. Bonhoeffer & J. Kruger*)   1
1. Introduction.   2
2. Conventional Analysis of Multi-Unit Recordings   3
    2.1 Representation of multi-unit spike trains   3
    2.2 Cross correlation analysis of pair interaction.   5
3. Gravitational Clustering   7
    3.1 Coherence as proximity: analysis of distance   8
    3.2 Stimulus-dependence of coherence   10
    3.3 Visualization of clustering: projection   12
4. The 'Dynamic Correlation Matrix': A Hebbian Algorithm.   15
    4.1 Logical wiring diagram and correlation matrix.   15
    4.2 Hebb's rule: 'learning' the correlation matrix   16
    4.3 Normalization: stationary activity   16
    4.4 Normalization: stimulus-induced nonstationary activity   19
        4.4.1 Synchrony matrix.   19
        4.4.2 PST-matrix   20
        4.4.3 Differential correlation matrix.   20
        4.4.4 Normalized correlation matrix.   21
    4.5 Results   22
5. Discussion   25
6. Acknowledgements   27
7. Appendix.   28
    7.1 Gravitational clustering.   28
    7.2 Dynamic correlation matrix.   29
    7.3 Relation between 'dynamic correlation matrix' and normal 'cross correlation'   30

HYPERCHAOS AND 1/F SPECTRA IN NONLINEAR DYNAMICS: THE BURIDANUS' DONKEY (*F. T. Arecchi*)   35

1. Introduction.   36
2. Noise Induced Trapping at the Boundary between Two Attractors.   37
3. Conclusion: Long Memory in Statistical Physics   42

THE SENSITIVITY METHOD: A STRUCTURAL APPROACH TO UNDERSTANDING SYSTEMS (*J. D. Becker & F. Vester*)   51

1. Problems.   51
2. Systems.   54
   2.1 The conventional approach: system dynamics.   56
   2.2 A structural approach: the sensitivity model.   59
3. Planning a Computer Based Implementation of the Sensitivity Method.   64
4. Conclusion.   66

WHY BUILD MODELS (*A. Borsellino*)   72

SOME PROBLEMS IN BRAIN SCIENCE AWAITING THEORETICAL TREATMENT (*V. Braitenberg*)   75

PHYSICS AND METAPHYSICS OF COGNITIVE PROCESSES (*R. Busa*)   79

Introducton.   81
1. Where to Start a Scientific Inquiry on Cognitive Processes?   83
2. The Heteroclytous Extremes of the Expressional Arc.   86
3. Mind as Interfacing: Intelligence — Reasoning   95
4. Appendices   99

PERCEPTION AS A DYNAMIC INSTABILITY: A QUALITATIVE MODEL (*G. Caglioti*)   102

1. Introduction.   102
2. The Model.   103
3. Conclusions.   107

AFTER THOUGHT (*E. R. Caianiello*)   113

A MODEL FOR COMMUNICATION OVER NOISY CHANNELS
(*R. M. Capocelli, L. Gargano & U. Vaccaro*)   117
1. Introduction. 117
2. Notations and Definitions 121
3. Unique Decipherability. 122
4. Decipherability with Finite Delay. 126
5. Synchronizability 134
6. Decoders with Initial State Invariance 140

DISSIPATIVE STRUCTURES EMERGING WITHIN AN
ABSTRACT MODEL OF NEURAL NETWORK (*G. M. Guazzo*)   163
1. Introduction. 163
2. The Model 164
3. Conclusions 168

THE RECOGNITION OF PATTERN FORMATION IN
SELFORGANIZING SYSTEMS BY MEANS OF INFORMATION
ENTROPY (*H. Haken*)   171
1. Selforganizing Systems — Synergetics 172
2. Pattern Formation close to Nonequilibrium
   Phase Transitions. 175
3. Information Entropy 176
4. Pattern Recognition. 178
5. Information Compression, Emergent Patterns, Semantics 182

THE INFORMATICS OF HIGH AND LOW CONTEXT SYSTEMS
(*E. T. Hall*)   185

A NON-DETERMINISTIC APPROACH TO ANALOGY, INVOLVING
THE ISING MODEL OF FERROMAGNETISM (*D. R. Hofstadter*)   209

CONSERVATION AND DISSIPATION IN NEURODYNAMICS
(*P. I. M. Johannesma & A. M. H. J. Aertsen*) 228
1. Dynamic Systems 229
2. Reactive and Creative Systems 230
3. Statistical Dynamics 233
4. Population Dynamics 234
5. Neurodynamics 237
6. Conservation and Dissipation 239
7. Simple C-D Systems 242
8. Three Examples 245
9. A Special Class of C-D Systems 251
10. Population Dynamics and Neurodynamics in C-D Form 253
11. Conclusion and Speculation 256

SELF-ORGANIZED SENSORY MAPS AND ASSOCIATIVE MEMORY
(*T. Kohonen*) 258
1. Sensory Maps in the Brain 258
2. The Basic Neural Network Model 260
3. Relaxation of the Activity 264
4. Self-organization of the Input Connections 266
5. Relationship of Sensory Maps to the Distributed Models of Associative Memory 269
6. Conclusions 272

A NEW, ALGEBRAIC, CALCULUS OF THE IDEAS IMMANENT IN NERVOUS ACTIVITY (*F. E. Lauria*) 274
0. Introduction 274
1. The Machine and the Associated Language 278
2. The Language's Meta-rules 285
3. The Product's Implementation as an Example 292
4. Conclusions 301

WHAT KIND OF UNCERTAINTY ATTACHES TO HUMAN ACTIONS (*D. M. MacKay*) 309
1. Stochastic Information-Processing 309
2. Systematic Unpredictability 310
3. 'Determinate' vs 'Predictable' vs 'Inevitable' 311

4.	Randomness	312
5.	I-story and Brain-story	313
6.	The Physical Costs of Knowing	314
7.	Logical Relativity	315
8.	Self-fulfilling Predictions?	316
9.	The Embodiment of Mind	317
10.	Openness vs Uncertainty	318
11.	The Retrospective View	319
12.	Manipulative Prediction	319
13.	Conclusion	320

THE BIOLOGICAL FOUNDATIONS OF SELF-CONSCIOUSNESS AND THE PHYSICAL DOMAIN OF EXISTENCE (*H. R. Maturana*)   324

1.	Purpose	324
2.	The Problem	324
3.	Nature of the Answer	325
4.	The Scientific Domain	326
	4.1 Scientific explanations	326
	4.2 Science	329
5.	Objectivity in Parenthesis	330
	5.1 An invitation	331
	5.2 Objectivity in parenthesis	331
	5.3 The universum versus the multiversa	332
6.	Basic Notions	332
	6.1 The observer	333
	6.2 Unities	333
	6.3 Simple and composite unities	333
	6.4 Organization and structure	334
	6.5 Structure determined systems	335
	6.6 Existence	337
	6.7 Structural coupling or adaptation	338
	6.8 Domain of existence	338
	6.9 Determinism	339
	6.10 Space	339
	6.11 Interactions	340
	6.12 Phenomenal Domains	340
	6.13 Medium, niche and environment	340
7.	Basis for the Answer: the Living System	341
	7.1 Science deals only with structure determined systems	341

	7.2	Regulation and control.	342
	7.3	Living systems are structure determined systems	342
	7.4	Determinism and prediction	342
	7.5	Ontogenic structural drift	344
	7.6	Structural intersection	347
	7.7	The living system.	349
		7.7.1 Implications.	349
		7.7.2 Consequences.	350
	7.8	Phylogenic structural drift.	352
	7.9	Ontogenic possibilities	354
	7.10	Selection	355
8.	The Answer		356
	8.1	Cognition.	356
	8.2	Language.	358
9.	Consequences.		361
	9.1	Existence entails cognition	361
	9.2	There are as many cognitive domains as there are domains of existence	361
	9.3	Language is the human cognitive domain	362
	9.4	Objectivity.	362
	9.5	Languaging: operation in a domain of structural coupling	363
	9.6	Language is a domain of descriptions.	364
	9.7	Self-consciousness arises with language	364
	9.8	History	365
	9.9	The nervous system expands the domain of states of the living system	365
	9.10	Observing takes place in languaging.	366
10.	The Domain of Physical Existence		369
11.	Reality		374
12.	Self-consciousness and Reality		375

ON ASSOCIATIVE MEMORIES (*G. Palm*)   380
1. Introduction. ... 380
2. The Basic Device ... 385
3. Methods for Pattern Mapping. ... 387
   3.1 Addressing by all possible inputs ... 387
   3.2 Serial search through list of actual (x, y) pairs. ... 388
   3.3 Hash coding of (x, y) pairs ... 389

	3.4	Parallel search using complex switches.	391
	3.5	Parallel search using simple switches	392
	3.6	Association between x and y for all (x, y) pairs	394
4.	Methods for Pattern Completion		395
	4.1	Serial search through list of all patterns	396
	4.2	Parallel matching using complex switches.	397
	4.3	Parallel matching using simple switches	398
	4.4	Auto-association of all patterns	398
5.	Comparison of the Various Methods		401
	5.1	Pattern mapping	402
	5.2	How to determine the dimensions of the matrix for the associative method	406
	5.3	Pattern completion	407
	5.4	How to determine the dimensions of the matrix for the associative method.	409
6.	Associative Memory and Sparse Coding		410
7.	Variations of the Basic Device		412
8.	Physical Realization of the Basic Device.		418

ON THE FUNCTION OF THE CAT'S VISUAL CORTEX
(W. von Seelen, H. P. Mallot & F. Giannakopoulos)   423

1.	Introduction.		423
2.	Dynamic Two-dimensional Spatial Networks		425
3.	Neuronal Feedback Systems		427
	3.1	Specific properties.	427
	3.2	Regularization by feedback	433
4.	Receptotopic Mapping		435
5.	Linear Theory of the Visual Cortex.		439
6.	Nonlinear Cortex Couplings.		443
	6.1	System characteristics	445
	6.2	Behavior with time-dependent stimuli	446

COHERENT ACTIVITY IN NEURONAL POPULATIONS: ANALYSIS AND INTERPRETATION

Ad Aertsen, Tobias Bonhoeffer

Max-Planck-Institut für Biologische Kybernetik
Spemannstrasse 38, D-7400 Tübingen, FRG

and

Jürgen Krüger

Neurologische Universitätsklinik
Hansastrasse 9, D-7800 Freiburg, FRG

ABSTRACT

To study neural interaction it is necessary to simultaneously record spike trains from a population of neurons under different experimental conditions. Evaluation of the data is normally done by crosscorrelating the spike trains of all possible pairs of neurons and inspecting the results for possible signs of coherence. With recording from relatively large populations (10-30) of neurons becoming experimentally feasible, this procedure starts to be prohibitively tedious. Moreover it does not really address the issue of interest: cooperativity in larger groups of neurons, possibly involved in so-called assemblies.

In this paper we discuss a recently developed technique for multi-unit analysis, explicitly designed to overcome these difficulties: 'gravitational clustering'. The basic idea is the following: each of the recorded N neurons is associated with a charged particle in a fictitious (N-)space. Due to the time varying charges, determined by the respective neurons' spike trains, the particles mutually exert forces which cause them to move: coherently firing neurons result in particles aggregating into clusters.

We also describe a slightly modified version of this approach: the 'dynamic correlation matrix'.

Both techniques will be illustrated with the results of their application to multi-electrode recordings from the cat's visual cortex.

1. INTRODUCTION

Much of our understanding of the functioning of the nervous system is based on electrophysiological recordings of the activity of individual neurons, the single unit spike trains. There is, however, ample evidence that the nervous system is more than simply a collection of independent elements. Especially at more central levels, it is thought to involve functional assemblies of neurons, acting together in a coherent manner, mediated by anatomical, presumably 'plastic' connections [20,12,6,31]. In order to gain understanding of processes within and between neuronal assemblies it is necessary to observe simultaneously and separably the activity of many neurons during appropriate experimental manipulation of the whole organism.

Within this context it is understandable that we are witnessing a growing interest in the study of neuronal interaction. Mainly due to recent developments in multi-neuron recording technology (multi-micro-electrodes, spike separation techniques), the simultaneous recording of spike trains from a population (10-30) of neurons has become a feasible enterprise (reviews are given in [15,27]).

The search for 'neuronal assemblies' raises an interesting sampling problem: What is the size of such an assembly and how large a fraction of an assembly does one have to observe in order to draw conclusions about the assembly as a whole? Since this issue remains largely unsettled, it is generally felt advisable to try to record from large groups of neurons. A relatively new technique which seems to be well suited to achieve this goal is the method of 'optical recording', which uses voltage sensitive fluorescent dyes [18,19]. Provided that this technique, when applied as a multi-unit recording device, reaches a spatial resolution which is in the single neuron range (the present resolution is - depending on the preparation - at best some 30 microns), it should be possible to record simultaneously and separably from some hundred neurons at a time. This means that a substantial part of the network is available for observation, as compared to relatively low fractions attainable with multi-electrodes. The advantage of this higher 'sampling' ratio for the analysis of network properties can hardly be overestimated.

Another potential advantage of the optical recording technique is that the neurons one is recording from, are literally visible. This in principle opens the way for a long overdue attempt to combine multi-neuron recording with a more thorough analysis of anatomical structure than currently possible with multi-electrodes 'poking in the dark'.

Facing the massive flow of data resulting from evermore sophisticated multi-neuron experiments, it must be granted that, unfortunately, theoretical developments regarding analysis and interpretation of these data have not been able to keep pace. The principal tool for analysis in use continues to be the cross correlation of pairs of spike trains (two neurons at a time), a method which was developed already in the sixties [34]. Although proven to be quite adequate for few-unit (2-3) recordings, this approach is being strained beyond extent in a situation where recording from some ten neurons simultaneously has become a relative routine in a number of laboratories. In this context the present paper discusses recently developed techniques of multi-unit analysis: *'gravitational clustering'* [17,14,1], and the related technique of *'dynamic correlation matrix analysis'*. These techniques will be described heuristically and illustrated with the results of their application to multi-electrode recordings from the cat's visual cortex [25].

2. CONVENTIONAL ANALYSIS OF MULTI-UNIT RECORDINGS

In the following we will look somewhat more closely at a typical multi-unit recording, in this case from Area 17 in the cat. In these experiments use was made of a linear array of twelve glass covered Pt-Ir electrodes, arranged in organpipe-like manner with a distance between electrode tips of 160 microns. The array was introduced into the left hemisphere of the visual cortex at an angle of 45° such that the electrode tips were situated approximately vertically underneath each other (for details see [25]). The actual recording sites could be reconstructed after the experiment on the basis of lesions made by some of the electrodes (generally 3 to 4). In the present example the array of recording sites indeed was shown to be approximately perpendicular to the cortical layers; individual sites were situated in the following layers: electrode number 1 in the white matter underneath layer VI, 2 in layer VI, 3 in VI, 4 in VI, 5 in V, 6 in IV, 7 in IV, 8 on the border of IV and III, 9 in II/III and, finally, 10 in I. The remaining two electrodes in this particular case did not yield a useful recording and are left out of consideration. Five out of ten electrodes (nrs. 2, 4, 5, 6 and 8) each gave a reliable single unit recording, the remaining five (nrs. 1, 3, 7, 9 and 10) were judged to be probably single unit, with possibly some additional spikes from a second unit.

The stimulus in this experiment consisted of a light bar (length 3°, width 14') moving at constant velocity in a direction perpendicular to its orientation (distance travelled 3°, duration 1.8 sec). At the end of the movement the bar remained stationary for 0.4 sec, after which it moved in the opposite direction (same distance and duration). Finally the bar was rotated over 22.5° (duration 1 sec) after which another, identical cycle of movement back and forth, perpendicular to the bar orientation started. This scheme was repeated periodically, so that after 8 cycles of 5 seconds the original direction of movement was reached again. The complete stimulus sequence, lasting 40 seconds, was presented repeatedly, in general some 20 to 30 times. In the particular example discussed below, the stimulus was presented to the right (contra-lateral) eye.

2.1 Representation of Multi-Unit Spike Trains

Spike trains recorded during 10 consecutive stimulus sequences are shown as dot displays in Fig. 1, with different neurons represented by different colours. Time runs horizontally, covering the complete stimulus sequence of 40 sec, vertical lines in the dot displays signify the different movement cycles of 5 sec each. Dot displays are arranged in the way the corresponding recording sites were located in the cortex: the pink dots at the top represent spikes from electrode 10 which occupied the most superficial position in the cortex (in layer I), the yellow ones at the bottom correspond to spikes from electrode 1 which was down in the white matter below layer VI. In the lowest part of Fig. 1 we have schematically indicated the stimulus program with arrows showing the direction of bar movement. From this Figure it becomes quite clear that not all directions are equally successful in eliciting a response; for instance the stimulus where the bar is moving to the upper left corner under an angle of 45° appears to be clearly favoured in the recordings 5 to 8. The more or less unanimous preference for direction of bar motion is consistent with the earlier mentioned vertical alignment of recording sites.

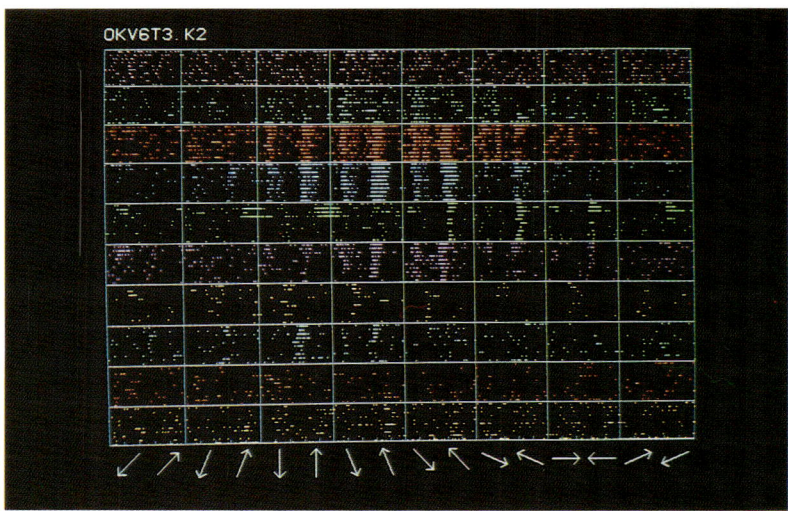

Figure 1. Dot displays of spike trains from a twelve electrode recording in the cat's visual cortex. Two electrodes were omitted because they did not yield a useful recording. Different colours represent different electrodes. The stimulus program (bar moving in different directions) is indicated schematically at the bottom. Further explanation is given in the text.

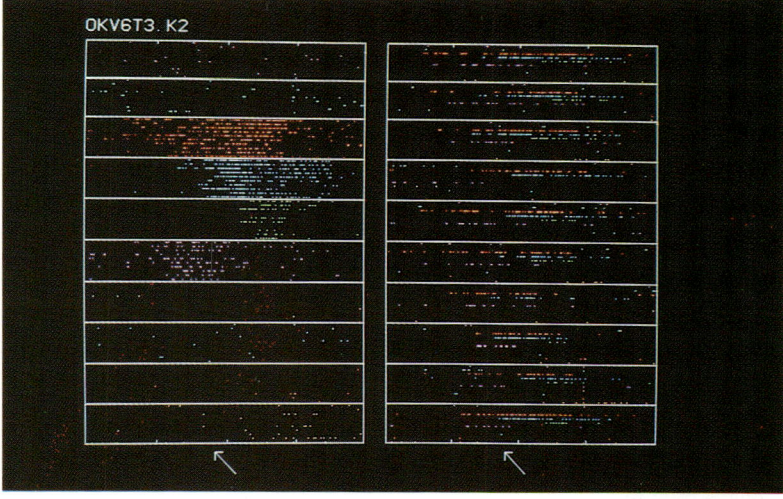

Figure 2. Dot display showing a selected part of the data from Fig. 1: one direction of bar movement indicated at the bottom of the Figure. At the left the data are displayed as in Fig. 1 (sweeps per unit), whereas at the right the adjacent traces in each box represent synchronous activity of the different neurons (units per sweep), different boxes corresponding to different stimulus presentations. The colour code is the same as in Fig. 1.

In order to study the effects of bar movement direction more closely, the data as displayed in Fig. 1 could be 'cut' into different vertical slices by selecting an appropriate time window, such that in every slice one has only one direction of stimulus movement. Using this procedure of editing the spike trains ('cut' and 'glue') we have artificially produced multi-unit recordings, in which the stimulus is a bar periodically moving in one and the same direction. In Fig. 2 we have displayed the response to the earlier mentioned preferred stimulus direction (45°, top left), by selecting a time window between 22 and 24 sec in the 40 sec sweep. On the left hand side the data are displayed in the same way as in the previous Figure (organized in sweeps per unit), whereas on the right hand side adjacent traces in each box represent the simultaneous activity of different units during one stimulus presentation (units per sweep), the different boxes corresponding to subsequent stimulus presentations. The latter representation, where colour now is essential to identify the different units ('Neurochrome' [10]), in fact is the more natural one, since it reflects most properly the activity as it happens to occur in the brain. Consequently, it is also the most appropriate one to look for possible time patterns in multi-unit activity. In Fig. 2 one observes how different neurons react differently to stimulation, both regarding the magnitude and the time course of the responses. Whereas the left hand part of Fig. 2 clearly shows the average dynamic properties of each single neuron's response to the ensemble of identical stimulus presentations, the right hand part emphasizes the intricate inter-relationships between the individual responses of different neurons during each single stimulus sweep. Note also the considerable variability of multi-unit firing patterns over subsequent presentations of the same stimulus (different boxes in right part of Fig.2). In this context we note that, when looking for temporal structure in multi-unit activity it should - apart from proper visualization - also seriously be considered to present the data to the investigator truly as signals in the time domain e.g. as identifiable, audible signals ('Neurophone' [2]).

2.2 Cross Correlation Analysis of Pair Interaction

The commonly used technique to investigate coherence among simultaneously recorded activity of different neurons is to select from the observed group two neurons at a time, and analyze this pair's firings for possible temporal relations. This analysis is based on the cross correlation function of the two simultaneously recorded spike trains [34]. Departures from background in the correlogram are taken as indicative of a *'functional connection'* between the two neurons, where this 'connection' may take different forms: excitatory or inhibitory connection, shared input, stimulus coupling [30]. We use here the term 'functional connection' as purely descriptive: for some, as yet not specified, reason there is a statistical relationship between the probabilities of firing of the two neurons.

An example of this type of analysis is shown in Fig. 3, which presents correlograms with different time resolutions for the pair of units 7 and 8, calculated for the activity evoked by presentation of the '45°, top left'-stimulus (cf. Fig. 2; for the correlograms we used 16 stimulus sweeps instead of only the 10 shown in the dot display). In the upper correlogram (Fig. 3a), the one with the largest time scale (-4 to 4 sec, 80 ms/bin), one clearly sees the dominating effect of the stimulus: periodical waxing and waning of synchrony of firing, with a period of 2 sec, corresponding to the period of stimulus presentation in our 'edited' spike trains. Even at this resolution, though, one also notices that the synchrony around zero time shift (middle peak: simultaneous activity) is slightly higher than in the adjacent peaks (corresponding to a time shift of one stimulus period). This is shown in more detail in the lower correlograms which have increased time resolution (Fig. 3b and 3c: 10 ms/bin, Fig. 3d and 3e: 1 ms/bin). In each of these pairs of correlograms the top one (3b resp. 3d) corresponds to zero time shift (simultaneous recording), whereas the lower one (3c resp. 3e) shows the synchrony for a time shift of one stimulus period (the so-called *'shift control'* [16]). Comparison of simultaneous correlogram and shift control, especially in the lower pair, clearly shows that, during this particular stimulus, the firings of units 7 and 8 exhibit a degree of synchrony which significantly goes beyond the much higher degree of synchrony induced by stimulation per se. The fact that this 'extra' synchrony has its peak at the origin suggests that at least the main contribution is due to shared neuronal input, whereby the rather irregular and unsymmetric form of the elevation does not exclude other neuronal sources of coherence.

Careful inspection of the shape of correlograms and the evaluation of quantitative characteristics, such as peak width, height, displacement from zero etc., are used to make inferences about possible connectivity in the network. An account of an analysis in these terms was given by Krüger [26]. It should be noted here that under conditions of relatively sparse firing, which is the normal case in cortex, the method of cross correlation of spike trains was shown to strongly favour the detection of excitatory connections (signified by an elevation in the correlogram) above inhibitory ones (a trough in the correlogram) [3].

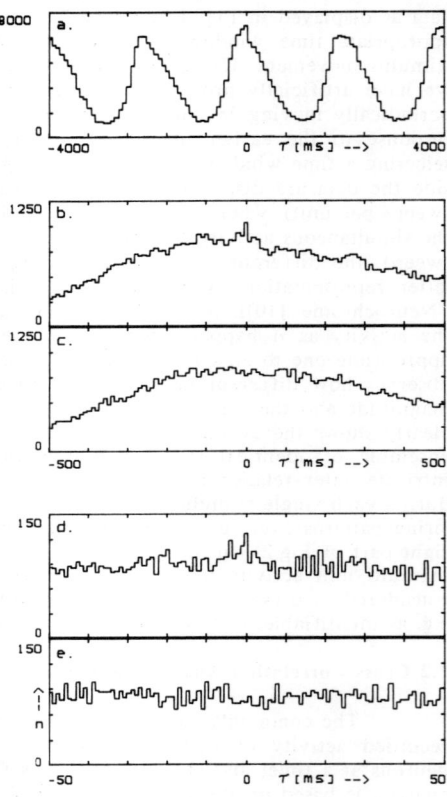

Figure 3. Cross correlograms of the spike trains from neurons 7 and 8 with three different time resolutions (a. 80 ms/bin; b. and c. 10 ms/bin; d. and e. 1 ms/bin). Data are the same as in Fig. 2 (only one direction of bar movement). Two correlograms (b. and d.) display the coherence between neurons 7 and 8 in a simultaneous recording, whereas c. and e. show their synchrony with a time shift over one stimulus-period ('shift control'). Further explanation is given in the text.

The problem about this approach is that the number of correlograms which have to be examined grows nonlinearly with the number of recorded neurons: for N neurons one has $N(N-1)/2$ different pairs, which in the present case of 10 neurons (cf. Figs. 1 and 2) implies 45 correlograms. Going to 30 neurons (this is about the maximum number of simultaneous recordings that came forward in the last few years, see e.g. [28]) the number of correlograms increases to 435. These numbers have to be multiplied still further, e.g. with the number of different stimuli (in our case, for instance, we have 16 directions of bar movement) and with the desired number of different time resolutions. After examination of all the individual pair correlations, the inference has to be recombined by the experimenter into a single conceptual picture. Quite evidently this procedure becomes prohibitively tedious for growing numbers of recorded neurons (imagine the pairwise analysis of a 100-unit recording from a fluorescent dye experiment!). Furthermore one may question whether the pairwise analysis of correlation addresses the real issue of interest: cooperativity in larger groups of neurons, possibly involved in so-called 'assemblies'.

3. 'GRAVITATIONAL CLUSTERING'

In order to overcome the difficulties discussed above, recently a new analysis technique was proposed: *'gravitational clustering'* [17]. A formal description of this approach has been given elsewhere [1,14,17]; we will restrict ourselves here to a short, rather more heuristic description.

The basic idea is as follows: with each of N neurons we associate a point-particle in a (fictitious) N-space. Initially these particles are located such that mutually they all have the same distance. This is achieved most easily by placing them on the vertices of an N-dimensional hypercube. Now each particle is given a time varying charge, which is determined by the respective neuron's spike train: each spike induces a charge increase which decays in the course of time (we have used an exponential decay with different time constants, e.g. 8 ms). Because we assume that in our fictitious space a kind of Coulomb law is valid (for equations see Appendix) the particles mutually exert forces, which cause them to move. As a consequence of temporal overlap of charge histories some of the particles will cluster, whereas others will on the average preserve their original positions.

The rules of the charge functions can be defined such that the forces lead to aggregation of those particles which correspond to neurons that tend to fire in synchrony (excitation) or, on the other hand, where firing in one is associated with silence in others (inhibition). Different clusters correspond to different assemblies. By this procedure we have translated the amount of temporal coherence among different spike trains into a distance measure in N-dimensional space: the higher the coherence among neurons, the smaller the distance between the corresponding particles and vice versa.

Since the gravitational representation translates synchrony into clustering, it is to be expected that the high degree in synchrony induced by stimulation per se (cf. Fig. 2 and 3) will lead to profound clustering of almost all particles, thereby effectively masking the much smaller amount of coherence of neuronal origin. Since at this point our main interest is in the latter contribution to multi-unit coherence, this purely stimulus-related component has to be taken care of. In a first order approximation this was accomplished by resorting to a differential type of analysis which may perhaps be characterized best as being analogous to subtracting the 'shift control' from the simultaneous correlograms in Fig. 3. We will return to this methodological issue in Sect. 4; a more extensive discussion on procedures to normalize for directly stimulus-induced nonstationarities c.q. synchrony will be presented elsewhere (Aertsen et al., in preparation). All results in the remainder of this Section were normalized for direct stimulus effects by this differential correction procedure.

3.1 Coherence as Proximity: Analysis of Distance

The original problem of analyzing temporal coherence now has been traded for the problem how to investigate the N-space for clusters of particles. Our first approach was to plot the different inter-pair distances as a function of time: downgoing curves will signify clustering particles while horizontal curves indicate that no net attraction occurred. This was done for the experiment described above. Again we selected the activity during the (45°, top left)-direction of bar movement (cf. Fig. 2) and subjected these spike trains to the gravitational analysis. The resulting pair distances, plotted against time are shown in Fig. 4, where colour was used to identify the pairs. Note that time runs horizontally from 0 to 32 sec: for the gravitation analysis we used 16 (instead of 10, cf. Fig. 2) stimulus presentations of 2 sec each. In this Figure four different bands of curves can be distinguished :

(1) The particles corresponding to neurons 5,7 and 8 cluster most strongly: their pair curves have the strongest negative slopes. Among these three especially the neurons 7 and 8 appear to have a high degree of coherence in their firing patterns (cf. Fig. 3). It should be noted here that in the gravitational algorithm, in order to avoid oscillations around a singularity, the attractive force between any two particles is 'switched off' the moment these particles have approached each other to a certain fraction (here 10%) of their initial distance. The interactions with the other particles, however, continue to be present so that eventually the clustered particles may drift apart again, whereupon their own pair force is switched on again. From the curves in Fig. 4 it can be observed that for these three particles this switching of pair forces actually took place several times.

(2) The three curves which lie closely together and slightly above the former three curves represent the somewhat smaller coherence between neuron 6 and the three previous ones.

(3) Still higher in the Figure one observes a yellowish band of curves which reflects the lower, but still significant degree of coherence between neurons 9 and 10 and the earlier four.

(4) Finally, the remaining particles (1, 2, 3 and 4) more or less preserve their mutual distances which indicates that the coherence among the corresponding neurons is negligible. The slight decrease in distance between the particles in this group on the one hand and those in groups (1), (2) and (3) on the other hand is an artifact: the distance between two particles may slightly decrease without the two themselves having anything to do with each other, simply because one of the two particles belongs to a certain cluster and the other one does not. A simple example in two dimensions may illustrate the point: imagine three particles on the corners of an equilateral triangle. If two of the particles are attracting each other and therefore decrease their mutual distance, this will also affect the distances between these two and the third, non-moving particle: these will also decrease slightly (to a minimum of about 87%). This decrease, however, does not reflect a genuine coherence. Small decreases in pair distances therefore should not be overinterpreted, in fact, they add to the obvious lower limit for effects to be significant, set by the diffusion type influence of the stochasticity of neuronal firing [13].

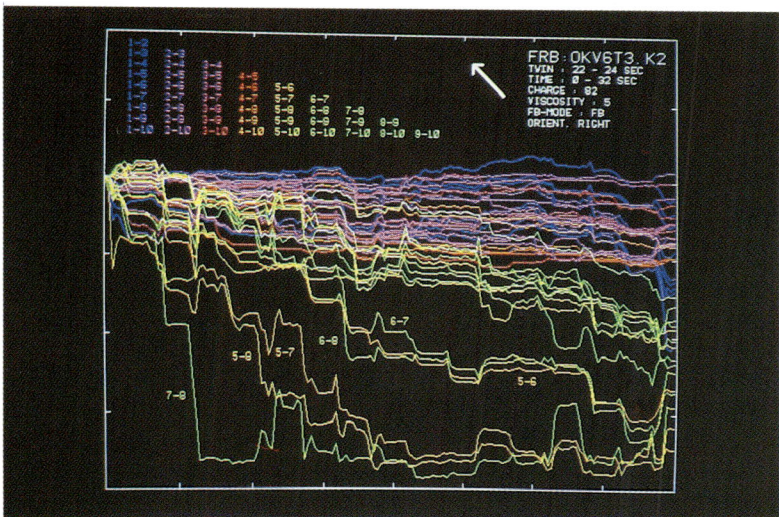

Figure 4. Gravitational clustering: inter-pair distance as a function of time for the 45 pairs of neurons in the 10-electrode recording of Fig. 2. The arrow indicates that the direction of motion of the stimulus was towards the upper left corner of the screen. The colour coding for the pairs is displayed in the upper left corner. For further explanation see text.

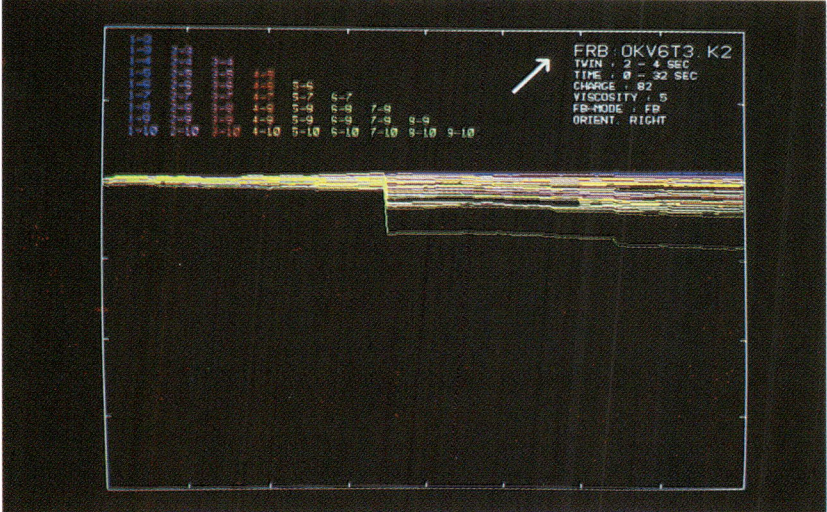

Figure 5. Gravitational clustering: distance plots for the 45 pairs of neurons. The arrow indicates that the direction of bar movement was towards the upper right corner of the screen. Further details as in Fig. 4.

Apart from the global behaviour of the various curves (overall shape), Fig. 4 also entails quite detailed information about the temporal structure of multi-unit coherence. Note, for instance, the temporal variations in slope along each individual curve and the temporal relationship between these variations among different curves. These reflect substantial variations in synchrony among spike trains as time proceeds and are obviously related to the stimulus induced time structure of the multi-unit spike trains, revealed in the 'Neurochrome' (cf. right hand part of Fig.2).

3.2 Stimulus-Dependence of Coherence

Being interested in a possible stimulus dependence of coherence among the multi-unit firings (remember we corrected for directly stimulus-induced coherence), we also analyzed the spike trains corresponding to the other directions of bar movement. Using again the spike train editing procedure, the data displayed in Fig. 1 were 'cut' into 16 vertical slices, each 2 seconds wide, such that each slice corresponds to one particular direction of bar movement; these data were submitted to the gravity procedure. Figure 5 shows the resulting distance curves for one of the other directions: 45° to the top right, which is perpendicular to the previous one. In this case one observes that, apart from some noise related 'diffusion', basically all particles remain their initial positions. Only the curve corresponding to neurons 8 and 9 shows a sudden, small decrease in particle distance after some 14 seconds of recording; this is caused by a coincident burst of firing early in the eighth stimulus sweep, which involved four neurons, with strongest contributions from numbers 8 and 9. On the whole Fig. 5 shows that for this particular direction, perpendicular to a more favoured direction (as far as single unit responses are concerned), at this point there is no evidence for coherence in firing, which goes beyond directly stimulus-induced effects.

In order to get a more complete picture of the influence of direction of movement on the degree of multi-unit coherence, one might plot all the distance curves corresponding to all the directions tested and compare these. Again we are facing a congestion of data, therefore we shall momentarily take a step back and go to a more global measure of coherence. To this end we disregard the dynamics of clustering for a moment and only look at the final result: the inter-pair distances after 16 stimulus presentations. More specifically, we have made a bar graph of the net decrease in pair distance: the higher the functional connectivity between a pair of neurons (i.e. the smaller the final distance between the corresponding particles), the higher we draw the bar for this pair.
In Fig. 6 three of these bar graphs are displayed. The arrow in the upper left corner of each graph represents the direction in which the stimulus moved. The upper graph corresponds to the direction 45°, top left (cf. Fig. 4), the middle graph to the movement in the obverse direction, and the bottom graph corresponds to a direction perpendicular to the other two (cf. Fig. 5). Comparing the upper two graphs it is apparent that the coherence among the earlier mentioned neurons in groups (1), (2) and (3) is comparably strong for movement back and forth along the negative diagonal of the screen, with the exception of neuron 6 which only 'joins in' during movement to the top left (in the middle graph the corresponding large bars for particle 6 are missing). In the orthogonal direction (lower graph) the coherence is generally quite weak, as we had already seen in Fig. 5.

Figure 7 summarizes the analysis of coherence for the whole spectrum of directions tested. In this Figure bar graphs were positioned at the tip of the vector which indicates the direction of motion of the stimulus, e.g. the graph for direction

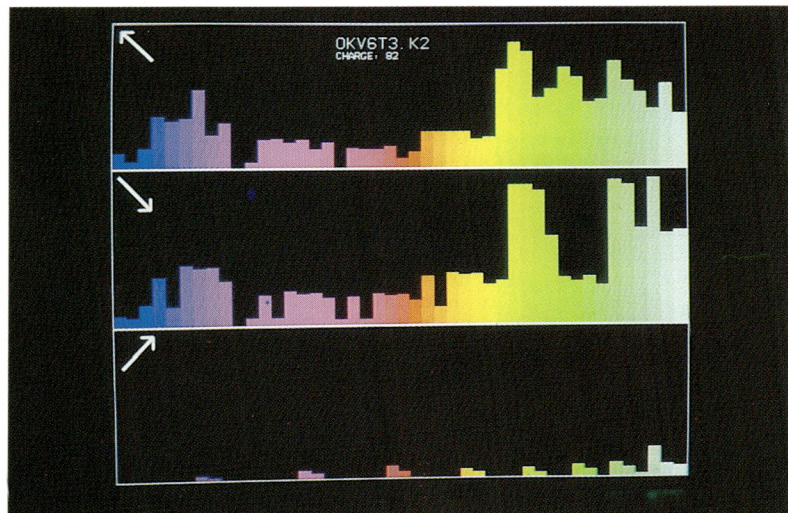

Figure 6. Stimulus dependence of coherence: three bargraphs for different directions of stimulus movement. Colour coding is the same as in Figure 4. The heigth of each bar corresponds to the degree of coherence in the respective pair of neurons. The direction of motion of the stimulus is displayed by the arrows in the upper left corner of each Figure. Further explanation in the text.

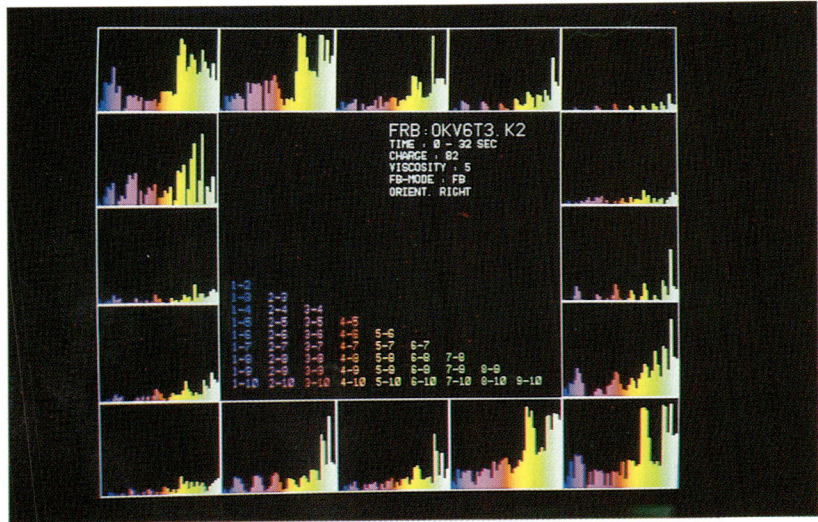

Figure 7. Stimulus dependence of coherence: bargraphs for all 16 directions of stimulus movement. Each graph is positioned on the tip of the vector indicating the direction of motion of the stimulus. Further details as in Fig. 6.

45°, top left (cf. upper graph in Fig. 6) is situated at 45°, top left in Fig. 7. The conclusion we drew from Fig. 6 is confirmed in Fig. 7. The degree of coherence between neurons is shown to be quite strongly stimulus dependent. At this point it should be emphasized once more that this is not simply due to the single neurons showing a stronger response to some directions of movement than to others; the directly stimulus-induced coherence was 'corrected' for by a differential procedure, thus Fig. 7 represents coherence that goes beyond the synchrony in spike trains which is merely due to the temporal correlation of single unit PST-histograms. This suggests that information about (the direction of) a stimulus may very well be represented also in the coherence among firings of different neurons and not only in single unit responses. Although we did not perform a quantitative analysis of direction tuning of single unit responses in our data yet, comparison with Fig. 1 suggests that in this particular experiment the directions with higher coherence globally coincide with the directions of stronger single unit responses. This might be taken to indicate that the differences between the individual bar graphs in Fig. 7 do not so much reflect a genuine stimulus dependence of coherence, but rather suggest that stimulus related rate variations were not adequately normalized for. Indeed it will be argued later (Sect. 4.4) that a true quantitative measure of coherence requires additional dynamic scaling for remaining stimulus effects. Application of this second stimulus 'correction', however, will be shown to confirm and further diversify the stimulus dependence suggested in Fig. 7, rather than reducing it to a mere artifact that can be eliminated by proper scaling.

Finally it should be emphasized that the abstraction of using the bar graphs, i.e. using only the final configuration and not the dynamics of clustering, provides a measure of the summated amount of coherence, integrated over time, and thus necessarily only presents a global picture. Temporal variations of coherence which might signify interesting dynamic reorganizations among different assemblies are integrated out; oscillations may even lead to cancellation and thus complete invisibility of substantial coherence. Furthermore the global picture may be distorted by the earlier mentioned saturation of attraction for very small distances (cf. Fig. 4) which will produce a negative bias for very strong coherence. Evidently for an analysis of temporal variations in multi-unit coherence and their dependence on dynamic properties of the stimulus ensemble one has to consult more detailed and dynamic measures, such as the distance curves (cf. Figs. 4 and 5) and/or other methods of visualization to be described in the next Section.

3.3 Visualization of Clustering: Projection

To envisage a process taking place in N dimensions, so far we have been using a one dimensional method: the analysis of pair distances as a function of time. We have seen that it results in curves which contain a wealth of quite detailed information. However, it may still be difficult for the investigator to imagine, from these curves, the actual particle trajectories in N-space. In order to enhance the spatial impression of the gravitational representation it may be worthwhile to extend our scope to two dimensions: project the N-dimensional particle trajectories on a plane, such that they can easily be visualized e.g. on a video screen.

For the data in our 45°, top left example (cf. Figs. 2 and 4) we have calculated 2-dimensional trajectories from the 10-dimensional ones, using standard geometric projection rules. When plotted on a video monitor this results in an animation movie with particles wandering around on the screen. From this movie six snapshots, taken at equal intervals of 6 sec, are shown in Fig. 8. The projection plane

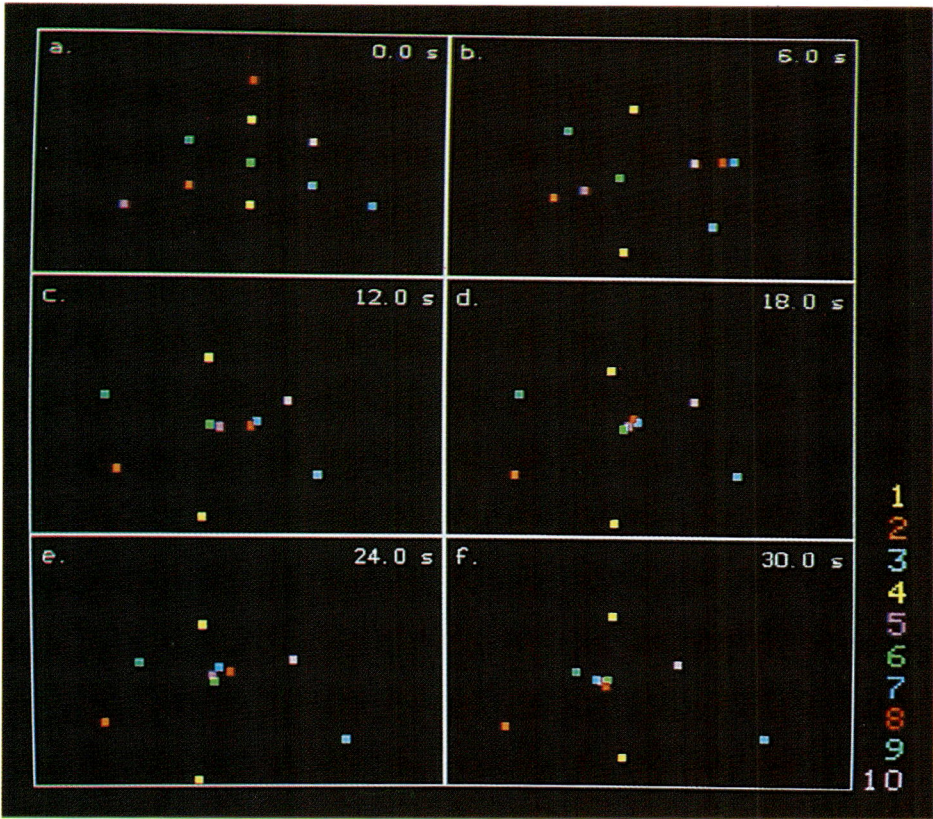

Figure 8. Geometrical projection: six 'snapshots' of the particle trajectories in N-space projected onto a 2-dimensional plane. The snapshots were taken at equal time intervals; time is indicated in the upper right corner of each Figure. Colours code for neurons (cf. Figs. 1 and 2). Further explanation is given in the text.

was chosen such that the initial N-dimensional particle configuration with equal mutual distances is preserved at least to a certain extent (cf. Fig. 8a) (obviously a 2-dimensional projection of 10-space still entails a considerable information loss). Observation of Fig. 8 shows that the information which had to be distilled from Fig. 4 by close inspection of the distance plots, now is much more evident: The particles corresponding to the neurons 5,6,7 and 8 are rather far apart in the beginning but already after (at most) 12 seconds (cf. Fig. 8c) they have formed a cluster at the center of the screen. The finding from Fig. 4 that the neurons 7 and 8 are very strongly connected is also clearly reflected in Fig. 8: already after (at most) 6 sec (cf.Fig 8b) they have formed a tight cluster. As time proceeds the cluster of four continues to collapse, moreover it starts to attract the particles 9 and 10, which apparently are less tightly connected to the former ones. The remaining particles on the average keep their mutual distances and wander around on the screen in a Brownian-motion like manner, which indicates that the corresponding neurons are firing incoherently. It may be noted that, although in the gravitational algorithm we used charge rules such that a distinction is made between the case of a neuron i driving a neuron j and the obverse case of j driving i [14], nevertheless the particles in Fig. 8 are attracting each other in a highly symmetrical way. This indicates that also the coherence among the neurons in our example is quite symmetrical, i.e. although there are clear associations of firing between certain neurons, there is no indication of a specific directionality in these associations. This in turn points to the coherence being the sign of shared neuronal input, rather than being the result of directed connections between members of the group. This observation is in accordance with our finding concerning the neurons 7 and 8 from inspection of their cross correlograms (cf. Fig. 3).

The results of projecting the particle trajectories onto a plane thus give a strong spatial impression of the aggregation process taking place in the N-space. This is especially so when actually witnessing the movements on a screen in the form of an animation movie; the snapshots in Fig. 8 in this respect give only a faint flavour of the dynamic experience. This spatial impression is much harder to be gained through conceptual reconstruction from the distance curves (such as Figs. 4 and 5) and even more cumbersome from analysis of all correlograms involved (such as those in Fig. 3) Especially the hierarchy in clustering, reflecting a hierarchy in coherence and pointing at a hierarchy in assembly organization, becomes quite evident from this type of visualization of the particle trajectories.

As already noted, mapping the N-space onto a 2-space clearly introduces a high information loss. Therefore the results strongly depend on the choice of the projection plane. An unfortunate choice may result in trajectories which are not of much help in analyzing the coherence of the investigated neuronal group. Evidently it is not a desirable situation when the outcome of an analysis technique is so highly dependent on an 'inspired guess' (or, in its absence, on a great deal of trial and error). A promising alternative would be to use the so-called 'structure preserving' nonlinear mapping introduced by Sammon [35] or modifications thereof [5,36]. A related possibility might be to use the self organizing mapping proposed by Kohonen [24]. These methods best preserve the shortest distances in the original N-dimensional distribution; therefore one would expect to see the same clusters as in the direct geometric projections shown here. It remains to be seen, however, how the relationships between clusters will be preserved in this type of approaches. There are a number of unsolved problems involved in this type of data dependent and iterative mappings, such as the dependence of the map on starting configuration, and the

distortion of trajectories. Moreover these methods generally involve heavy computation.
The utility of projection schemes is to provide visual feedback to the investigator in a readily interpretable form involving information about the entire assembly of neurons. An alternative to this type of visualization would be to develop methods that describe the evolution of clusters and their shapes directly in the N-space. An undistorted description in terms of e.g. hyperspheres and hypertubes in N-space might be more accurate and provide more insight than projections. Visualization could then be obtained by (mentally) projecting such gestalts (rather than the individual points) to a two- or three-space [37].

4. THE 'DYNAMIC CORRELATION MATRIX' : A HEBBIAN ALGORITHM

We have also investigated a slightly modified version of the gravitational representation, based on the evaluation of a dynamically modifiable connectivity (correlation) matrix. In this approach also the formal relation which exists between gravitational clustering and cross correlation analysis becomes more explicit. To this end we now shortly return to the conventional approach of analyzing the coherence in a group of neurons by cross correlating pairs of spike trains, two at a time.

4.1 Logical Wiring Diagram and Correlation Matrix

In Sect. 2.2 it was already observed that, when applied to the simultaneously recorded activity of larger groups of neurons, the pairwise analysis of cross correlation inevitably leads to considerable numbers of correlograms. It was soon realized that there is the need for a compact representation of results, in order to avoid a state of conceptual plethora when faced with such numbers of correlograms. One possible way, inspired by graph theory and electronics schematics, is to present 'abbreviated' correlograms of the whole group in the form of a so-called *'logical wiring diagram'* [13], with the width (colour, length) of the connecting wires between any two nodes representing a specific aspect of the correlation of the corresponding neurons (e.g. strength or latency). The 'logical wiring diagram' thus aims at giving a global visualization of the 'functional connectivity' within the observed group of neurons. Note that, unlike in normal electronics diagrams, the connection between any two nodes in a logical wiring diagram may consist of two directional 'wires': the 'connection' between neuron i and neuron j needs not be symmetrical (neither does the cross correlogram).

Another compact way to present the correlations within a larger group of neurons, isomorphic to the previous one, is related to linear algebra: the *'functional connectivity matrix'*, in which the value at a particular entry (i,j) denotes a specific aspect (e.g. strength) of the functional connection from i to j. Again, in general the correlation matrix will not be symmetrical: $C(i,j) \neq C(j,i)$. A visual representation of the 'connectivity matrix' is obtained by displaying the matrix values using a grey- or pseudo-colour coding. A question of considerable interest arises when the connectivity matrix is relatively sparse, i.e. when it contains a considerable number of zero (or very small) elements. In that case one may try to reorganize the matrix into some desirable form by means of an appropriate reordering algorithm [7]. One may hope to arrive in this way at a more or less 'natural' form of the connectivity matrix, which by its appearance allows direct inference regarding the global properties of the connectivity structure (e.g. a band, a block triangular or a block diagonal form).

4.2 Hebb's Rule: 'Learning' the Correlation Matrix

We have adopted the format of the connectivity matrix in a modified realization of the 'gravitational clustering algorithm'. It was observed [1] that the equations defining the gravitational representation bear a close resemblance to the formalism describing plasticity of synaptic connections between neurons, the so called *Hebb's rule* [20]. According to this rule, the principal drive for 'learning', i.e. changing synaptic weight, resides in pair interactions of participating neurons (coherent pre- and post-synaptic firing). The image of particles moving around in N-space, subject to mutual attractive and repellent forces as a consequence of dynamically varying charges, by only minor modifications can be translated into a Hebbian-type learning algorithm (formal equations are given in the Appendix). The functional connections (i.e. correlations) between N neurons are represented in a NxN correlation matrix. Initially the matrix is specified to be uniformly zero, which corresponds to our complete lack of knowledge at the outset of the analysis. With every near-coincident firing of two neurons the value of the corresponding entry in the matrix is increased slightly, completely analogous to the charge interaction between any two particles in the gravity representation: Each spike generates a local influence ('charge') which diminishes in time, this influence is 'sensed' by another neuron, provided it has a spike, the influence of which overlaps in time with that of the first one. The increase in 'synaptic' weight for that particular pair then is proportional to the amount of overlap of the influence (charge) histories. In the course of the analysis the spike trains are processed step by step; at each time step the connective weights in the correlation matrix are updated. Thereby the investigator is able to observe how the matrix gradually 'learns' the functional connectivities underlying the observed spike trains. This learning is based on Hebb's rule which mirrors a supposedly similar development in real neuronal networks.

The updating of matrix entries is defined such that it distinguishes between a spike from neuron i, closely followed by a spike from neuron j (entry i,j), and the opposite case: a spike from neuron j, closely followed by a spike from neuron i (entry j,i) (analogous to the 'two-charge rule' in the gravitational algorithm, see [14]). The resulting connectivity matrix needs not be symmetric, which allows for inference of directionality of the 'connections'. Analogously to gravitational clustering, the time constants in the influence ('charge') histories can be manipulated and thus provide the means to focus on specific temporal configurations in spike patterns. Through the selection of the time course of the elementary (i.e. single spike) charge function, the experimenter effectively makes an operational definition for the 'coherence' being evaluated in that particular analysis run. Throughout the work presented here we have used an exponential waveform, with a time constant of 8 ms.

4.3 Normalization: Stationary Activity

Even in the case of completely incoherent spike trains the algorithm sketched above would result in positive matrix values for all pairs of neurons, simply because in any two spike trains one would a priori expect a number of near-coincidences for purely statistical reasons. For stationary Poisson spike trains this can be corrected for by requiring that each neuron's 'charge' history has a total value of zero, when integrated over time [17]. One possible way is to define the elementary charge function such that it has zero mean value (bi- or poly-phasic waveform); another way, in fact the one we used here, is to subtract from the various charge histories the corresponding time average, which amounts to a simple DC-shift. For uncorrelated, stationary Poisson spike sequences, this zero-mean-normalization of the charge histories will, in expectation, result in a value of zero for the respective entry in the correlation matrix. A relative abundance of spike-spike and silence-silence combinations will lead to a positive 'connection' ('excitation'), whereas a net surplus of spike-silence combinations will give rise to a negative value ('inhibition').

In order to illustrate the working of this correlation matrix algorithm, we have applied it to artificial spike trains generated by a simulated network of 10 neurons, which was analyzed extensively in the original 'gravity' papers [17,14,1]. The network is shown schematically in Fig. 9a, each of the neurons fires spontaneously according to a Poisson statistic with a spontaneous firing rate of about 10 spikes per sec. The connections in the network all have a strength of 0.35 (on a linear scale from 0 to 1), a latency of 1 ms and a width of 4 ms (for details of the simulator see [3]. Fig. 9b shows the correlation matrices at 4 different moments during the 'learning' analysis: after 1, 2, 4 and 8 seconds. Values in the i-th column of the matrix (i numbered from left to right) represent weight factors of 'outgoing' connections, i.e. originating in neuron i; values in the j-th row (j numbered from bottom to top) represent weight factors of 'incoming' connections, i.e. ending in neuron j. Values along the diagonal (entries i,i) have not been updated and thus represent zero correlation. The strength of the correlation is coded in grey: the larger the value, the darker the grey. The four matrices in Fig. 9b have been drawn with identical scaling in order to enable a direct comparison of matrix values and to emphasize the 'growth' aspect of the analysis.

Comparison of Figs. 9a and 9b shows that already after 2 seconds of 'recording' (i.e. after a mere 20 spikes of each neuron) the connectivity structure starts to show up clearly in the two bottom rows, with additional weak correlations scattered over the entire matrix. These background correlations either represent partial coherence (indirect correlation, e.g. entry 7,8) or are due to purely statistical effects. As time progresses the values in the two bottom rows, corresponding to 'real' connections in Fig. 9a, keep increasing steadily. The other locations in the matrix show this growing behaviour to a clearly smaller extent and rather more erratic. Effectively this implies that, as time proceeds the algorithm, through 'learning', gathers evidence for 'true' correlations, and improves its 'signal to noise ratio' with respect to statistical background correlations. This increase of reliability with time is emphasized in Fig. 9c, where the same matrices from Fig. 9b are shown, but now each matrix was scaled individually, according to its own range of values. Again we notice the dominating 'connections' in the two bottom rows. In this case, however, these values no longer grow (i.e. become darker) in time, a direct consequence of the dynamic scaling applied. In contrast with the relatively stable bottom row correlations, the background correlations appear to be diminishing in time: the noisy background in the matrix gradually cools down to a comfortably quiet landscape, against which the 'true' correlations are gradually gaining significance. It should be noted that in this representation already after one second of recording (only some 10 spikes per neuron, cf. Fig. 9c, matrix at top left) the structure of the network starts to be visible. This makes the dynamic correlation matrix, like the gravitational clustering, a very sensitive tool to analyze coherence in multi-unit spike trains.

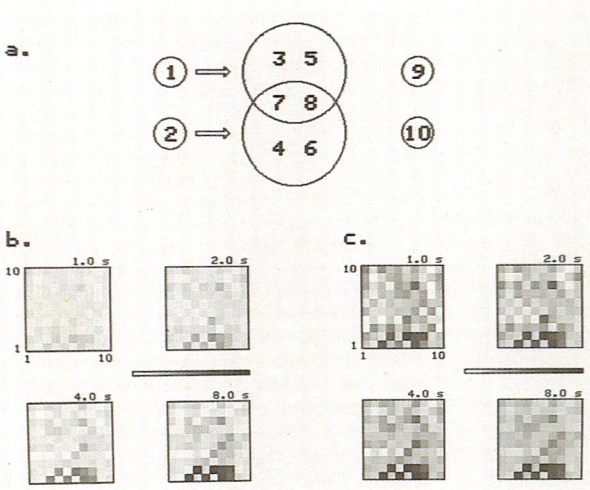

Figure 9. Time evolution of a correlation matrix for a simulated network of neurons. In a. the 'wiring diagram' of the network is given; b. and c. show the evolution of the correlation matrix at different moments in time (indicated at top right). In b. all four matrices are shown with one scaling based on the overall range of values, whereas in c. each matrix is scaled individually according to its own range of values. For further explanation see text.

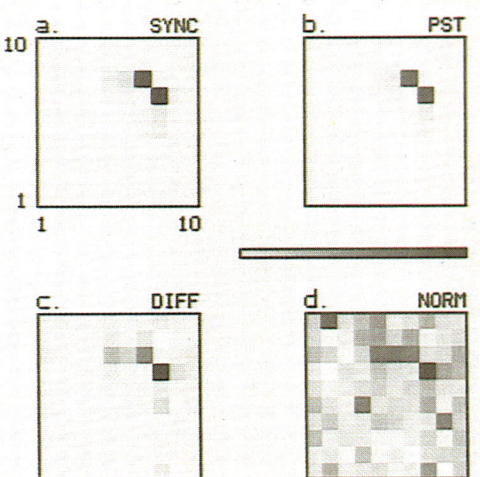

Figure 10. Four versions of 'dynamic correlation matrix' analysis, applied to the spike trains from Fig. 2: a. synchrony matrix; b. PST-matrix; c. differential correlation matrix (first 'correction' for stimulus effects); d. normalized correlation matrix (second 'correction' for stimulus effects). Further explanation is given in the text.

4.4 Normalization: Stimulus-Induced Nonstationary Activity

It is clear from the dot displays in Figs. 1 and 2 that multi-unit activity recorded from the visual cortex under dynamic stimulus conditions does not qualify as a stationary Poisson process. The firing rates of the different neurons show clear variations in time, which appear to be connected to the stimulus: more or less reproducible behaviour over subsequent stimulus presentations. Moreover the time course of these variations appears to be linked to the nature of the stimulus: movement of the bar in different directions leads to different variations of firing rate. This, of course, is all very well known, and in fact is used as the common experimental paradigm to study single neuron response properties. At the same time, however, this stimulus-induced nonstationarity of firing probability presents a nontrivial problem in the analysis of coherence among multi-unit spike trains. First there is a formal problem: the mere fact that spike trains under stimulation become nonstationary causes severe theoretical problems in a field where the mathematical formalism of stationary Poisson processes remains the only mathematical tool to describe spike trains that was developed to any useful extent. Secondly, and more importantly, there is a substantial problem: stimulus-induced nonstationarities provide an additional, strong source of synchrony among spike trains from different neurons (see also our remarks in connection to Fig. 3). In contrast to the simulated example in Fig. 9, neurons may not only fire in close coincidence because they are connected or share a common neural source of input, they may also show coherence in their firing patterns because they are affected in synchrony by some (not necessarily the same) properties of the stimulus (the shared preference for the direction of bar movement among a number of neurons in Figs. 1 and 2 provides a clear example). At this point one may proceed along different, not mutually exclusive ways.

4.4.1 Synchrony Matrix

The first way is simply to treat the spike trains for what they are: time varying signals, transferred from neuron to neuron in a massively interconnected network. Information is thought to be carried both by the time variation of single neuron firing rates as well as by the time varying degree of coherence among spike trains from different neurons. In this view, coherence is considered as a 'code' which signals sensory configurations in the 'outer world' and which at the same time reflects the instantaneous activity state and functional connectivity structure of the neuronal network. This implies that it is worthwhile to look for synchrony of firing, irrespective of what possibly brought it about (stimulation and/or connectivity). This also more or less reflects the way the nervous system is confronted with the problem: how to handle incoming information with time varying characteristics, having no independent knowledge of the external world, or, to put it more strongly: how to 'construct' an external world with its stationary and nonstationary properties, without actually knowing what these are, and only the time varying spike trains to go on.

Based on these considerations, we have analyzed the spike trains from our 45°, top left example (cf. Fig. 2) for synchrony per se, quite irrespective of what caused it (we did use the stationary, zero charge correction, though). The resulting *synchrony matrix* is shown in Fig. 10a, values were coded using a linear grey scale. From this Fig. it is clear that the high synchrony between neurons 7 and 8 (cf. Fig. 3) completely dominates the other pairs. Apart from the synchrony between these two, one also observes a fair amount of co-firing for the pairs (6,8), (5,8), and, although hardly discernible, (5,7) and (6,7). Note that the synchrony of firing for these pairs is quite symmetrical (cf. Fig. 2). A slightly 'negative' synchrony, i.e. the co-occurrence of firing in one spike train and silence in another one, can be observed (although hardly discernible) for the pair (8,9). This can be understood from the suppression of firing in neuron 9 and the simultaneous activation of neuron 8 (cf. Fig. 2). From the comparison of Figs. 2 and 10a it is also clear that the visibility of synchrony using this kind of measure is strongly influenced by the numbers of spikes involved: the higher the firing rate of a neuron, the larger (in general) the matrix entries involving that neuron. It should therefore be considered seriously whether some kind of scaling procedure, e.g. division by the product of the numbers of spikes of the neurons involved, to arrive at something like a 'synchrony index' might not be the more appropriate way to proceed [9].

4.4.2 PST-Matrix

The second approach to the question of stimulus-induced nonstationarities also involves scaling, but rather for more substantial reasons. This approach interprets coincident firing in different neurons as the sign of 'functional connectivity' and aims at recovering the 'logical wiring diagram' from coherence analysis of the multi-unit spike trains. From this point of view the synchrony brought about by stimulation is considered a nuisance, which masks the effect of primary interest (see also our remarks in connection to the correlograms in Fig. 3); consequently one needs a 'correction procedure' which 'unmasks' the connectivity-related coherence. To this end we developed a procedure, which will be described here heuristically; formal expressions are given in the Appendix, a more fundamental discussion on this methodological issue will be presented elsewhere (Aertsen et al., in preparation).

As a first step we want to measure the synchrony induced by purely stimulus-related modulations of neuronal firing rate. What is asked for, in fact, is an explanatory model of the relation between sensory stimulation and firing probability of the neurons under observation. This in general not being available, we have to suffice with a statistical model, based on the data at hand: different realizations of a multi-variate point process with time varying rates. The best estimate for the stimulus-related nonstationarity that can be obtained from these data is simply the average: for each unit the spike trains recorded during consecutive stimulus presentations are mapped onto a single time interval, covering one stimulus sweep. By this procedure we obtain a new multi-unit spike train, each single train being roughly M times as dense as the corresponding original spike train, the latter of course being M times as long, with M the number of stimulus presentations in the experiment. This new train we call the *multi-unit PST-train*: the time varying density of each single unit component is estimated by the well known peri-stimulus time histogram (PSTH). The multi-unit PST-train presents our best statistical estimate of the relation between stimulation and firing probability. By subjecting this multi-unit PST-train to our correlation matrix analysis we can estimate the purely stimulus-induced component of coherence. The resulting *PST-matrix*, being scaled for the number of stimulus sweeps, for our particular example is shown in Fig. 10b, using the same scaling as in the synchrony matrix (Fig. 10a).

4.4.3 Differential Correlation Matrix

In order to assess the neuronal component of multi-unit coherence one has to compare the synchrony matrix (Fig. 10a) with the PST-matrix (Fig. 10b): any coherence in the multi-unit spike trains which goes beyond purely stimulus-induced coherence is taken to reflect functional connectivity in the network. Note that this is a one-way statement: from the absence of any difference between the two matrices one cannot infer the absence of functional connectivity; the effects of the latter might simply be buried under (generally much larger) stimulus effects. Not surprisingly (cf. Fig. 3), in the present case the synchrony matrix (Fig. 10a), looks conspicuously like the PST-matrix (Fig. 10b), which once more emphasizes the dominating role played by stimulation. In order to quantify the dissimilarity between synchrony matrix and PST-matrix we chose to calculate the algebraic difference of the two: the *'differential correlation matrix'* equals the synchrony matrix *minus* the PST-matrix (analogous to the subtraction of the 'shift control' in Fig. 3). The result is shown in Fig. 10c, grey scale being adjusted to the range of values in the differential matrix. For the present example this matrix represents the first order estimate of the connectivity contribution to multi-unit coherence. The conclusions from this Figure are completely in register with those from gravitational analysis (cf. Fig. 4), where in fact the same differential procedure was used to get rid of purely stimulus-induced coherence.

4.4.4 Normalized Correlation Matrix

It may be illustrative to note that the actual calculation of these differential measures took a somewhat different path, although numerically the outcome is completely equivalent. In the stationary case (cf. Sect. 4.3) the gravitational representation and the dynamic correlation matrix were calculated on the basis of the respective charge histories, each one corrected for overall rate by undergoing a DC-shift to obtain zero total charge. In the nonstationary case this DC-term, representing the statistical estimate of the time invariant firing rate, was replaced by its time varying analogon: the (properly scaled) charge history of the (periodically continued) PST-train. In other words: the differential correlation matrix measures the coherence between time varying signals which have been *shifted* dynamically to obtain a constant expectation value equal to zero. This, of course, is quite a common procedure in the correlation analysis of ensembles of time varying analog signals (so-called 'cross covariance function' [4,33]. As a matter of fact, this analogy can be pushed one step further: in order to arrive at a *normalized* cross covariance function, i.e. with values between -1 and +1, in conventional signal analysis an additional *scaling* with auto-covariance is applied, equivalent to a transformation to ensembles with zero mean and unit variance. Also in the present case such normalization appears mandatory. First of all, our measure for coherence so far is only a relative measure, not an absolute one. Moreover, in the differential measure we still have rate effects playing an important role for the obvious reason that when the expected firing rate is high, then also the magnitude of the deviation of actual firing rate from expected rate is expected to be high. This implies that, although stimulus induced modulations of firing rates were subtracted, the differential correlation matrix still presents a distorted picture of neuronal coherence, with strongly firing neurons dominating through the sheer amount of their output, and hence, the expected large variability over subsequent stimulus sweeps (compare Fig. 10c with Fig. 2). The same statement holds for the 'differential' gravitational results (cf. Figs. 4-7)

In order to overcome these difficulties, and inspired by the analogy with conventional signal analysis, we have applied additional dynamic scaling to arrive at the *'normalized correlation matrix'*, with matrix values at any moment by definition restricted to the range between -1 and +1. The precise form of the scaling procedure is given in the Appendix, it amounts to a dynamical scaling of entries in the differential correlation matrix with the appropriate time dependent auto covariance of the (shifted) charge histories (spike trains) involved. The resulting normalized matrix for the present example is shown in Fig. 10d, values range between -0.03 (white) and +0.14 (black). From this Fig. one observes that the normalized matrix indeed differs significantly from the differential matrix (Fig. 10c): the picture now is rather more diverse: the coherence for the original group remains present, but is not dominant anymore, in addition one observes connections entering the scene which so far escaped from being noticed, especially involving the lower electrode numbers. Coherence on the whole assumes only moderate values, with a maximum of 0.14.

This Figure also points to a problem of the normalized correlation matrix: its sensitivity to noise. Due to the scaling procedure, which in fact amplifies the influence of weakly and/or regularly firing neurons at the expense of strongly and/or irregularly firing neurons, certain matrix entries may be blown up for possibly the wrong reason, e.g. because the respective neuron(s) hardly fired. This is for instance the case in the pairs (2,10) and (9,4), where the matrix indicates relatively high coherence for spike trains containing only very few spikes (cf. Fig. 2). Although this may surely point at a genuine albeit only sporadically detectable coherence, it will be hard to establish statistical significance. On the one hand the normalized correlation matrix is asserted to be our best theoretical approximation to the functional connectivity of the group, on the other hand, however, as a statistical estimate it is clearly suffering from a high variance. Results therefore have to be interpreted with care and dot displays of the actual spikes trains should always be within reach; it is also obvious that the analysis of variance of the estimator and the development of criteria for significance are important topics for further research.

4.5 Results

The results of application of these different correlation matrix techniques to the spectrum of all directions of bar movement tested in the experiment (cf. Fig. 1) are summarized in Figs. 11 to 13. Figure 11 shows the synchrony matrices, Fig. 12 the differential correlation matrices and Fig. 13 the normalized correlation matrices. The layout of these Figures is analogous to Fig. 7: each matrix is positioned at the tip of the vector which indicates the corresponding direction of bar movement. In each Figure the 16 matrices were plotted using one scale, covering the range of matrix values in that particular Figure (ranges are indicated above the grey scale in the Fig.).

In Fig. 11 one observes how larger values in the 'synchrony matrices' are basically restricted to a few directions, associated with movement along or close to the negative diagonal, with the pair (7,8) very much dominating. The PST-matrices (not shown here) give practically the same picture. This once more emphasizes the strong influence of stimulation on multi-unit synchrony, simply through the single unit's dependencies on stimulus properties (in this case direction of bar movement, cf. Fig. 1). Subtraction of direct stimulus influence renders the 'differential correlation matrices' (Fig. 12) clearly more sensitive to smaller values of coherence so that the picture becomes more diverse (more neurons involved); still, basically these results are in register with the synchrony matrices: high values for certain neurons in the case of movement along or close to the negative diagonal, for other directions virtually no signs of coherence (see also the bar graphs in Fig. 7).
Additional scaling leads to the 'normalized correlation matrices' in Fig. 13; these present a much more differentiated picture of coherence and especially of its stimulus-dependence. Whereas the earlier noted preference of certain neurons to cohere for movements along the negative diagonal remains present, albeit less clear,

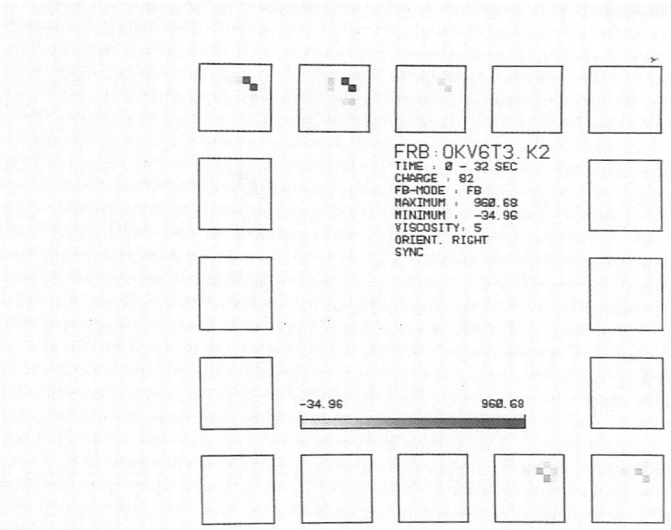

Figure 11. Synchrony matrices for all 16 directions of bar movement (spike trains from Fig. 1). Matrices are positioned at the tip of the vector indicating the direction of motion of the stimulus. All matrices have been plotted according to the same grey scale. Further explanation in the text.

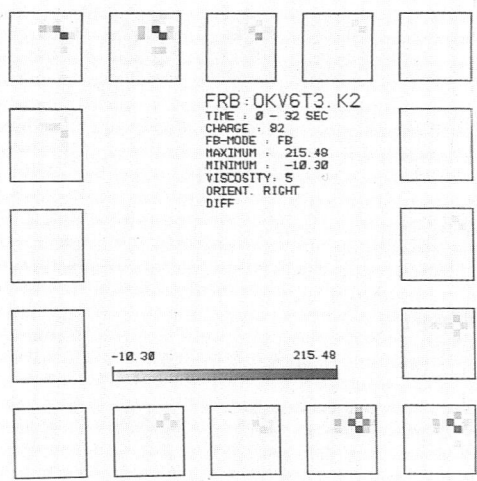

Figure 12. Differential correlation matrices (first stimulus 'correction') for all 16 directions of bar movement (spike trains from Fig. 1). Further details as in Fig. 11.

now also movements in other directions, especially those associated with the positive diagonal, appear to result in increased coherence. Purely horizontal or vertical stimulus movements clearly are less effective in eliciting coherence. In the present example coherence effects are essentially restricted to the electrodes numbered between 5 and 10 (corresponding to recording sites in cortical layers V to I), the actual distribution of coherence varying with direction (e.g. 7 and 8 co-vary strongly along the negative diagonal, 9 becomes involved more strongly with movement along the positive diagonal). Values of the coherence in this experiment attain relatively small values, between -0.05 and +0.20. The asymmetry of this range of values with respect to the value zero may reflect a true asymmetry in the connectivity, although we rather suspect that (a substantial part of) it is due to a preferential sensitivity of our analysis technique for 'excitatory' connections above 'inhibitory' ones (see a discussion on this issue in the context of cross correlograms in [3]. Finally, the matrices in Fig. 13 as compared to those in Fig. 12 also show an increased tendency of not being symmetrical with respect to the diagonal of the matrix.

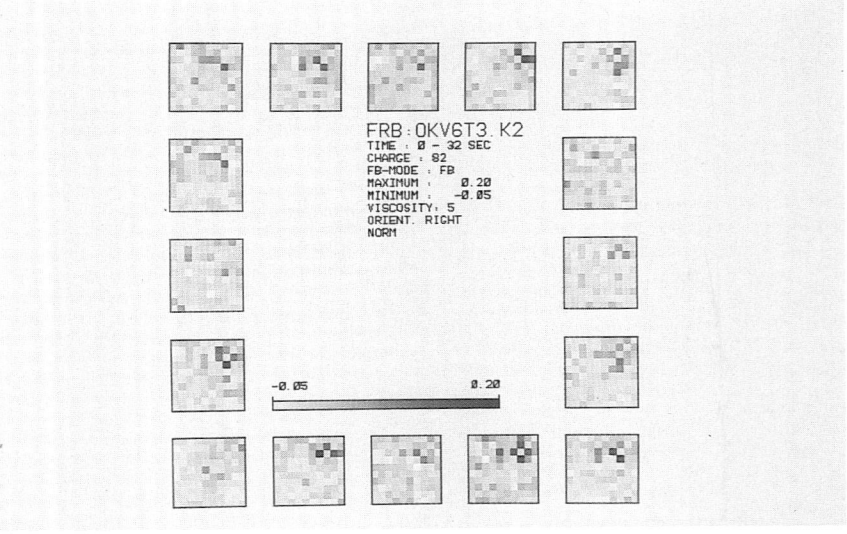

Figure 13. Normalized correlation matrices (second stimulus 'correction') for all 16 directions of bar movement (spike trains from Fig. 1). Further details as in Fig. 11.

So far our discussion of results from dynamic correlation matrix analysis (Figs. 11-13) has focused on more global properties, such as the degree of (dis)similarity of matrices for different stimuli. Like gravitational clustering, this technique provides a comprehensive and readily digestible overview of the distribution of coherence among larger groups of neurons, and its properties such as dependence on stimulus and time. It should be born in mind, however, that in fact the time development of each individual grey pixel in each of the matrices represents highly condensed information regarding the coherent behaviour of a specific pair of neurons under certain experimental conditions (comparable to the information contained in the individual distance curves as in Fig. 4). The more detailed analysis of relatively subtle effects like variations of individual matrix entries (c.q. pair distances or projections) is more comparable to conventional analysis of cross correlograms (cf. Fig. 3). An adequate interpretation of results from multi-unit experiments regarding the functional and physiological significance of neuronal interaction requires the continuous interplay of these different levels of analysis.This type of more substantial interpretation of these findings, together with detailed analysis of other multi-unit recordings from similar experiments is subject of current work; preliminary findings seem to be that in general the coherence in these experiments is clearly stimulus-dependent, is confined to relatively low values (roughly between -0.1 and +0.3) and is not so much pointing at small, local circuits but rather at relatively diffuse distributions of coherence, involving rather large fractions of the neurons in the multi-unit recording.

In how far these findings are related to, for instance, the experimental program and the electrode arrangement (linear array, tip distances 160 microns, recording sites positioned vertically with respect to cortical layers) remains subject to further investigation.

5. DISCUSSION

In this paper we have described the recently developed multi-unit analysis technique of 'gravitational clustering' and a technique derived from it: dynamic correlation matrix analysis. These techniques were applied to multi-electrode recordings from the cat's visual cortex, in order to investigate these recordings for possible coherence among the activity patterns of different neurons. This coherence is assumed to reflect the functional connectivity in the neural network recorded from. In the course of this analysis it became clear that the presentation of stimuli provides a strong, if not fully dominating effect on multi-unit coherence, simply through the large, stimulus-induced modulations of single unit firing rates. This evidently poses severe problems to a naive approach of identifying synchrony of firing with neuronal connectivity. Different approaches to this problem were discussed.

The first approach is to apply the measures for coherence simply to the spike trains as they are, thereby measuring the 'synchrony' or 'co-variation' of firing, irrespective of what caused it (stimulation and/or connectivity): the 'synchrony matrix'. In this view multi-unit coherence conveys information to the nervous system about what is going on 'outside' and, at the same time, reflects the instantaneous activity and connectivity status 'inside'. This interpretation of coherence is more or less in line with the proposal that the functional order of a neuronal net may be defined through the order induced by the cross correlations of signals carried by the neurons [22,23]. It is also related to the notions of 'syntactic' and 'semantic' aspects of activity patterns in neuronal populations, as expressed by Johannesma et al. [21].

In the second approach one is mainly interested in the multi-unit coherence as far as it goes beyond purely stimulus driven coherence. The aim is to artificially separate 'outside' from 'inside' by pealing off the strong stimulus-induced coherence. Thereby one hopes to obtain the purely neuronal component of coherence, which may be read as a 'logical wiring diagram', specifying the instantaneous 'functional connectivity' state of the network. In this approach we have argued for a 'stimulus correction' procedure consisting of two stages: subtraction, followed by scaling.

These two approaches are certainly not meant to be mutually exclusive, we rather view them as complementary and in their interrelation providing a conceptual framework for understanding the activity of an entire population of neurons as an entity, rather than as a collection of single elements. A central issue here is to arrive at a quantitative notion of appropriate 'state variables' to describe the neuronal population, analogous to the phase space representation of complex systems, used, for example, in statistical physics. Related to this type of description in terms of microvariables would be the introduction of the appropriate macro-variables (like e.g. pressure and temperature in thermodynamics) to describe the global properties of the system, disregarding the precise micro-states.

Moreover, the dichotomy between purely stimulus-induced coherence on the one hand and coherence due to 'true' connectivity on the other hand is somewhat artificial to begin with, insofar as, firstly, such a separation requires 'extra-neuronal' knowledge about the experiment, which only the experimenter has, and, secondly, there is assumed to be a strong interaction between the two: connectivity shaping the stimulus-induced coherence (for structural reasons) and stimulus-induced coherence shaping the connectivity (presumably by a Hebb-type mechanism).

Related to this is the paradoxical observation that in order to separate the 'connectivity-coherence' from 'stimulus-coherence', one has to rely on the 'noise' in the system: the more reproducibly the stimulus affects the neurons' firing probabilities, the less room is left for connectivity to make itself known. This is especially clear in the case of highly 'adequate' stimuli, where subsequent presentations evoke strongly driven and virtually identical spike trains: the correlation of multi-unit firing becomes identical to the expected correlation, determined on the basis of PST-trains, and consequently no inference can be drawn regarding possibly underlying connectivity. In other words: if we suppose that a Hebb-type mechanism is operative in shaping the neuronal connectivity, then this implies that the stronger and more reliable it has done so, causing the network to react to external input in a relatively deterministic fashion, the harder it is for an external observer to find out about it. The investigation of functional connectivity by 'pealing off' the 'purely stimulus-induced coherence' through a 'shift control'-type procedure then effectively reduces to a case of 'throwing out the baby with the bath water'.

Throughout this discussion the various approaches were illustrated by results of their application to simulated spike trains and to multi-unit recordings from the cat's visual cortex during presentation of a stimulus ensemble, consisting of a bar moving in different directions. These results show that, like gravitational clustering, the dynamic correlation matrix is truly a very sensitive device to detect possible coherence in multi-neuron activity. This high sensitivity, combined with the multi-dimensionality and the dynamic character inherent to this approach makes these methods a promising tool, which in principle should also be able to detect changes in coherence among spike trains with time constants in the range of one to a few seconds, such as are induced by manipulation of the stimulus and which in our experimental data indeed were shown to be present.

In the present paper we have used the term 'functional connectivity' as purely descriptive: there exists a statistical relationship between the firing of different neurons. This notion of 'functional connectivity' (correlation) is used here as opposed to 'structural' or anatomical connectivity of the neuronal net, involving mono- and poly-synaptic connections. Although, obviously, these two notions are intricately connected, they should not be confused. Functional connectivity may change dramatically on a short time scale, such as induced by manipulation of the sensory environment, as was demonstrated in the present paper (for other experimental evidence see e.g. [8,13]. It has been proposed that synaptic links in the central nervous system have to be modifiable on the fast time scale of fractions of a second ('synaptic modulation' [29]). Simulation studies, on the other hand, suggest that fast changes in functional connectivity are not necessarily associated with corresponding changes in structural (i.e. synaptic) connectivity [11]. At any rate, the dynamic connectivity structure of neuronal networks has certainly become an element in recent developments in brain theory, and further investigation of the relation between structural and functional connectivity, both theoretically as well as experimentally, appears to be essential.

The formal relations between the techniques of 'gravitational clustering' and the analysis of the 'dynamic correlation matrix' are strong and relatively direct (see also Appendix); the same goes for the kind of information yielded by their application to experimental data, as illustrated by our examples. There are also differences, the main one related to the geometrical nature of the gravitational representation: the influences of different particles (neurons) on another particle (neuron) are combined by vector addition into a single driving force, whereas the correlation matrix keeps pair interactions separated into their respective, modifiable matrix entries. Both approaches have their own virtues, an extensive discussion of which is beyond the scope of the present paper (an example is the quite straightforward relation which can be derived between the dynamic correlation matrix and the 'normal' cross correlation function, see Appendix). Both approaches essentially are highly parallelized calculations of correlations between spike trains. Although so far all calculations were performed on conventional, i.e. serial computer configurations, this parallel nature of both approaches makes them natural candidates to be implemented on a truly parallel machine. This would enable a much faster evaluation of coherence in multi-unit spike trains, with the possibility of having (preliminary) results on-line during the actual experiment; the advantage for the experimenter of having an appropriate feedback cannot be overstressed. In our lab we are currently investigating the possibility of implementing this type of multi-unit analysis techniques on a parallel, hardware realization of an associative network [32]. This would have the additional advantage of closing the conceptual circle: for the analysis of experimental data from real, neuronal networks one uses a machine which in itself represents a model of such a neuronal network.

6. ACKNOWLEDGEMENTS

Calculations were performed using a Fortran software package developed by one of the authors (AA), which was integrated with the software package MATFUN, developed at the Dept. of Medical Physics and Biophysics of the University of Nijmegen, The Netherlands, and implemented on the VAX 750 at the MPI-Tübingen. The latter package was also used for graphics display of the results.
The authors wish to thank Peter Johannesma and Günther Palm for stimulating discussions, Manfred Caeser for his work on the evaluation of histological data, Volker Staiger for skillful assistance in generating the Figures and Shirley Würth for helping out in the final processing of the manuscript.

7. APPENDIX

7.1 Gravitational Clustering

With each one of a group of N neurons we associate a particle in N-space, with initial position $\vec{r_i}(0)$ (i=1,...,N) such that all particles are positioned on the vertices of an N-dimensional hypercube; therefore all initial mutual distances are equal to some constant d_0. After the system starts to evolve, the position of the i-th particle is at any time t given by the N-dimensional vector $\vec{r_i}(t)$.

Let $z_i(t)$ be the spike train from neuron i. The charge function associated with particle i, used to generate inter-particle forces is given by a filtered version of the spike train:

$$Q_i(t) = \int ds\, q(s)\, z_i(t-s) \tag{A1}$$

In the 'two charge model' [14], designed to allow inference regarding the *direction* of interaction, with each particle we associate two charges: an *effector* charge which generates the propulsive field:

$$Q_{e,i}(t) = \int ds\, q_e(s)\, z_i(t-s) \tag{A1a}$$

and an *acceptor* charge, used to calculate the force on that particular particle:

$$Q_{a,i}(t) = \int ds\, q_a(s)\, z_i(t-s) \tag{A1b}$$

In the present paper we have used for the elementary effector charge $q_e(t)$ a decaying exponential starting at and following the spike, and for the elementary acceptor charge $q_a(t)$ a rising exponential that terminates at the spike.

The propulsive field $\vec{E}_{ij}$ at position $\vec{r_i}$ generated by the particle j at position $\vec{r_j}$ is given by

$$\vec{E}_{ij} = Q_{e,j}\, \hat{r}_{ij} \tag{A2}$$

with the unit vector $\hat{r}_{ij}$ given by

$$\hat{r}_{ij} = \frac{\vec{r}_{ij}}{r_{ij}} = \frac{\vec{r_j}-\vec{r_i}}{\|\vec{r_j}-\vec{r_i}\|} \tag{A3}$$

From (A2) one notes that the field was chosen to be independent of distance. The force exerted on particle i by particle j is then given by

$$\vec{F}_{ij} = Q_{a,i}\, \vec{E}_{ij} = Q_{a,i}\, Q_{e,j}\, \hat{r}_{ij} \tag{A4}$$

The total force on i is given by the vector sum of all the pair forces:

$$\vec{F}_i = \sum_{j\neq i} \vec{F}_{ij} = \sum_{j\neq i} Q_{a,i}\, Q_{e,j}\, \hat{r}_{ij} \tag{A5}$$

As a consequence of this force the particles will move. The dynamic equation is defined by

$$\mu \dot{\vec{r}}_i = \vec{F}_i \qquad (A6)$$

with μ denoting the 'viscosity' of the medium in N-space. Note that the acceleration term is lacking in (A6): we assume high viscosity and therefore velocity is proportional to force. The displacement, finally, is calculated by numeric integration (Euler) using a time step δ :

$$\vec{r}_i(t+\delta) = \vec{r}_i(t) + \frac{\delta}{\mu} \vec{F}_i(t) \qquad (A7)$$

7.2 Dynamic Correlation Matrix

The formalism of the dynamic correlation matrix is a simple modification of the equations describing 'gravitational clustering'. With N neurons we associate a NxN-correlation matrix C, with the entry C_{ij} denoting the correlation ('functional connection') *from* neuron j *to* neuron i, i.e. related to the probability of having an i-event *after* a j-event. Note that this implies that C needs not be symmetrical: direction of connectivity is preserved. Initially the matrix C is set to be uniformly zero, signifying our complete lack of knowledge at the outset of the experiment. In the course of time the matrix values are gradually updated, according to a Hebb-type rule for synaptic modification, which is directly borrowed from the gravitational representation: the pair force F_{ij} (Eq. A4) is interpreted as the drive for modification of the 'connection', very much the same as the dynamic equation for the particle motion (Eq. A6):

$$\mu \dot{C}_{ij} = Q_{a,i} Q_{e,j} \qquad (A8)$$

In our correlation matrix algorithm we have implemented a somewhat more general formulation which also incorporates a leakage term:

$$\mu \dot{C}_{ij} = Q_{a,i} Q_{e,j} - \alpha C_{ij} \qquad (A9)$$

This extra term, which signifies 'forgetting' of earlier 'experience', is quite common in theoretical work on synaptic plasticity. In the present paper, though, we have always used Eq. (A8) (i.e. $\alpha = 0$), to enable a better comparison with the gravitational algorithm.

The main difference between the gravitational representation and the dynamic correlation matrix is the vector addition (Eq. A5), which is missing in the latter approach. In the gravitational representation the influence of different particles (neurons) on another particle (neuron) is combined into a single force, the correlation matrix keeps all the pairs separated in the respective, modifiable matrix entries. Both approaches have their own advantages, one advantage of the correlation matrix is that certain relations can be derived in a straightforward manner, which, in the gravitational approach, is more cumbersome because of the dynamic interaction of force vectors.

7.3 Relation between 'dynamic correlation matrix' and normal 'cross correlation'

As an example of such a relation it may be illustrative to show the explicit relation between the dynamic correlation matrix and the 'normal' cross correlation function of spike trains. The net 'change' of a connection C_{ij} over some period of time, say from t_1 to t_2, is given by

$$\mu(C_{ij}(t_2) - C_{ij}(t_1)) = \int_{t_1}^{t_2} dt\, Q_{a,i}(t)\, Q_{e,j}(t) \qquad (A10)$$

Substitution of the expressions for the charge functions (Eqs. A1a and A1b) leads to

$$\mu(C_{ij}(t_2) - C_{ij}(t_1)) = \int ds_1 q_a(s_1) \int ds_2 q_e(s_2) \int_{t_1}^{t_2} dt\, z_i(t-s_1) z_j(t-s_2) \qquad (A11)$$

On the right hand side one easily recognizes a term which also appears in the time dependent cross correlation function:

$$R_{ij}(\tau_1, \tau_2) = E(z_i(\tau_1) z_j(\tau_2)) \qquad (A12)$$

A considerable simplification is obtained when the correlation is *not* time dependent, i.e. for stationary processes z_i and z_j:

$$R_{ij}(\tau_1, \tau_2) = R_{ij}(\tau_1 - \tau_2) \qquad (A13)$$

This implies that, for *stationary* spike trains, Eq. (A11) can be rewritten as

$$\mu(C_{ij}(t_2) - C_{ij}(t_1)) = \int ds_1 q_a(s_1) \int ds_2 q_e(s_2)\, R_{ij}(s_1-s_2; t_1, t_2) \qquad (A14)$$

where

$$R_{ij}(s_1-s_2; t_1, t_2) = \int_{t_1}^{t_2} dt\, z_i(t-s_1)\, z_j(t-s_2) \qquad (A15)$$

is the estimate of the cross correlation between i and j, obtained by *time averaging* over the interval between t_1 and t_2.

Equation (A14) in fact states that, for stationary spike trains, the net change of the 'connection' C_{ij} over the internal (t_1, t_2) is proportional to the (t_1, t_2)-estimate of the cross correlation function between neurons i and j, 'smoothed' by $q_e(t)$ and 'weighted' by $q_a(t)$. From this relation it becomes clear how, by appropriate choice of q_e and q_a one can obtain directional information from the cross correlogram.

The zero total charge normalization, applied in the case of stationary spike trains (Sect. 4.3), amounts to a DC-shift in the charge functions, equal to the (time averaged) estimate of firing rates $\bar{z}_i$ resp. $\bar{z}_j$:

$$\mu(C_{ij}(t_2) - C_{ij}(t_1)) =$$

$$= \int ds_1 q_a(s_1) \int ds_2 q_e(s_2) \int_{t_1}^{t_2} dt \ (z_i(t-s_1)-\bar{z}_i)(z_j(t-s_2)-\bar{z}_j) \quad (A16)$$

This DC-shift effectively implies a translation from cross correlation to cross covariance: the departure of the cross correlation from its background level due to the non-zero expectations $\bar{z}_i$ and $\bar{z}_j$.

The *'differential correlation matrix'* applied for *nonstationary* spike trains as a first approximation to 'functional connectivity', with direct stimulus effects subtracted (Sect. 4.4.3), uses precisely the same formalism as given in (A16), with the difference that now the expected firing rates are time dependent, and in the case of periodic stimulation can be estimated through the *PST-trains* (Sect.4.4.2):

$$M \ \bar{z}_i(t) = \sum_{m=1}^{M} z_{i,m}(t) \quad (A17)$$

with M the number of stimulus presentations, and t ranging from 0 to T, T being the duration of one stimulus sweep. It is clear that in this case (A16) is directly related to the time dependent cross covariance function of z_i and z_j (e.g. [4]). Furthermore, one easily sees that, with $\bar{Q}(t)$ denoting the charge function for the PST-trains, scaled for the number of sweeps M, it holds that

$$\mu(C_{ij}(MT) - C_{ij}(0)) = \int_0^{MT} dt \ (Q_{a,i}(t) - \bar{Q}_{a,i}(t))(Q_{e,j}(t) - \bar{Q}_{e,j}(t)) =$$

$$= \int_0^{MT} dt \ Q_{a,i}(t) \ Q_{e,j}(t) - \int_0^{MT} dt \bar{Q}_{a,i}(t)\bar{Q}_{e,j}(t) \quad (A18)$$

or: the charge-shift procedure to calculate the differential correlation matrix, is equivalent to subtraction of the *'PST-matrix'* (Sect. 4.4.2) from the *'synchrony matrix'* (Sect. 4.4.1).

The *'normalized correlation matrix'* (Sect. 4.4.4) finally is obtained by additional scaling for auto covariances:

$$C'_{ij}(t_2) - C'_{ij}(t_1) = \frac{\int_{t_1}^{t_2} dt \, \{Q_{a,i}(t) - \overline{Q}_{a,i}(t)\} \{Q_{e,j}(t) - \overline{Q}_{e,j}(t)\}}{\left\{ \int_{t_1}^{t_2} dt \, \{Q_{a,i}(t) - \overline{Q}_{a,i}(t)\}^2 \cdot \int_{t_1}^{t_2} dt \, \{Q_{e,j}(t) - \overline{Q}_{e,j}(t)\}^2 \right\}^{\frac{1}{2}}} \quad (A19)$$

It is easily shown that, because of the Cauchy-Schwarz inequality, the scaling in (A19) effectively results in matrix values restricted to the range between -1 and +1, i.e. the procedure in (A19) is truly a normalization.

8. REFERENCES

1. Aertsen, A., Gerstein, G. and Johannesma, P., From neuron to assembly: Neuronal organization and stimulus representation. In: G. Palm and A. Aertsen (eds.), Brain Theory, pp. 7-24, Springer, Berlin (1986).
2. Aertsen, A.M.H.J., Erb, M. and Johannesma, P.I.M., The 'Neurophone': Acoustic representation of neural activity patterns (in preparation).
3. Aertsen, A.M.H.J. and Gerstein, G.L., Evaluation of neuronal connectivity: Sensitivity of cross correlation. Brain Res, **340**, 341-354 (1985).
4. Bendat, J.S. and Piersol, A.G., Random data: analysis and measurement procedures. Wiley-Interscience, New York, (1971).
5. Biswas, G., Jain, A.K. and Dubes, R.C., Evaluation of projection algorithms. IEEE Trans Pattern Anal Machine Intell, **PAMI-3**, 701-708. (1981).
6. Braitenberg, V., Cell assemblies in the cerebral cortex. In: R. Heim and G. Palm (eds.), Theoretical approaches to complex systems. Lecture Notes in Biomathematics, pp. 171-188, Springer, Berlin (1978).
7. Duff, I.S., A survey of sparse matrix research. Proc IEEE, **65**, 500-535 (1977).
8. Eggermont, J.J., Epping, W.J.M. and Aertsen, A.M.H.J., Stimulus dependent neural correlations in the auditory midbrain of the grassfrog (Rana temporaria L.). Biol Cybern, **47**, 103-117 (1983).
9. Epping, W.J.M., Auditory information processing in the midbrain of the grassfrog. Ph.D. Dissertation. University of Nijmegen, The Netherlands (1985).

10 Epping, W., Boogard, H. van den, Aertsen, A., Eggermont, J. and Johannesma, P., The Neurochrome: An identity preserving representation of activity patterns from neural populations. Biol Cybern, **50**, 235-240 (1984).
11 Erb, M., Palm, G., Aertsen, A. and Bonhoeffer, T., Functional versus structural connectivity in neuronal nets. In: Proceedings 9. Kybernetik-Kongress (DKG), p. 23 (abstract). Göttingen, FRG (1986).
12 Gerstein, G., Functional association of neurons: detection and interpretation. In: F.O. Schmitt (ed.), The Neurosciences: Second Study Program. pp. 648-661, Rockefeller University Press, New York (1970).
13 Gerstein, G., Aertsen, A., Bloom, M., Espinosa, I., Evanczuk, S. and Turner, M., Multi-neuron experiments: Observation of state in neural nets. In: H. Haken (ed.), Complex Systems - Operational Approaches, pp. 58-70, Springer, Berlin (1985).
14 Gerstein, G.L. and Aertsen, A.M.H.J., Representation of cooperative firing activity among simultaneously recorded neurons. J Neurophysiol, **54**, 1513-1527 (1985).
15 Gerstein, G.L., Bloom, M.J., Espinosa, I.E., Evanczuk, S. and Turner, M.R., Design of a laboratory for multi-neuron studies. IEEE Trans Systems, Man and Cybernetics, **SMC-13**, 668-676 (1983).
16 Gerstein, G.L. and Perkel, D.H., Mutual temporal relationships among neuronal spike trains. Biophys J, **12**, 453-473 (1972).
17 Gerstein, G.L., Perkel, D.H. and Dayhoff, J.E., Cooperative firing activity in simultaneously recorded populations of neurons: detection and measurement. J Neurosci, **5**, 881-889 (1985).
18 Grinvald, A., Real-time optical imaging of neuronal activity. TINS, **7**, 143-150 (1984).
19 Grinvald, A., Real-time optical mapping of neuronal activity: From single growth cones to the intact mammalian brain. Ann Rev Neurosci, **8**, 263-305 (1985).
20 Hebb, D.O., The organization of behaviour. A neuropsychological theory. Wiley, New York (1949).
21 Johannesma, P., Aertsen, A., Boogard, H. van den, Eggermont, J. and Epping, W. From synchrony to harmony: Ideas on the function of neural assemblies and on the interpretation of neural synchrony. In: G. Palm and A. Aertsen (eds.), Brain theory, pp. 25-47, Springer, Berlin (1986).
22 Koenderink, J.J., Simultaneous order in nervous nets from a functional standpoint. Biol Cybern, **50**, 35-41 (1984).
23 Koenderink, J.J., Geometrical structures determined by the functional order in nervous nets. Biol Cybern, **50**, 43-50 (1984).
24 Kohonen, T., Self-organized formation of topologically correct feature maps. Biol Cybern, **43**, 59-69 (1982).
25 Krüger, J., A 12-fold microelectrode for recording from vertically aligned cortical neurones. J Neurosci Methods, **6**, 347-350 (1982).
26 Krüger, J., Investigation of a small volume of neocortex with multiple microelectrodes: evidence for principles of self-organization. In: H. Haken (ed.), Complex Systems - Operational Approaches. pp. 71-80, Springer, Berlin (1985).
27 Krüger, J., Simultaneous individual recordings from many cerebral neurons: Techniques and results. Rev Physiol Biochem Pharmacol, **98**, 177-233 (1983).
28 Krüger, J. and Bach, M., Independent systems of orientation columns in upper and lower layers of monkey visual cortex. Neurosci Lett, **31**, 225-230 (1982).
29 Malsburg, C. von der, Nervous structures with dynamical links. Ber Bunsenges Phys Chem, **89**, 703-710 (1985).
30 Moore, G.P., Segundo, J.P., Perkel, D.H. and Levitan, H., Statistical signs of synaptic interaction in neurons. Biophys J, **10**, 876-900 (1970).

31 Palm, G., Neural assemblies: An alternative approach to artificial intelligence Springer, Berlin, (1982).
32 Palm, G. and Bonhoeffer, T., Parallel processing for associative and neuronal nets. Biol Cybern, **51**, 201-204 (1984).
33 Papoulis, A., Probability, random variables and stochastic processes McGraw-Hill Kogakusha, Tokyo, (1965).
34 Perkel, D.H., Gerstein, G.L. and Moore, G.P., Neuronal spike trains and stochastic point processes. II. Simultaneous spike trains. Biophys J, 7, 419-440 (1967).
35 Sammon, J.W., Jr., A nonlinear mapping for data structure analysis. IEEE Trans Computers, **C-18**, 401-409 (1969).
36 Terekhina, A.Yu., Methods of multidimensional data scaling and visualization (survey). Avtom Telemekh, 7, 80-94 (1973).
37 Zahn, C.T., Graph-theoretical methods for detecting an describing Gestalt clusters. IEEE Trans Computers, **C-20**, 68-86 (1971).

HYPERCHAOS AND 1/F SPECTRA IN NONLINEAR DYNAMICS: THE BURIDANUS'DONKEY

F.T. Arecchi

Istituto Nazionale di Ottica

Largo E. Fermi 6, 50125 Firenze (Italy)

Abstract

The power spectrum of a nonlinear dynamical system with more than one basin of attraction becomes $1/f$ in the presence of noise, provided the basin boundary be fractal. The result is numerically proved by a cellular automaton. This phenomenon resembles the random-random walk in a one dimensional lattice, which undergoes a subdiffusive motion with the same type of spectra. Experimental evidence is shown in systems including nonlinear electronic oscillators and lasers. The relevance to biochemical evolution is discussed.

1. Introduction

In the middle 1300 the following problem has been attributed to Johannes Buridanus, a philosopher at the University of Paris. Suppose a donkey is just halfway between two equivalent choices (e.g. two food baskets that we call F1 and F2). What will be its decision? In the solution attributed to Buridanus the donkey dies, having no elements to decide for either solution. The current modern solution, upon which most of statistical physics is built, is more optimistic. The initial condition between the two choices is an unstable one, like the maximum $x=0$ in a quartic potential well $V(x)=-ax^2+bx^4$ ($a,b > 0$) and it would be left immediately once the donkey (taken as a material point initially at $x=0$) is coupled to the rest of the Universe, which provides for a thermal bath including fluctuations (even at zero temperature there would be quantum flunctuations).

Let us model the fluctuations as an additive white noise (no memory) source. If we use a discrete time approach and introduce an uncertainty Δx per step (the donkey's feet have a finite size), there is a single time scale τ, that corresponding to the average first passage time through Δx. Afterwards, because of the uniqueness theorem for the solution of a differential problem, noise will not play any extrarole and the donkey will go either to F1 or F2. The time scale τ provides an exponential decay of correlations, that is, of the memory of the initial uncertainty and an associated Lorentzian power spectrum

$$G(\omega) \simeq \frac{1/\tau}{\omega^2 + 1/\tau^2} \qquad (1)$$

As well known a log-log of (1) has two asymptotic straight lines, a high frequency one with a slope -2 (20 db/decade) and a horizontal one for low frequency, corresponding to lack of correlations (white spectrum). The two lines cross at $\omega = 1/\tau$. The long time lack of correlation is the basis of all Markoffian approaches to statistical physics.

Buridanus' solution would be then wrong, since the donkey does not die but it performs a decision with a definite time scale.

These considerations where the basis of an approach to decay of unstable states motivated by an early experiment on a transient laser /1/ and then formalized in a general procedure /2/.

If, however, still with the same two-valley potential (bistable solution) and in the presence of a white noise, we increase the number of degrees of freedom up to 3 in order to allow for a chaotic dynamics, then we observe experimentally, the possibility of jumps back and forth from a decision to the other one.

Aim of this paper is to show that this is equivalent to provide Buridanus'donkey with a fractal boundary between the two choices. Indeed an irregular rugged boundary can be crossed from several directions, and it will provide a large number of length scales rather than a single one Δx, and hence a large number of time scales. This will be equivalent to the superposition of many Lorentzians spectra as (1), thus providing a power spectrum utterly different from a Lorentzian one. The donkey will keep a long memory of the initial uncertainty and it might die as expected by Buridanus.

In the following we approach the problem with reference to chaotic dynamics, offering a solution in terms of an elementary model.

2. Noise induced trapping at the boundary between two attractors

Addition of random noise in a nonlinear dynamical system with more than one attractor may lead to 1/f spectra, provided that the basin boundary be fractal /3/. Combining the features leading to deterministic chaos with a random noise is somewhat equivalent to a double randomness and we call "hyperchaos" such a situation. Indeed random-random walks in ordinary space, as diffusion in disordered systems, have shown a 1/f behavior /4-5/. Thus, hyperchaos here introduced is a random-random walk in phase space, where in fact one of the two sources of complex behavior is due to the fractal structure arising from deterministic dynamics.

To evaluate the impact of the following arguments, I premise some historical remarks on 1/f spectra in nonlinear dynamics.

Some years ago it was discovered /6/ that in a nonlinear dynamical system with more than one attractor, introduction of random noise induces a hopping between different basins of attraction, giving rise to a low frequency spectral divergence, resembling the 1/f noise well known in many areas of physics (Fig. 1). Such a discovery was confirmed by a laser experiment implying two coexisting attractors /7/ (Fig. 2), and later the effect was observed in other areas as e.g. Josephson tunnel junctions /8/.

The effect was questioned with two objections:
a) a noise induced jump across a boundary leads to a telegraph signal, hence to a single Lorentzian spectrum /9a/;
b) a computer experiment yielded a power law only over a limited spectral range /9b/.

The questions were answered /9c/ with a statement of the empirical conditions under which the 1/f spectra appeared, namely:
i) coexistence of at least two attractors (so called " generalized multistability" /7/),
ii) presence of noise,
iii) some "strangeness" in the attractors.

As a matter of fact this third condition was rather vague. To make it more precise, two theoretical models were explored, namely, a one dimensional cubic iteration map with noise /10/ and a forced Duffing equation with noise /11/. Both these papers disclose interesting features, bringing more light on the above assumption iii). Fig. 2 of Ref. 10 shows that the size of the 1/f spectral region increases with the r.m.s. of the applied noise, that is, with the probability of crossing the basin boundary by a noise-induced jump.

The numerical evaluation of Ref. 11 showed that for some control parameters the boundary between basins of attraction was an intricated set of points, through which it was impossible to draw a simple line. In such cases the noise was most effective in yielding low frequency spectra 1/f-like.

On the other hand a fundamental logical approach to the 1/f problem was based on the composition of a large number of Lorentzians (or elementary Markov processes with exponential decay) whose weights are log-normally distributed /12/, thus fulfilling the relation

$$\int_{\gamma_1}^{\gamma_2} \frac{\gamma}{\omega^2 + \gamma^2} p(\gamma) d\gamma \simeq const \times \frac{1}{\omega}, \qquad (2)$$

provided $p(\gamma) \sim 1/\gamma$, and for the frequency range $\gamma_1 \ll \omega \ll \gamma_2$.

Motivated by the rate processes considerations, which yielded a single Lorentzian for two attractors, we developed a kinetic model /10/ based on a single transition rate for each pair of attractors. In the case of M attractors, this yielded M-1 Lorentzians. To approximate the integral (2) by a sum (5% accuracy in fitting a 1/f law would require about one pole per decade) a large number $M \gg 2$ of attractors is necessary and hence the integral of eq. (2) would be replaced by the sum over the M-1 Lorentzians corresponding to the eigenvalues of the kinetic model, however there is no reason to weigh the Lorentzians according to their reciprocal widths, hence no satisfactory reconstruction of a 1/f spectrum was possible. In fact, an experiment on a forced and noisy Duffing oscillator with an increasing number of attractors /13/ did not offer a clear evidence of the expected scaling of the spectral exponent with the number of attractors. On the contrary, Ref. 11 showed that the boundary region between just two attractors was sufficient to yield 1/f-like spectra, at variance with the many-attractor model. Thus, this suggested that the boundary structure was the real responsible for a large number of decay constants (possibly log-normally distributed).

In the meantime, the fractal, structure of a basin boundary was explored in some examples /14/. This means the following. As the phase point wanders within one basin of attraction, if we draw a sphere around the point defining its distance from the other basin of attraction, the radii of these spheres are distributed with all scale lengths, according to the self similar structure of the fractal boundary.

Based on the above considerations, we have built an elementary cellular automaton which models the motion of the phase point within a fractal basin boundary under the presence of random noise. We model the boundary region of two basins of attraction A and B as two adjacent one-dimensional lattices of sites. Suppose we start from site i. At each discrete time step, if i belongs to A (i = i_A) it moves one step forward on the same lattice ($i_A \to (i_A + 1)$) and if it belongs to B it goes one step backward ($i_B \to (i_B - 1)$). In the absence of noise, once the motion has started on one basin, it will remain on it forever. In the presence of noise, at each time step there is a finite probability of a "cross" jump at the same lattice site, from stripe A to B: $i_A \to i_B$.

We call L the maximum size of the boundary region and $\ell_i \leq L$ any of the possible sizes of the fractal set. At each time step, the probabilities of permanence and jump are respectively

$$P_{AA} = P_{BB} = \ell_i/L$$

$$P_{AB} = P_{BA} = 1 - \ell_i/L \quad . \tag{3}$$

To build a self-similar structure we allow ℓ_{i_k} to scale as $\frac{\ell_{i_k}}{L} = (1/2)^{V(i_k)}$ where $V(i_k)$ is a natural number sorted randomly. for each site i_k (i=$-\infty$ to ∞, k = A,B). To deal with a real numerical experiment we consider finite sequences of N sites (e.g. N = 10^3) and we truncate the fractality by imposing $0 \leq V(i_k) < F$. Here, F is a finite integer denoting the maximum partitioning $(1/2)^{F-1}$, that is, the ultimate resolution of the measuring device in appreciating the fractal structure of our set. With all this in mind, for each evolution we extract a double sequence of N integers randomly distributed between 0 and F-1, and denote each site i_k by the corresponding number $V(i_k)$. This means that we have attributed to each site an "area of respect", that is, a specific separation ℓ_{i_k} from the other attractor, with ℓ_{i_k} depending on $V(i_k)$ as shown above. We start, e.g. on the basin A from i_A =N/2.

At this step, to account for a suitable noise yielding the permanence and jump probabilities (3), we generate a random number y uniformly distributed between 0 and 1. If $y \leq (1/2)^{V(i_A)}$, then at the next time the point goes to i_A + 1 on attractor A; if $y > (1/2)^{V(i_A)}$, then the point jumps instanteneously to site i_B and at the next time it goes to i_B -1 on attractor B.

By measuring the position coordinate, taking the Fourier transform and squaring it, we can build the power spectra, that is, the transforms of the position correlation functions.

In Fig. 3 we show two power spectra for F=4, and 14 respectively. In fact, we have measured spectra for all integer values of F between 4 and 14, but we just report two samples over slightly more than three frequency decades. The sequence shows that, as the fractality increases, the slope of the log-log plot goes from about 2 (single Lorentzian) to about 1 (1/f spectrum). This appears better in Fig. 4, where the slope α of the $f^{-\alpha}$ spectral law is plotted versus the fractality F. The Lorentzian (α = 2) of the random telegraph model is easily recovered for F=1, thus showing that noise induced jumps across a regular line boundary fulfill the intuitive expectation of a single decay rate. An analogy with the random-random walk /4,5/ is easily drawn. Indeed our motion is bound with an r.m.s. deviation going from about $\sqrt{t}$ to $|\log t|^2$ as the fractality F increases from 4 to 14, according to Sinai.

For comparison we mention other approaches leading to 1/f or anyway nonLorentzian low frequency spectra:
i) Pomeaeu-Manneville type-3 intermittency corresponds to slowly

diverging trajectories with a 1/f power spectrum /15,16/. This behavior is intrinsic to the dynamics, hence it occurs without noise.

ii) A deterministic diffusion process may occur beyond "crisis" /17/ when two otherwise disjoint attractors merge into a single one. Here again no noise is required, and a comparison of this behavior with noise induced jumps was given in Ref. 11.

iii) Another comparison of intrinsic versus noise -induced intermittency was carried on for a damped driven pendulum, which models a Josephson juction /18/. This last papers offers numerical evaluations of spectra, showing an 1/f region extending over two decades, but to our knowledge nobody has tried so far to analyze the role of fractality and draw a comparison with Sinai subdiffusive motion.

Among other things, the results of this paper may strongly affect our current understanding of Optical Bistability (OB) phenomena. OB is described in terms of two fixed point attractors, which however are the result of a collective dynamics implying many degrees of freedom. There are no exhaustive analyses of the structure of the basin boundary, thus possible fractal structures may appear if the dynamics is evaluated in detail. On the other hand, in order to reduce the signal power necessary to drive the OB device from one state to the other, the system is usually set very near to the boundary. Thus, unavoidable random noise might induce low frequency spectra of the type above described.

3. Conclusion: long memory in statistical physics

Let me conclude with a speculation on the role of the long time terms in nonequilibrium statistical mechanics.

We have shown that, whenever in non linear dynamics more than one attractor is present, there are two distinct power spectra:

i) a high frequency one, corresponding to the decay of correlations within one attractor;

ii) a low frequency one, corresponding to noise induced jumps.

Based upon i), the usual transport coefficients for macroscopic equations of evolution have been built. Effect ii) has been overlooked so far. Here, I wish to consider an example showing the relevance of ii) with respect to multiphoton molecular excitation.

Let me consider a molecule with two isomeric states (cis and trans) of almost equal energy, separated by an energy barrier, say, of 1 eV (e.g. rodhopsin molecule in the retina of vertebrates).

We know that an IR laser such as a CO_2 laser (λ =10 μm, $h\nu \simeq$ 0.1 eV) may give rise to a multiphoton absorption process if it is powerful enough to provide 10 photons within one coherence time of the "cis" valley, so that 10 small photons pile up to 1 eV excitation. (We are considering a molecule large enough so that the barrier is already a classical one, and so quantum tunneling is possible). We know that a vibrational mode IR active decays by intra-molecular relaxations toward the thermal bath of all other modes, in a time of order of 10^{-12} s =1 ps. In order to have a multiphoton isomeric transition, we should have a laser power of

10 photons/1ps $\simeq 10^{-7}$ watt

over a cross-section of $\sim (1\text{Å})^2 = 10^{-16}$ cm^2, and thus a laser intensity of 10^9 W/cm. But this was a Markovian point of view, based on a memory time of 1ps related to the high frequency spectral broadening. A double potential Valley dynamics is described by a Duffing equation (see Ref. 6 and 11) and the presence of an IR laser illumination as a forcing term yields a motion on an attractor not necessarily confined in one valley, even for very low intensities (see Fig. 1).

Such a chaotic motion may pass near to the boundary, hence requiring an activation energy much less than the barrier of 1 eV to be introduced into an Arrhenius type law. For instance, in Fig. 2 we have seen a high frequency spectrum around 100 KHz, and the corresponding low frequency jump spectrum at 1 Hz (5 decades below). By the same reasoning, we might expect that an intensity 5 or 6 decades lower (that is, 10 photons/μs or just 10^3 W/cm) might be sufficient for a multiphoton isomerization process.

If we could use such a large enhancement factor in most activation processes of biochemical relevance, the consequence would be that the times necessary for biochemical evolution on Earth could be correspondingly reduced.

This is just a guess, to show how the introduction of the long memory processes here described for the first time may open new routes in the physics of complex systems.

REFERENCES

/1/ F.T. Arecchi, V. Degiorgio and B. Querzola, Phys. Rev. Lett. 19, 1168 (1967).

F.T. Arecchi and V. Degiorgio, Phys. Rev. A3, 1108 (1971).

/2/ F.T. Arecchi and A. Politi, Phys. Rev. Lett. 45, 1215 (1980)

F.T. Arecchi, A. Politi and L. Ulivi, Nuovo Cimento 71B, 119 (1982).

/3/ F.T. Arecchi and A. Califano, Europhysics Letters 2, (1986).

/4/ Ia. G. Sinai, in Proc. Berlin Conf. on Math. Problems in Theoretical Physics, R.S. Schrader et al. eds., Springer, 1982, p. 12.

/5/ E. Marinari, G. Parisi, D. Ruelle and P. Windey, Phys. Rev. Lett. 50, 1223 (1983).

/6/ F.T. Arecchi and F. Lisi, Phys. Rev. Lett. 49, 94 (1982).

/7/ F.T. Arecchi, R. Meucci, G.P. Puccioni and J.R. Tredicce, Phys. Rev. Lett. 49, 1217 (1982).

/8/ R.F. Miracky, J. Clarke, and R.H. Koch, Phys. Rev. Letters 50, 856 (1983).

/9a)/ M.R. Beasley, D.D' Humieres and B. A. Huberman, Phys. Rev. Lett. 50, 1328 (1983).

b) R. Voss, Phys. Rev. Lett. 50, 1329 (1983).

c) F.T. Arecchi and F. Lisi, Phys. Rev. Lett. 50, 1330 (1983).

/10/ F.T. Arecchi, R. Badii and A. Politi, Phys. Rev. A29, 1006 (1984).

/11a) F.T. Arecchi, R. Badii and A. Politi, Phys. Lett. A103, 3 (1984).

b) F.T. Arecchi, R. Badii and A. Politi, Phys. Rev. A32, 402 (1985).

/12/ E.W. Montroll and M.F. Shlesinger, Proc. Nat. Aca. Sci. USA, 79, 3380 (1982).

/13/ F.T. Arecchi and A. Califano, Phys. Lett. A101, 443 (1984).

/14/ G. Grebogi, E. Ott, and J.A. Yorke, Phys. Rev. Lett. 50, 935 (1983).

S.M. Mc Donald, C. Grebogi, E. Ott and J.A. Yorke, Physica 17D,

125 (1985).
/15/ Y. Pomeau and P. Manneville, Comm. Math. Phys. $\underline{74}$, 189 (1980).
/16/ I. Procaccia and H. Schuster, Phys. Rev. $\underline{A28}$, 1210 (1983).
/17/ T. Geisel and S. Thomae, Phys. Rev. Lett. $\underline{52}$, 1936 (1984); T. Geisel, J. Nierwetberg and A. Zachere, Phys. Rev. Lett. $\underline{54}$, 616 (1985).
/18/ E.G. Gwinn and R.M. Westervelt, Phys. Rev. Lett. $\underline{54}$, 1613 (1985).

FIGURE CAPTIONS

Fig. 1 Electronic non linear forced oscillator obeying the law $\ddot{x} + \gamma \dot{x} - ax + bx^3 = A\cos(2\pi ft)$. Hopping between two attractors and associated 1/f spectrum in the purely bistable case. (a) Symmetric phase-space plots; (b) log-log spectrum showing the low frequency divergence, a broadened f/8 line, and a narrow f/4 line; (c) a sample of the x(t) plot.

Fig. 2 Bistability in a CO_2 laser with loss modulation. (a,b) coexistence of two attractors (period 3 and 4 respectively) high frequency spectrum around 100KHz, (c) comparison between the low frequency cut-off when the two attractors are stable (dashed line) and the low frequency divergence when noise is added (solid line).

Fig. 3 Power spectra (vertical) versus frequency (horizontal) in log-log scale. Wavy lines: measured spectra, straight lines: best fits, whose slopes α are reported in the next figure. The two samples shown refer to F = 4, and 14, respectively.

Fig. 4 Exponents α of the power law $f^{-\alpha}$ versus fractality F.

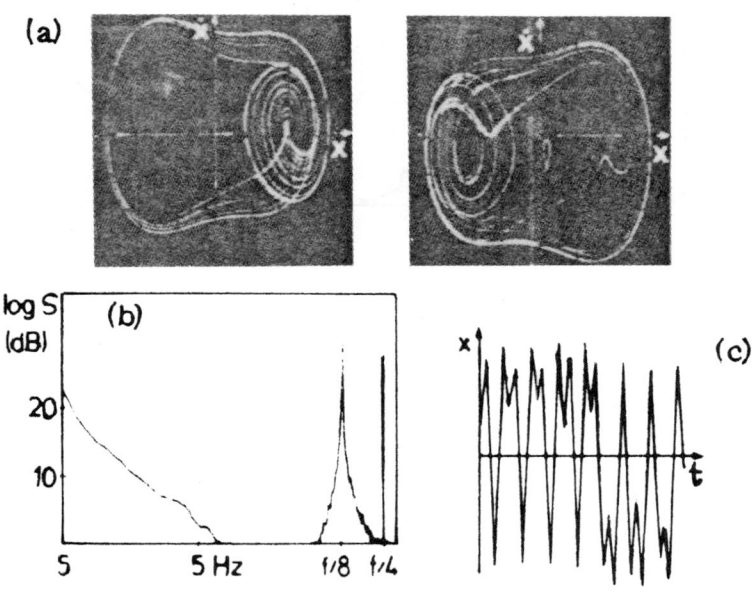

Fig. 1

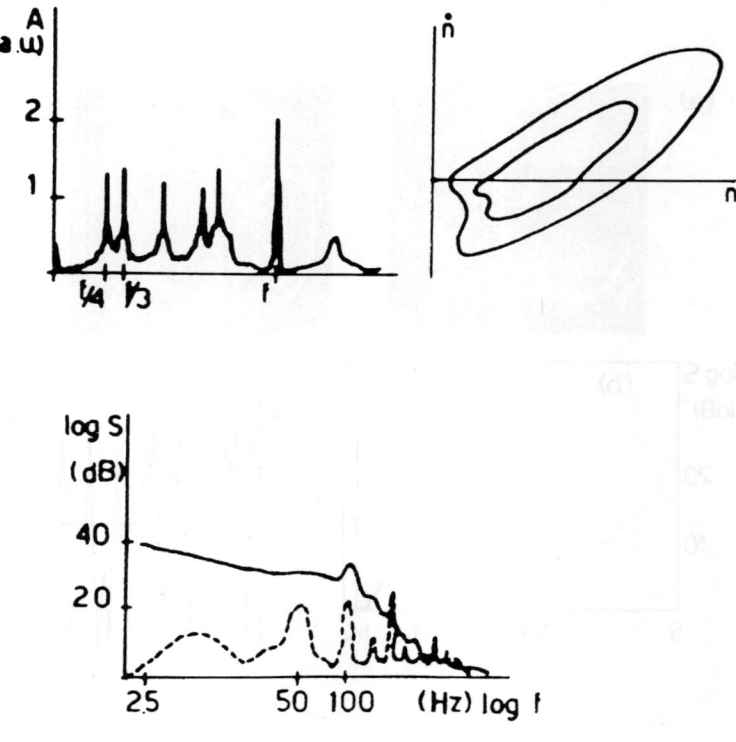

Fig. 2

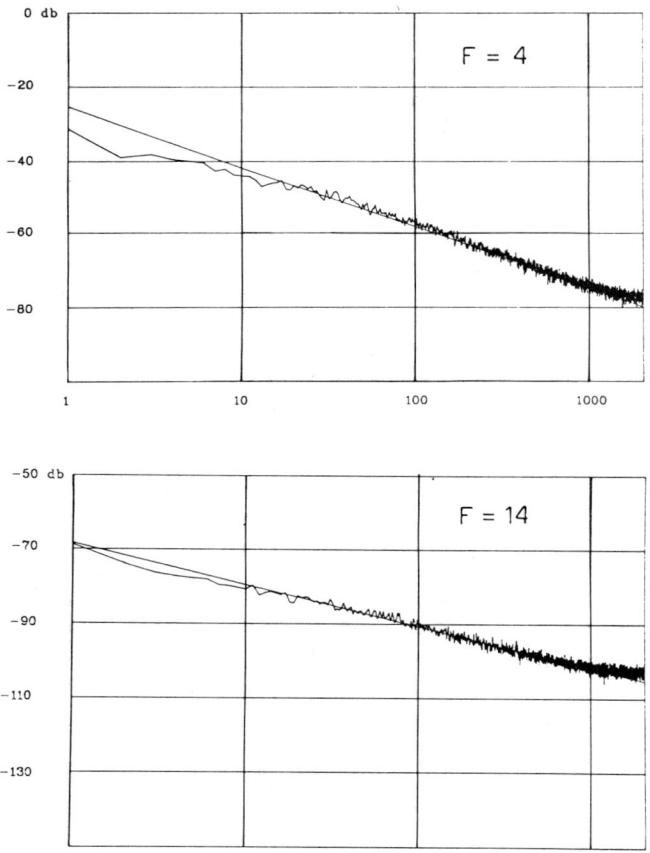

Fig. 3

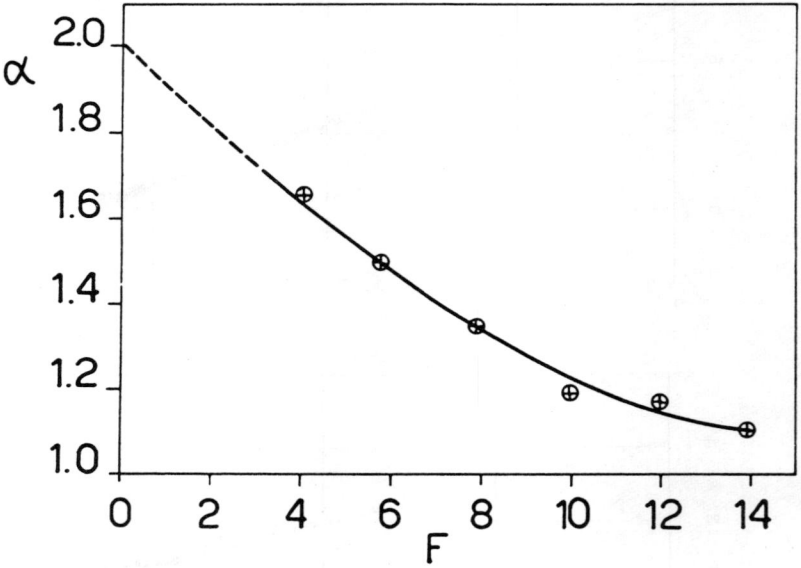

Fig. 4

THE SENSITIVITY METHOD:
A STRUCTURAL APPROACH TO UNDERSTANDING SYSTEMS

Jörg D. Becker

Universität der Bundeswehr München, Fakultät für Elektrotechnik
D-8014 Neubiberg

and

Frederic Vester

Universität der Bundeswehr München, Fakultät für Sozialwissenschaften
D-8014 Neubiberg; and

Studiengruppe für Biologie und Umwelt GmbH, Nußbaumstraße 14
D-8000 München 2

ABSTRACT

The Sensitivity Method provides the means for investigating and planning in complex open systems for which systems dynamics alone is insufficient. We discuss the method and suggest an implementation as an intelligent, evolutionary computer system.

1. PROBLEMS

According to K. POPPER[1], the main difference between living organisms and lifeless things is that organisms have <u>problems</u>. "I conjecture that the origin of <u>life</u> and the origin of <u>problems</u> coincide", he wrote.

We may define the term "problem" as a deviation of an actual state S_α from a desired state S_ω; we may call $R_{\alpha\omega}$ a solution of the problem if it transforms S_α into S_ω; and we may call happiness a situation in which S_α and S_ω coincide.

We should like to stress that here the term "state" is not restricted to its meaning in physics; it may refer to an operating mode, to a structure, to a pattern of interactions, to a set of abilities, etc.

From the definition it should be clear that values play an important rôle, because they suggest us which states are desirable.

Surprisingly enough a general study of problems (classification, strategies for solving) has started to develop only in recent years, mainly in the fields of cognitive psychology and artificial intelligence. In cognitive psychology one has to mention in particular the work of D. DÖRNER[2] and M. OSTERLOH[3]. Following M. OSTERLOH we give a classification of problems in TABLE 1.

If S_α, S_ω, and $R_{\alpha\omega}$ are all known then the only problem left is to carry out $R_{\alpha\omega}$ (which is a matter of our laziness, of the speed of a computer, of an industrial, social, or political norm; in this case we may have to enlarge the system under consideration, thus probably changing the problem class). Hence, we call this rather a task than a problem. Fast Fourier Transform would be an example for a task; you may even show that there is no faster algorithm for its purpose.

Interpolation problems, for which we don't know the solution, but we do know primitive rules from which it is to be constructed, form the typical case for the application of PROLOG. (see APPENDIX for an example).

Note that the problem class usually depends on time and on people concerned with the problem. Thus, to make gold from lead was a synthesis problem in the Middle Ages, and it still is a synthesis problem for most people; but to a nuclear physicist it is simply a task.

For most problems, however, and in particular for the most important ones, S_α, S_ω, and $R_{\alpha\omega}$ are unknown. Think of "quality of life" this is certainly an important problem, but we don't know exactly what it means, nor do we know how to get there, nor do we know suf-

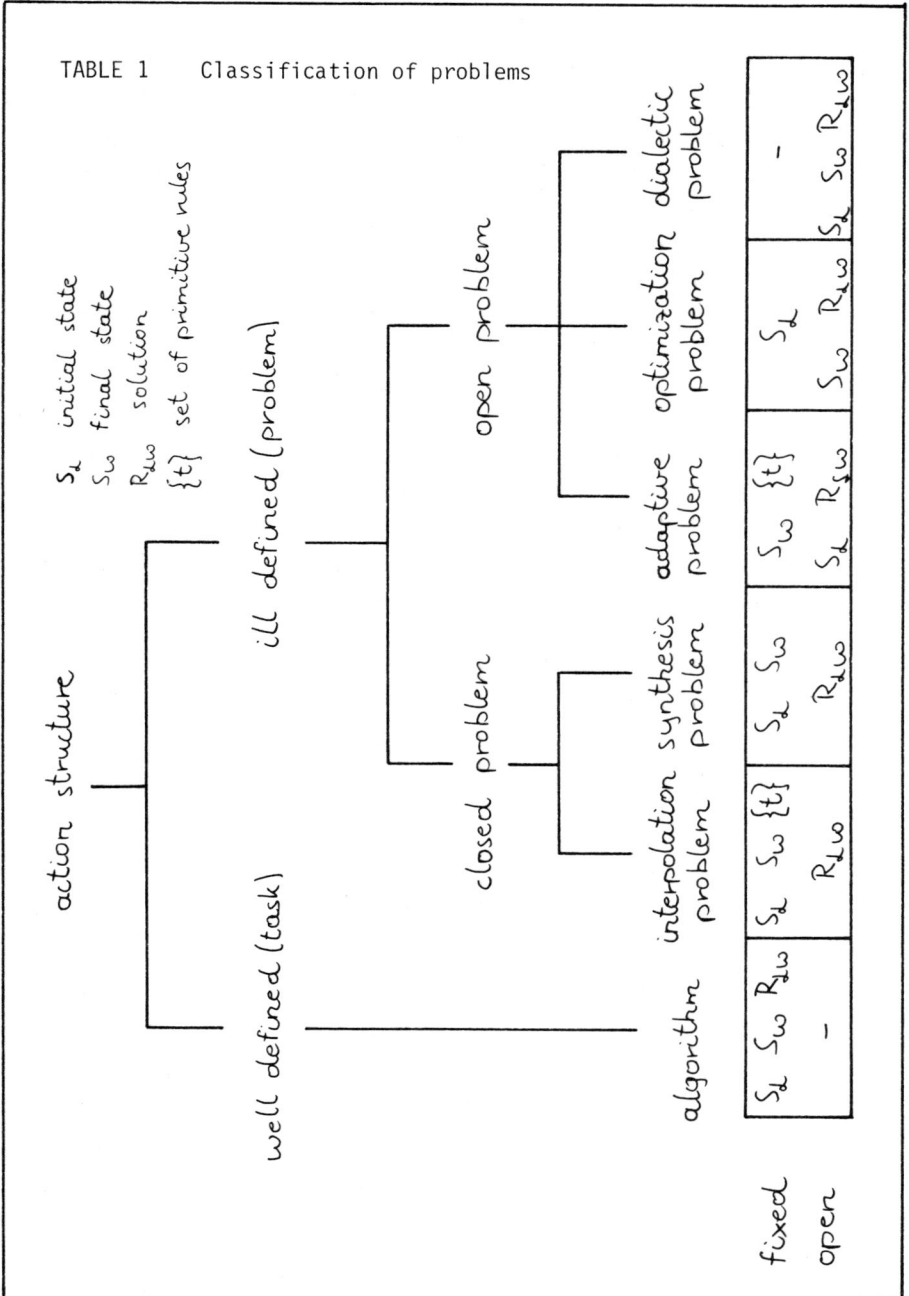

TABLE 1 Classification of problems

ficiently well the present state.

Hence, methods are needed which help us to deal with problems of this type. Such methods will have to include the subject, and they will have to be recursive. As A.v. MÜLLER[4] has pointed out, classical logic and the notion of causality will no longer apply in this case.

2. SYSTEMS

Since our labour resources are limited, and since too many details as well as too large a state space may obscure the important features of a problem, we usually cut a piece out of reality which we call a system. Whether something is considered to be a system depends on the problem. A carbon atom is a system if we are interested in the shape of the wave functions, but it is not a system if we want to talk about the specific heat of diamond (here, the whole crystal is the system); and if we are interested in the psychosomatic rôle of adrenalin we have to choose a still larger piece as a system, including the brain and the immune system.

A system should be roughly seperable from the outside world; i.e. the flows across the system boundaries (flows of matter, of energy, of information) ought to be minimal in some sense.

Furthermore we require that a system be composed from smaller elements between which there are relations; i.e. it should have a structure.

Thus, we arrive at the following tentative definition:

A <u>system</u> is a structured piece of reality with boundaries across which flow is minimal. (Note that various types of flows may have different minimal surfaces.)

In order to solve complex problems we have to understand systems to a certain extent. That this view is not generally shared may be seen from two examples. The first one concerns the energy consumption in the Federal Republic of Germany (cf. fig. 1, data from SÜDDEUTSCHE ZEITUNG[5]); it is pretty clear that the "predictions" are not biased

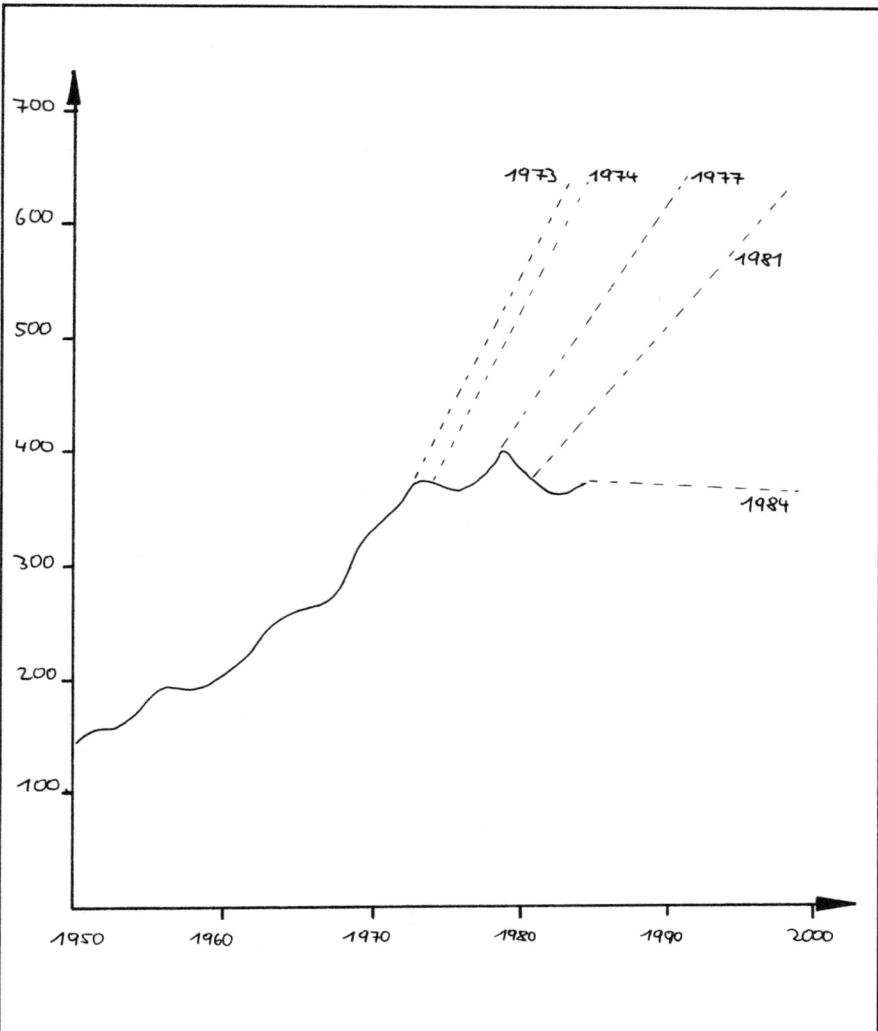

Fig. 1 Energy consumption and predictions (dashed) in the Federal Republic of Germany

by any system analysis. The second example concerns biological pest control. Rats like sugar cane; so if you grow sugar cane you may order some mungos who are pleasant companions and are very efficient at eating rats. When the rats are finished the mungos get hungry, and they might decide to eat your chickens. Still worse, they also like lizards, with the effect that certain insects loose their natural enemies. Thus, in the end you have bugs instead of rats eating your sugar cane[6].

We conclude that system modelling makes an important part of problem solving.

2.1 The Conventional Approach: System Dynamics

The central goal in system dynamics is to model the system with a set of differential equations, ordinary or partial, linear or nonlinear, deterministic or stochastic. The solution of these equations, subject to suitable initial and boundary conditions, is then interpreted as a "prediction". Even if such an approach has its merits, as we shall discuss below, we have to acknowledge that it cannot meet its claim. (see also M. CLARKE[7]).

We shall briefly discuss some of the difficulties of system dynamics. As an example, let us take the population dynamics of a predator-prey system. In many cases such systems show periodic oscillations of the populations in nature. Following VOLTERRA we may model the system in the following way (for a detailed discussion see, e.g. the book by J. HOFBAUER and K. SIGMUND[8]; for a cybernetic discussion of population dynamics see F. VESTER[9]). Let x denote the population of the prey, y the population of the predator. The Lotka-Volterra equations then read:

$$\frac{\dot{x}}{x} = a - by$$

$$\frac{\dot{y}}{y} = -c + dx$$

This means that the growth rate of the prey population is a constant (due to propagation) minus a term proportional to the predator population (the more predators the bigger the loss). The growth rate of the predator population is a negative constant (if there is not prey they starve) plus a term proportional to the prey population. The Lotka-Volterra equations allow for an exact solution. It may be represented by closed concentric trajectories in the x-y-plane; in the middle there is a fixed point. Time goes around counterclockwise (See fig. 2). At the first sight it seems that the Lotka-Volterra equations describe reality rather well. However, there are two drawbacks. First, if there are not predators the prey population will grow exponentially. Second, neither the trajectories nor the fixed point are asymptotically stable; if we add fluctuation terms to the Lotka-Volterra equations there is a finite chance that at least one of the species will die out in a given time.

We may try and cure the first drawback by introducing logistic growth limiting terms in the equations:

$$\frac{\dot{x}}{x} = a - ex - by$$

$$\frac{\dot{y}}{y} = - c + dx - fy;$$

but now there are no closed trajectories any more; we only get a stable fixed point.

In order to obtain a stable limit cycle (i.e., a periodic attractor) there are two simple possibilities. Either we add a term proportional to $(x^2 + y^2)$ to both equations; but there is not obvious interpretation of such a term. Or we let three species interact; but this seems rather arbitrary, apart from the fact that a set of three coupled Lotka-Volterra type equations exhibits easily chaotic solutions.

We see that nonlinear differential equations may react very sensitively to arbitrary small changes of their structure.

Furthermore, we should not forget that we have neglected time

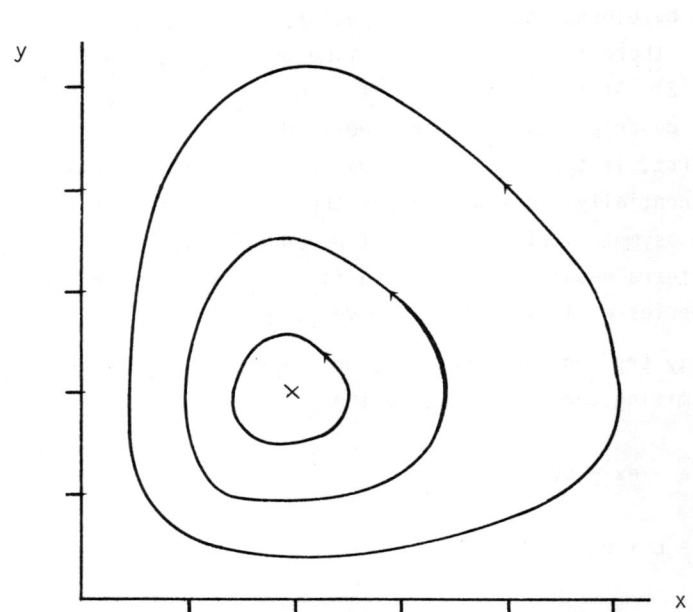

Fig. 2 Solutions to the Lotka-Volterra equation.
x: fixed point; →: time arrow.

lags (e.g., animals breed only once or twice a year) and spatial variations (territorial behaviour of predators, retreat areas of the prey), and these may be important for understanding the behaviour of the system. The inclusion of these effects leads to partial nonlinear integro-differential equations, and it becomes even more difficult to make plausible the connection with reality.

To summarize these and other unpleasant features of system dynamics we may say that the temporal behaviour of solutions is strongly model dependent; that it is difficult to say whether one has choosen a relevant set of variables, and that the form of the couplings is not at all evident; that initial conditions are rarely well known, and that chaotic solutions destroy every possibility to make "predictions"; that non-measurable quantities may be important, and that differential equations are an excellent tool to hide any bias of the system analyzer (even to himself).

Clearly, system dynamics also has its merits. It helps with the formation of concepts; it reveals possible scenarios and tendencies; it gives a classification of dynamical behaviour; it shows very clearly the "non-logic", the counter-intuitive behaviour of complex systems; and, of course, there are also systems for which this approach is appropriate (circuits, chemical reactions). Furthermore, general results are very useful (e.g., catastrophe theory, Hopf bifurcation, slaving principle, scaling laws, dissipation-fluctuation theorem, etc.).

2.2 A Structural Approach: The Sensitivity Model

Problem solving requires understanding of the system. But of complex systems we have only a limited knowledge; and system dynamics is only a tool to investigate system behaviour in general, without giving us the necessary answers in the individual case. We may now ask two questions:

How can we extract relevant information about the system?

And: Who could give us good criteria for system design?

A possible answer to the first question is: from the structure of the system. Thus, we are not so much interested in the detailed temporal behaviour of the system, nor in all of its elements, as in its global cybernetic properties, such as its adaptive and self-regulative abilities. Without knowing the detailed dynamics we may get an overall picture of the system from its structural properties, e.g. the ratios

$$\frac{\text{number of connections}}{\text{number of elements}} \quad ; \quad \frac{\text{number of negative feedbacks}}{\text{number of positive feedbacks}} \quad ;$$

or the classification of elements with respect to their rôles in the system (active, passive, critical etc.), as it may be obtained from an influence matrix of interdependencies.

A possible candidate as an answer to the second question (who gives us criteria for system design?) is Nature. With over four billion years of experience, with an enormous variety of species of a most varied kind, with her gigantic processing of materials, energy, and information, with her ability to do all this without having problems with raw materials or waste, with her ability to cope with environmental changes, Nature may well serve as a paradigm for system design.

Thus, the idea is not to look for defects and malfunctions of a system and to cure them ("repair service behaviour"), but to design systems in a way that malfunctions do not occur.

Admittedly a structural approach would be rather dangerous without the guidance of Nature. Roughly one could say that interfering with the elemtns of a system may cause a temporal damage, but interfering with the structure of a system may endanger the system much more if we don't do it in accordance with the laws of biocybernetics.

On the basis of these considerations, one of the authors (F.V.) has developed a method for analyzing and designing complex systems. It has been called "Sensitivity Model" (see e.g. A. v. HESLER, F. VESTER[10]; F. VESTER[11]; MESSERLI[12]; Z. NAVEH, A.S. LIEBERMANN[13]).

We shall briefly review the method; for details we refer to the literature just mentioned.

Information, as MARUYAMA[14] says, is about objects, about relations, and about relevance. We may add a fourth quality, change. Hence, we first need information about the objects, i.e. we need a relevant set of variables to describe the elements of the system. Second, we have to find relations between the elements, which gives us the structure of the system. Third, we have to interpret and to evaluate the system structure; this is done according to general stability theory (mathematics, thermodynamics, synergetics) and to biocybernetics (see below). Fourth, we have to consider change; change in state (i.e., system dynamics, but used as a tool mainly to get a _feeling_ for the system), change in structure, and change in values (this is achieved by the feedbacks which are built into the method). The sensitivity method provides a frame for the whole procedure, without pinning the planner down to specific models (he may try out his own ones). The various steps are summarized in TABLE 2.

For the evaluation of the system structure, one of the authors (F.V.) has extracted eight biocybernetic design rules from Nature. They seem to govern living systems on every scale, from biochemistry to ecology. In TABLE 3 we list the eight rules in a brief form.

The goals of the sensitivity method are not so much directed towards a detailed understanding of the system, or to detailed recipies to cure some evil. The method rather aims at a global understanding of the system and its behaviour, at an evaluation in the light of biocybernetics, and at an eventual improvement of the system structure in order to meet better the objections of Nature. For instance, the question whether energy consumption should go up or down[+] is of secondary rank to the global value of _quality of life_ which includes e.g. survival, living standards, clean air, minimal social stress,

---

[+] In the course of evolution, species with a low energy consumption relative to the mass of the body have shown to have a better chance of survival.

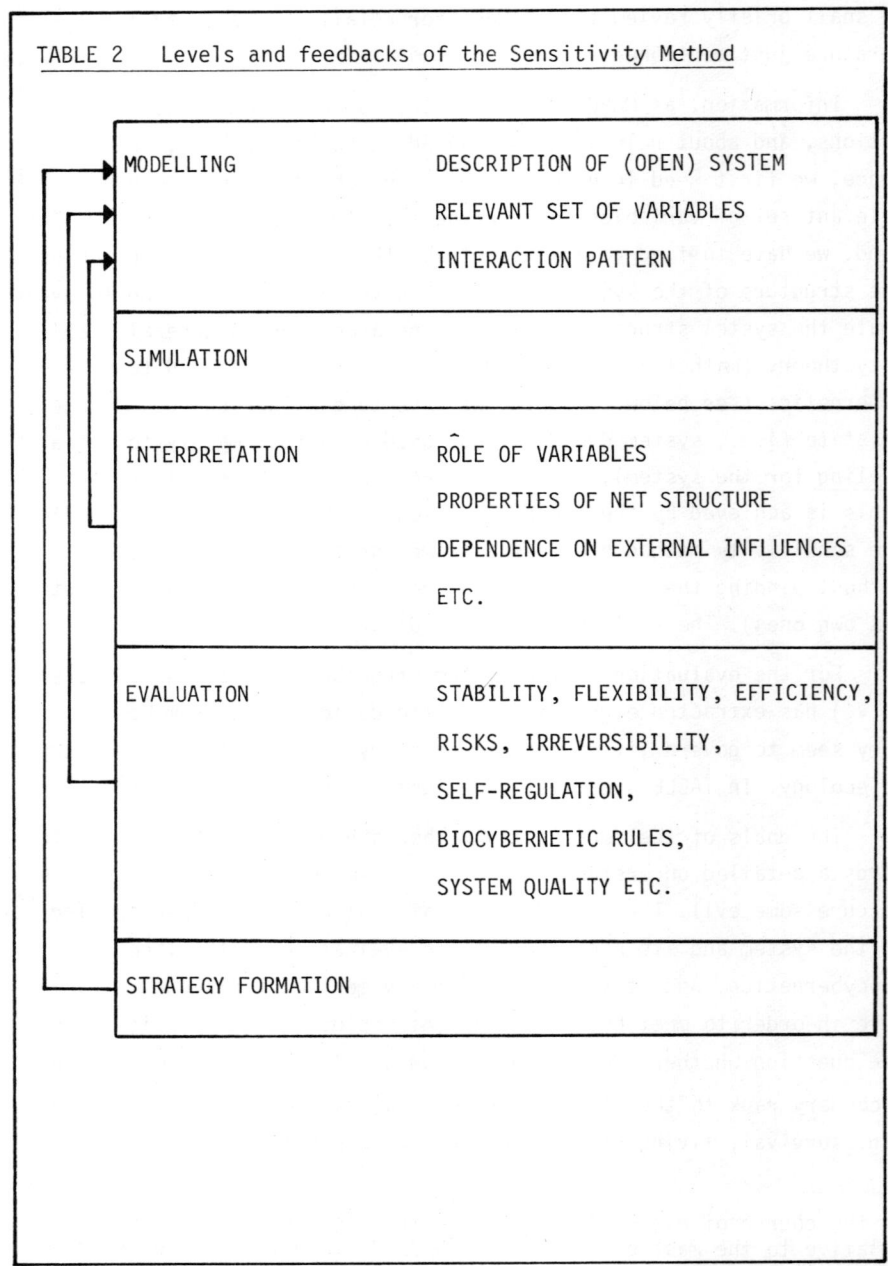

TABLE 2 Levels and feedbacks of the Sensitivity Method

MODELLING	DESCRIPTION OF (OPEN) SYSTEM RELEVANT SET OF VARIABLES INTERACTION PATTERN
SIMULATION	
INTERPRETATION	ROLE OF VARIABLES PROPERTIES OF NET STRUCTURE DEPENDENCE ON EXTERNAL INFLUENCES ETC.
EVALUATION	STABILITY, FLEXIBILITY, EFFICIENCY, RISKS, IRREVERSIBILITY, SELF-REGULATION, BIOCYBERNETIC RULES, SYSTEM QUALITY ETC.
STRATEGY FORMATION	

TABLE 3   Eight rules of biocybernetics

- NEGATIVE FEEDBACK DOMINATES OVER POSITIVE FEEDBACK

- SYSTEM FUNCTION INDEPENDENT FROM GROWTH

- SYSTEM FUNCTION ORIENTED (NOT PRODUCT ORIENTED)

- USE OF EXISTING FORCES (JIU-JITSU PRINCIPLE)

- MULTIPLE USE OF PRODUCTS, FUNCTIONS, AND STRUCTURES

- RECYCLING

- SYMBIOSIS

- BIOLOGICAL DESIGN OF PRODUCTS

Taking any of these values seperately will necessarily lead to unsatisfactory or even dangerous "solutions".

Let us now list some properties of the sensitivity method:
- It is a complete method, including modelling, dynamics, interpretation, evaluation, and strategy formation.
- It is a flexible method, encouraging the planner to try out his own models.
- It is an expert system, both with respect to the problem field and to biocybernetics.
- It is highly structured (net structure, level structure, feedback etc.).
- It is a transparent method.

- It gives an overall picture of the system, and it's aimed at qualitative improvements.
- It is an interactive method, treating the designer as part of the system.
- It uses a complex notion of time.

From these properties we conclude that the method is based on a non-classical, but consistent paradigm of time and logic in the sense of A. v. MÜLLER[4], and that it is ideally suited for an implementation as a computer aided planning system.(Independently also M.F. WOLTERS[15] arrived at the last conclusion.)

Before turning to a sketch of a possible computer implementation we shall briefly report on previous experiences with the sensitivity method. It has been applied to regional planning (e.g., the lower Main region, as part of the UNESCO programme "Man And Biosphere", and the Upper Rhine Valley, a three-country project), to development aid, to management, to large scale architecture, to ecology, to agriculture, etc. Up to now there are some 30 applications which have demonstrated the functioning of the method; (some references are given under[16]).

Some empirical facts found in these applications still lack a theoretical understanding; e.g. the equivalence of various sets of variables, provided they represent all important types; or the fact that very few variables may be used as indicators for the functioning of the system. A closer contact between the sensitivity method and the field of synergetics may be advantageous for a better understanding of these phenomena.

3. PLANNING A COMPUTER BASED IMPLEMENTATION OF THE SENSITIVITY METHOD

The rich structure of the sensitivity method suggests an immediate implementation as a computer system (i.e. a realisation as a code in a specific software and hardware environment); but it will have some unconventional features. It will be a kind of expert system (in various senses, as stated above), but it will be a dynamical, evolutionary system both with respect to the system under consideration and with respect to itself. It will even forbid a clear distinction between the outside world and itself. It will in general make no distinction between objects and subjects; it will treat the designer as well as the decision maker as parts of the overall system. The structure of the method guarantees transparency such that the decision maker may share the view of the system. As an output it will rather produce new structures, patterns, instead of endless columns of numbers. (For a more detailed discussion of the concept of structural design see a forthcoming paper[17] of one of the authors (J.B.).)

In order to make use of these features a number of computer aided facilities has to be added. We think of programming aids, automatic communication between users (know-how exchange and distribution), automatic evaluation of experiences, and automatic documentation.

We realize that giving a computer system the chance to change itself is considered dangerous; but such a feature is an inevitable consequence of the whole philosophy. Automatic documentation will allow for going back to previous states if necessary. Furthermore one ought to make sure that changes in the computer system should happen on a longer time scale as compared to the time of everyday work with it.

With the software and hardware available today it will be possible to realize such an implementation. Since we want to define and process structures we could think of LISP or PROLOG as possible languages. Whereas PROLOG is more suitable for representing facts, LISP is more suitable for representing procedural knowledge. Since also numerical simulations have to be carried out an interface to an algorithmic language (e.g., C) is needed. Similar properties are to be expected from the Intellicorp packages KEE and SIMKIT.

New languages may also unify the features of LISP, PROLOG, and conventional algorithmic languages. Thus, F. MÜNDEMANN[18] has defined and implemented such a language (called LANGUAGE) which uses a PASCAL-like notation. (Actually, F. MÜNDEMANN views a language as a formal descriptive system, and he sees the main problem in the implementation, possibly on a parallel machine, by means of an intelligent translating mechanism.)

As for the hardware, one would think of a SYMBOLICS, a SUN, or a similar machine; and for the interconnection between the users one could use existing or planned data nets (DATEX, ISDN). For documentation etc. it would be useful to have a mainframe computer as a host somewhere in the net.

TABLE 4 shows the problem structure of a possible implementation.

4. CONCLUSION

Hardware and software of available computing systems allow for the implementation of a dynamical, interactive expert system for investigation and planning in complex open system. Such a dynamical expert system should be based on the sensitivity method, which is a structural approach to systems and which uses biocybernetic principles for evaluation. It has successfully been applied in various fields in which system dynamics alone is not sufficient.

Such a computer based method would be of vital importance to a better understanding of systems, and hence to many problems of our time (e.g., quality of life) which are of dialectic type, i.e., we have only a limited knowledge about actual state, desired state, and resolvent transformation.

We have discussed the structure and the properties of the sensitivity method; we have seen that it uses a non-classical, but consistent picture of time and logic; we have looked into the possibilities of a structured implementation as a computer system; and we have argued that such an implementation of the sensitivity method can be achieved with existing means.

The best method for designing such an implementation is, of course, the sensitivity method itself; <u>reflexivity</u> is, after all, a natural property of the method.

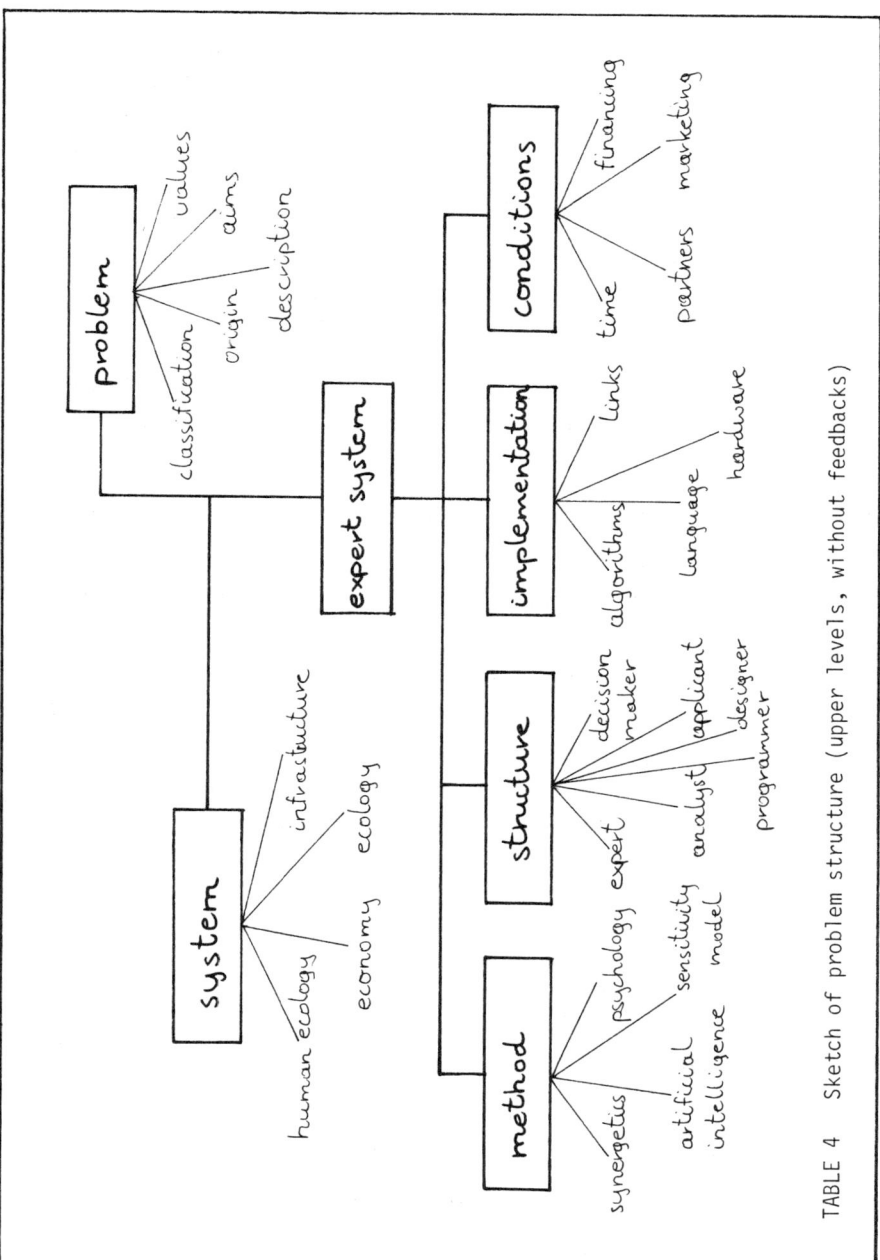

TABLE 4   Sketch of problem structure (upper levels, without feedbacks)

Since an implementation of the sensitivity method would greatly facilitate its usability it could be of great value not only for practical applications but also for basic research in complex systems, for instance, in the fields of brain research and artificial intelligence.

ACKNOWLEDGEMENTS

For one of the authors (J.B.) it is a great pleasure to express his warmest thanks to Prof. E.R. CAIANIELLO for his invitation to present this paper and for his marvellous hospitality.

The same author is also very much indebted to U. DOMPKE and F. MÜNDEMANN for most helpful discussions concerning the use of languages and machines.

APPENDIX: INTERPOLATION PROBLEMS AND PROLOG

Here we give a brief example for the use of PROLOG. In PROLOG one does not write programmes (in the sense of algorithms). Instead one specifies rules (corresponding to a set $\{t\}$ of primitive rules, cf. TABLE 1) and facts (corresponding to an initial state S_α). One may then ask questions, or set goals (corresponding to a final state S_ω). The PROLOG interpreter then tries to find a resolvent $R_{\alpha\omega}$, composed from elements of $\{t\}$, by working itself through the logics. Hence, PROLOG is a device for solving interpolation problems. Example:

rules: GRANDFATHER (x,z) ← FATHER (x,y) AND PARENT (y,z).
PARENT (x,y) ← FATHER (x,y).
PARENT (x,y) ← MOTHER (x,y).
facts: FATHER (EDUARDO, EVA).
MOTHER (EVA, NATALIA).

(Such a set of rules and facts specifies an expert system.) We may now ask questions to the system, e.g.,

goal: ? GRANDFATHER (NATALIA, EDUARDO)

In this case the system would tell us that it did not find a solution, but we have to find out ourselves that we should have asked a different question.

This simple example should not make us believe that programming in PROLOG is an easy task. On the contrary, one rather tends to make more mistakes than in other languages. Also problems which intuitively look rather simple may turn out quite difficult for a PROLOG programmer. Thus, modelling the kinship relations of bees[19] in PROLOG is much more difficult than it is for those of human beings (F. MÜNDEMANN, private communication).

For further details concerning PROLOG we refer to the literature[20].

REFERENCES

1) Popper, K., Unended Quest. Fontana 1976. (See in particular chs. 29 and 37).
2) Dörner, D., Problemlösen als Informationsverarbeitung. Kohlhammer 1976.
3) Osterloh, M., Handlungsspielräume und Informationsverarbeitung. Huber 1983.
4) v. Müller, A., Zeit und Logik. Wolfgang Baur 1983; and: v. Müller A., On the Way Towards a Complex Notion of Time. In (22).
5) Abkehr von der Atomenergie? Folge 1. SZ vom 7./8.6.1986.
6) Vester, F., Neuland des Denkens. dva 1980. (See in particular ch. 9, passage "Nutzung biologischer Systeme".)
7) Clarke, M., The Development of an Environmental Simulation Game. Working Paper 208, School of Geography, University of Leeds (1977).
8) Hofbauer, J. and Sigmund, K., Evolutionstheorie und dynamische Systeme. Parey 1984.
9) Vester, F., Ansätze zur Erfassung der Umwelt als System; and: Vester, F., Zukunftsprognosen, Modelle, Strategien. In: Buchwald and Engelhardt (Eds.), Handbuch für Planung, Gestaltung und Schutz der Umwelt (vols. 3 resp. 4). BLV Verlagsgesellschaft.

10) v. Hesler, A. and Vester, F., Sensitivitätsmodell/Sensitivity Model (German/English). Regionale Planungsgemeinschaft Untermain, Frankfurt 1980.
11) Vester, F., Cybernetic Principles of Organization. In (21).
12) Messerli, P., Memorandum der Programmleitung über Modellarbeiten in MAB-Programm. Geogr. Inst. d. Univ. Bern 1980.
13) Naveh, Z. and Liebermann, A.S., Landscape Ecology - Theory and Applications. Springer 1983.
14) Maruyama, M., Metaorganisation of Information. Cybernetica $\underline{4}$ (1965).
15) Wolters, M.F., Intelligenztechnologie - "Artificial Intelligence" Methoden - Eine Chance für den Industriepraktiker? In: M. Giloi, M. Schulz-Vorberg (Eds.), Intelligenztechnologie. Teubner 1983.
16) Vester, F. et al, Systemstudie Ökoland (4 volumes, 1983-1986). Arbeitsberichte aus dem Institut für Interdependenz von Technik und Gesellschaft. Fakultät für Sozialwissenschaften, Universität der Bundeswehr München;
Müller, H.R., Tourismus in Berggemeinden: Nutzen und Schaden. Schlußberichte zum Schweizerischen MAB-Programm Nr. 19 (1986); and J. Combe, Ch. Frei, Die Bewirtschaftung des Bergwaldes. Schlußberichte zum Schweizerischen MAB-Programm Nr. 22 (1986). Bundesamt für Umweltschutz, Bibliothek Bern.
17) Becker, J.D., Structure and Parallel Processing. In (22).
18) Mündemann, F., Erzeugen von dynamischem Maschinencode für Datenflußrechner mit Hilfe attributiver Grammatiken. In: Conference Report 9. Jahrestagung der Österreichischen Gesellschaft für Informatik. Berichte aus den Informatik-Instituten. Fakultät für Mathematik und Informatik, Universität Passau 1986; and:
Mündemann, F., internal memorandum. Fakultät für Informatik, Universität der Bundeswehr München.
19) Wickler, W. and Seibt, U., Das Prinzip Eigennutz. dtv 1983.
20) Clocksin, W.F. and Mellish, C.S., Programming in PROLOG. Springer 1981. And:

International Symposium on Logic Programming. IEEE Computer Society 1984.
21) Becker, J.D. and Eisele, I. (Eds.), WOPPLOT 83. Proceedings Springer 1984.
22) Becker, J.D. and Eisele, I. (Eds.), WOPPLOT 86. Proceedings to be published by Springer.

WHY BUILD MODELS

A.Borsellino
SISSA-International School for Advanced Studies
ICTP-International Center for Theoretical Physics
34014 - TRIESTE - Italy

As an introduction to the general theme of the present discussion I will briefly describe reasons and objectives of the practise of model building, in physics as well in other fields of science. Very probably what I am going to say will not appear new to you, but it is some time good to say explicitely, to pronounce rules and principles that are followed in our everyday activity. Furthermore we can show how to express quantitatively our intuitive judgements on the goodness of a model.

The relevant concepts and extimates are better expressed if we adopt the general terms of the theory of information. Let first take notice that the external world, from which we derive most of the informations in which we are interested (and that are necessary for our survival) is full of redundancies: they are identified in various forms, as permanence of objects, continuity of line and surfaces, invariance of properties or, more generally, as correlations. As a consequence, the messages that we collect to construct the "internal" map, necessary to guide our activity, are not encoded in the most efficient way. The symbols, forming the messages, will not carry the maximal possible information. We can therefore state that the average information per symbol $H < H_{max}$ and the redundancy $\mathcal{R} = 1 - H / H_{max}$ is often very large.

To make a specific example, in which we are certainly interested in a very direct way, suppose that we are making a measure and that as a result of the measurement we succeded to reduce the uncertainty of the measured quantity from the previous value, say within the range X, to the residual value given by the experimental error S. Then the information we gain with the measure is given by $I_1 = \log_2 (X / S)$ (Brillouin, 1956).

When we make n measurements, the information obtained with them is the sum only in the case that they are statistically independent. In all other cases the total information is less than the sum, being reduced by existing correlations between the data, by a "law", a regularity, etc. What the model is supposed to do is to make explicit the statistical dependence of the data, in such a way to get rid of the redundancies and represent the available informations in the most efficient code. The new code is using the "parameter space", the parameters that are used to fix the model. It is clear that a model is efficient or powerful if it can represent the available data with a small number of parameters. Naturally this imply that we are using statistically independent parameters.

We can illustrate what has been said with a very simple example. Suppose that we have measured n values of a dependent quantity y and that they are $y_1, y_2, ...y_n$, corresponding to the n values of the independent variable $x_1, x_2..$, x_n. We now try a representation of our data that will appear quite silly and such that nobody would consider it as a useful model. To make things clear we choose the simplest representation, a polynomial of degree n - 1:

$$y = f(x) = a_0 + a_1 x + a_2 x^2 + ... + a_{n-1} x^{n-1}$$

The n coefficients $a_0, a_1, ..., a_{n-1}$ of the polynomial can now be determined in such a way that the function $f(x)$ will pass exacly through the n points of coordinates (x_i, y_i). What we have accomplished with this operation? We have only substituted the n original data y_i with the n values of the coefficients a_j. We changed the representation of data, transcripting them from the code "y" to the code "a". We did not make any improvement or any tentative in "data reduction".

The operation "model building" starts as soon as we try to represent the data using a polynomial of degree m smaller than n - 1. We have now to take in account (or better to make use of) the fact that the data have an experimental uncertainty. Therefore if the "fitting " function cannot be adjusted to pass exactly through the "points", it is enough if it passes through the "windows" defined by the experimental errors. The operation "data reduction" is more successful if we can fit the data with a smaller polynomial degree: it indicate that the polynomial code "a" is efficient and can eliminate a large portion of the redundancy from the original data "y".

To make a rough extimate: suppose we have 10 values y and we are satisfied in representing them with a quadratic parabola, it means using 3 values of the "a". Without involving in the discussion of error propagation, we can extimate that the redundancy of the data was 70%. If the coefficient "a" are independent we can conclude that the data in the new code are mostly free of redundancy.

In exploiting the art of model building we very often are prone to take risk being thrifty, pleading for the most economic of the models. The risk represented by a poor fitting must be balanced against the benefits we obtain or expect. Remember the twenty years that Newton let pass before he got the new measure of the Earth diameter that fitted at his satisfaction the gravitational model he was developing.

The "cost" of the imperfect fitting can be extimated by some global discrepancy function like the χ^2 and the balance between benefit and cost can be evaluated for example with the ratio

$$\Delta \mathcal{R} / \chi^2$$

or using more conventional extimate like the Fisher function $F_{n,k}$.

Once a model has been introduced,it can start its way from the more modest labeling of "phenomenological " one , towards the more exacting status of "theory " . Along the route it will be pushed by its ability to represent new data or by its "predictive " value ,a test that is very important from the psychological point of view.

The psychology of discovery enter the art of model building in the decision process by which we choose to "exclude" (cancel,forget) some results as due to "spurious" disturbancies (a conservative attitude) , while in other case we are ready to shout aloud the "aha aha" alarm,at the risk to end up with a paper accepted by the "Journal of Irreproducible Results".

The conflict between the two tendencies confirms that our search for models of increasing power towards a well extablished theory has its roots in the complexity of our subjectif psychology.

References

The literature on this topics is quite vast : I limit the citations only to a few references that take the same informational approach :

Brillouin L. , Science and Information Theory ,Acad.Press (1956) .
Rothstein J., Communication,Organization,and Science,The Falcon's Wing
    Press (1956) .
Borsellino A. , L'Informazione ed il Progresso della Scienza,in :Il Concetto di
    Progresso nella Scienza (E.Agazzi,ed.), Feltrinelli,Milano (1976).

SOME PROBLEMS IN BRAIN SCIENCE AWAITING
THEORETICAL TREATMENT

Valentino Braitenberg

Max-Planck-Institut für Biologische Kybernetik,

Spemannstrasse 38, D-7400 Tübingen

Theoretical brain science includes activities ranging from the development of new mathematical techniques for discrete neural nets to philosophical speculations on the nature of consciousness. These all have one thing in common, namely that they are met by the majority of laboratory dwellers with the feeling that such works may be amusing to the theoretically inclined but rather unnecessary in practice.

This is often the attitude of people who refuse to call their own thinking theory, while insisting on calling other people's theory speculation. What can be learned from them is, however, a distinction between "primary" theories arising in the heads of an experimenter and "secondary" theories which are elaborations of the primary ones by people who have no first hand acquaintance with reality. There is a difficulty here. The pure theorist may tend to look down on the experimenter, whose thinking he considers amorphous. Nevertheless, he is dependent on the abstractions produced by such amorphous theorizing since these (and not the experimental findings) are the starting point for his own thinking. It is important to distinguish between the two kinds of theory, primary and secondary, for though each has its weaknesses, a combination of the two is even worse. It allows facts to turn into abstractions and vice versa and is no doubt the main reason why brain theory is met with suspicion in certain circles. I believe that the essential role of theory in brain science will be recognized as soon as theoreticians learn to produce papers which experimentalists (a) are unable or unwilling to write themselves, (b) are able and willing to read and (c) recognize as necessary steps for progress. I should like to make a few suggestions on the nature of such papers.

First, the entire project of theoretical neurosciences hinges on the acquisition of the data as near as possible to their origin. Whenever possible, theoreticians should be present in the laboratories during experiments, where they are no doubt welcome, provided a certain code of conduct is respected. This includes refraining from extemporaneous interference with the experiment, but does not forbid careful checks on the entire data gathering procedure. Such checks are important in microelectrode electrophysiology, a technique which relies on surgical procedures, therefore implying many on-line decisions on the side of the experimenter. Such decisions are known to introduce subjective expectation into the sampling. The resulting bias must be estimated by a critical observer versed in epistemology, such as a theoretician should be.

With or without direct insight into the laboratory, the theoretician should acquaint himself with original experimental reports in the largest possible measure. As it is, critical reviews of a field in neuroscience, mostly commissioned by journals, are invariably written by experimental scientists, and often carry the imprint of their unwillingness to "steal time" from their proper occupation in the laboratory. The bias (which makes a review readable) is one generated by the ambitions of competing laboratories. Such a bias may not be compatible with a balanced appraisal of the available facts and does not help the reader much either. I suggest that one of the most important tasks for theoreticians in neuroscience could be the systematic analysis of the experimental literature and the writing of review papers which assemble the facts around conceptual models. This does not guarantee a balanced appraisal of facts either, but the bias in this case is more interesting and likely to generate theoretical discussions and experimental questions. Such reviews would surely be met with gratitude on the side of the experimentalists, who would find themselves freed from a chore and would enjoy seeing their labours contributing to an intellectual discussion. The inclination to read and quote other people's papers being related to the pleasure of finding one's name among the references, the fame of theoreticians would undoubtedly profit from their own thorough digestion of the experimental literature. They could thus also counteract the criticism that they work on the experimentalists' abstractions, rather than on their facts.

Next I would suggest to the theoreticians some well known problems of brain science in which the lack of intuitive, naive solutions is apparent even to those experimentalists who otherwise have a high opinion of their own unaided thinking.

The problem of serial order in behaviour, pointed out by Lashley many years ago and as fresh today as ever. This problem, although arising in the study of all animal behaviour, is particularly striking when one considers certain human performances. Rats, cats, monkeys etc are all able to produce fairly rigid sequences of motor acts lasting a few seconds. Humans (and perhaps whales) produce highly ordered sequences lasting many minutes, such as playing a piece of music by heart or reciting a long poem. There are no moving parts in the brain, no tapes or disks as carriers of information allowing a spatial arrangement to be decoded by means of mechanical movement into a temporal sequence. The sequence of internal states in the brain must be entirely enacted by the rules of succession of sets of active neurons and these rules of course are embodied in the pattern of their synaptic connections. The high reliability of this system is astonishing: errors are fairly rare in speaking and especially rare in musical performance. We had a demonstration of this at our meeting: a full length piano recital without a single detectable mistake. It is likely that such sequences are governed by a few square centimeters of cortex, and there is no indication of time being represented there by a spatial coordinate: no one-dimensional progression in the succession of active neurons. On the contrary, the same neurons are presumably activated over and over again in different combinations and the state following a given state of activation must be determined by the activity of sets of neurons. All of this in a network composed of notoriously unreliable elements in which signals are exchanged which, as far as we know, are mostly bursts of action potentials with no obvious temporal order. It is difficult to say more. The theoretician who proposes an explanation which does not run counter to any established neurophysiology and anatomy will undoubtedly earn great respect.

The problem of shifting coordinate frames. Again, this is a serious problem, mainly due to the lack of mechanically movable carriers of information in the brain. The observations to be explained are of this nature: there are many instances of topographical correspondence of points in the brain with points of some sensory (visual, tactile etc.) or motor (the array of muscles in the body, the array of points to be reached by movements of the limbs) spaces. In many cases the internal representations of outside space in the brain have been anatomically well studied and usually appear to preserve the neighbourhood relations, if not the metric of the original array. Now, much of the activity of the brain involves corresponding points of different maps: an object seen at a certain position on the visual map may become the target of a movement directed toward the corresponding position on the motor map; a source of sound localized (through well-known mechanisms) on the acoustic map is immediately identified with a certain object occupying the corresponding position on the visual map; the movement of the entire visual panorama due to a change in the position of the head is identified with the corresponding movement recorded in the "vestibular map" (subserved by the inner ear) and the "proprioceptive map" (subserved by the sensors in muscles, joints and skin). The peculiar performance of the nervous system to be explained here is the adjustment of the point-to-point relations of the different maps which are necessary to compensate for certain deformations of the body. Rotation of the head on the neck requires a shift of the map governing movements with respect to the visual map, a feat which is accomplished with such ease that we are never aware of it. Similarly, rotation of the eyes in their sockets produces a visual input which is indistinguishable at the retinal level from that produced by rotation of the whole body, but does not produce any sensation of rotation. Apparently the expected visual input is somehow subtracted from the actual one, so that no conflict with the vestibular input is generated. Generally the visual and vestibular input cooperate with great accuracy, and discrepancies between the two (for instance when during a rotation of the body the eyes fixate an object so that no shift of the retinal image is produced) are always interpreted correctly. Again we are not aware of the automatic adjustments between the visual and vestibular maps, except in extreme situations when we meet the limits of the system's performance, sometimes experiencing unpleasant illusions such as dizziness.

Here the amateur theorist is at a loss. Rough qualitative models are insufficient to explain the phenomena at the level of neural performance. The theorist should seize the opportunity to theorize without any preconceptions of a dubious nature.

Statistics and geometry of neuronal connections. This is the field where in my own research I frequently feel the need for (and sometimes was lucky to avail myself of) professional theorizing. Apart from biochemical diversification, which is a different matter, the neurons in the neuropil, i.e. in the densely packed synaptic tissue, have widely different shapes. Since the signal processing seems to be basically very similar in all of them, we must assume that the properties of neural networks are somehow connected to the distribution of dendritic and axonal cell processes in the tissue. Both show great variation in density. In different regions of the cerebral cortex we find the main neuronal type, the pyramidal cell sometimes with dendrites forming a thick brush in the vicinity of the cell body, and sometimes with very long, sparse dendrites. Even more varied is the distribution of axonal ramification, where the density (percent contribution of any one neuron to the axonal network in it vicinity) varies by several orders of magnitude. The shape of dendritic and axonal trees also varies a great deal, being sometimes almost precisely globular in outline, sometimes very irregular and sometimes, as in the pyramidal cells, regularly composed of two portions with different geometry. We get the feeling that the genetic message which an individual neuron encodes takes the form of a probability distribution describing the location of a few thousand afferent synapses (on the dendrites) and as many efferent synapses (on the axon) in a coordinate frame anchored at the location of the cell body. Note that the total density of synapses in the neuropil is fairly constant; it is only the contribution of individual neurons to this density that varies. According to this picture, every neuron is a device which computes the convolution of the excitation in the tissue with the geometry of its own dendritic tree, transforms the result by some function related to its membrane properties and then contributes excitation (or inhibition, roughly negative excitation) to the original distribution according to the geometry of its axonal ramification.

This situation is appealing to those who like computer simulation, provided they have the necessary computer space at their disposal. But it also poses many questions to be tackled by analytical methods. What geometries of dendritic ramification are compatible with what distributions of cell bodies if we want to fill the space compactly, or to fill it taking into account also the space occupied by the axonal ramifications of the same and of other neurons? What is the average length of the fibers which connect all points of a folded surface to each other, or conversely (and more concretely), given the volume and density of the fibers of the white substance in the cerebral hemispheres, with what pattern of internal cortico-cortical fiber connections is this compatible?

These and many other such questions may not be aesthetically appealing to the mathematician, but they go beyond the possibilities of the practicing neuroanatomist and they are certainly relevant.

Further problems. It would not be difficult to produce a much longer list, but the theorists, once they get into the habit of reading the original literature, will undoubtedly get ideas of their own. One should warn them against involving themselves in theories for which an adequate experimental basis is simply not available. How is the geographical map which governs the navigation of the migrating bird encoded in the brain? In what form are the programs laid down in the brain which underlie the complex sequence of motor and perceptive acts which occur in mating or aggressive behaviour? And what is the mechanism which is present in human brains so that they can learn to speak, but which is apparently absent in animals?

Strangely enough, the "higher cognitive functions" may well be understood before such down-to-earth mechanisms as mating or even walking can be approached. The principle of associative memory is well understood and takes care, it seems, of most of the semantic difficulties that for some time were quoted as the real obstacle in any model of language and of thinking.

Dedicated to *Prof. E.R. Caianiello* on his 65th birthday: a token of our old friendship.

ROBERTO   BUSA   SI

PHYSICS AND METAPHYSICS OF COGNITIVE PROCESSES

Summary

Introduction :   1 -
             :   2 -

Chapter 1   :   §1 - The starting point: linguistic analysis
            :   §2 - my credentials
            :   §3 - kinds of done researches.

Chapter 2   :   The two ends of the expression operational arc:
            :   §1 - speaker - hearer
            :   §2 - thinking - talking
            :   a̲ - method
            :   b̲ - what's "concept"
            :   A - mind and text are heteroclytous
                      - the parameters of "intelligence"
            :   B - its hardware the "I"
            :   C - linguistic symptoms for the A diagnosis
            :   D - appendix: about "spirit".

Chapter 3 : Intelligence - Reason Interfacing
: §1 - in language: the message - the carrier
: §2 - the language carrier is the "first" information
: §3 - personalising "ontology" generative of all acting, thinking, talking
: §4 - in all lexica specific words and universal words
: §5 - the latter are basic in the former
: §6 - this inner intelligence connects logically the known-present-particular-visible to the unknown-hidden-possible-invisible-whole, pointing to the more, all, always, and to the new, beautiful
: §7 - this inner logic is the starter, the undeniable first.
: §8 - the other face is reasoning, deducting, computing
: §9 - consequently physics-metaphysics round-trip.

Chapter 4 : Appendices
: §1 - first intelligence cannot be delegated to an instrument; reasoning can
: §2 - the only reasonable confuting
: §3 - some epistemology from a Dante's metaphor and from .... Nietzsche.

INTRODUCTION

1. Trying to use a 'common' vocabulary.

2. The theme has three elements: cognitive processes, physics, metaphysics.

3. I define physics roughly, as the science of everything experimental and measurable.

4. Metaphysics is a science behind-after the science of physics.

   Metaphysics is a science of what is behind-beyond-after the experimental.

5. Both physics and metaphysics are activities only of the animal which we call man.

6. 1 billion human beings (average 60 Kg.) would fill a cube with side of 390 m; 4.5 billion would fill a cube with side 646 m.

7. Physics - Metaphysics: flags of contending academic weltanschaungen.

8. In all activity
 — Exercise of general energy,
 — with a specific content
   = a specified target

9. — Starting from something ⟨ source / transmitter
   — Arriving to ⟨ itself / another

Active : passive = sender : receiver = producer : product.

10. Metaphysics ⟨ Another cognitive process, different from physics?
                 Or same cognitive process on a different object?

11. Exit to : Knowledge ⟨ Sensitive
                         Intellectual ⟨ Superior intelligence
                                        Reasoning

12. Physics ──────── = Cognitive processes
    Metaphysics ────

    Our theme : Cognitive process of cognitive processes
              = Physics of physics
                         of metaphysics

              = Metaphysics of physics
                            of metaphysics

13. Γνῶθι σαυτόν

14. Two parts: 1. Cognitive process - physics - metaphysics: round trip
              2. Inner interface: personal "metaphysics".

15. <u>Chapter One:</u> Where to start a scientific inquiry on cognitive processes?

16. §1 <u>Linguistic analysis as a starting point</u>

17. Study the talking before studying the knowledge.

    An inquiry on inner cognitive processes must start by inquiring into the speech and texts of human discourse.

    How large is the mass of these physical events speech and text?

18. Words = strings of phonemes and graphemes =
    = physical entities and events: experimentable
                                              measurable
                                              publicly documentable.

19. Computerised linguistic analysis: A: Count and classify by computer all elements of human discourse as physical entities only.

    B: Then count and classify them semantically:

20.   i.e. parametrically and progressively from the bottom = the minimal, common, first and proper lexical unit signified by a word.

This basic nucleus of meaning is taken from "the context"= i.e. human speech or text = a. the text itself b. the vocabularies of that language c. the personal mastery of the same language.

21. Chapter One: §2 'My credentials'

2M	text lines	⎫ processed by computer
15M	words	⎭

- 2M text lines ⎫
- 15M words ⎬ processed by computer

16 languages   arminian, greek, latin, catalan, english, finnish, french, german, italian, portuguese, russian, spanish, arabic, aramaic, hebrew, nabataean

8 alphabets   arabic, arminian, cyrilic, gothic, greek, hebrew, latin, phonetic

450M records in i/o, 80 to 500 bytes each

21M text lines   photocomposed in 70,000 pages

of the 56 volumes of the I.T.:

- 50 statistical tables in 10 volumes

- 4 concordances in 49 volumes:
  2.5 + 6 + 0.6 + 1.2 M key-words
  = contexts

- in 7 volumes the republished Opera Omnia.

22. Chapter one: §3 "main types of completed or current researches"

In each or some texts:

- pre-editing
  - the graphics system
  - the typologies
- digitising and many cycles of correcting
- analysing each line into text elements
- enumerating and recapitulating all graphic types of text elements
- lemmatising (minimally)
- thematising
- formatting the 2/3 levels of the lexicological system of the text
- statistical morphematic segmentation
- statistical classification by semanticity types
- grouping by elementary semantic correlations
- searching the "syntagmas"
- censuses of latin endings in grammar, Forcellini, I.T.
- census of all flexional categories, of all classical latin lemmas
- etc.

23. Chapter two: the heteroclytous extremes of the expressional arc

24. two arcs: speaker-hearer and thinking-talking

25. § 1. speaker - hearer dialoguing

26. a very complex process

27. the paradoxical activity of symbolising = assuming and subsuming "signs" = "signifiers" to express himself:

    $\underline{A}$ to $\underline{B}$ via the sign $\underline{C}$ communicates an information $\underline{D}$:

    $\underline{C}$ is neither $\underline{A}$ nor $\underline{B}$ nor $\underline{D}$ but "carries" $\underline{D}$ following an instinctive vitally evolutive social pact.

28. Transferring information man-to-man is not mechanicistic.

    It is not imposing but proposing.

    Each one can examine and criticise all received information before accepting it and personalising it.

    (note on the contrary the imperative force of any idea which he personally has found...)

29. Chapter two: §2. The arc thinking - talking

   the arc thinking-talking = knowing-expressing = concept-word = mind-language ... operates within both the speaker and the hearer.

30. a. Methodologically: can the structures of the concept (source) be inferred from the structures of the expression (target)?

   Can we know some of the specific features of an author having only his works to hand?

31. b. Good question: what's a concept?
   information, idea, thought ...?

   everyone knows what it is to know, everyone knows what a concept is ...
   here "to know" =   to use, to exercise

   but to define it = to byte your teeth with your own teeth, to catch your hand using only it, to look into your own eyes directly without having recourse to a mirror....

32. to myself I describe an idea as follows:

   -   "in me an of mine operative formula of another"

   -   or "that aspect which I find in two objects because of which I affirm their similarity"

   -   how much more than a photo?

33. -   (note that not all knowledge is a concept or an information... in additional there is the "experience" = "touch"
        - mutual-activities - of the presence of another = of yourself too:
        you and the other are not "ideas"!

        all presence is accompanied by some concepts, but it is not them ...
        concept are still in me when the presence is gone.

34. <u>A. hints on mind and text as heteroclitous</u>

35. Expression is not a train arriving as it departed ...
    Expression is generative = engendering = author ---> work

36. text is granular
             sequential
             linear

                              mind is intuitive
                                      multi-dimensional
                                      co-present

                         diffusely centred everywhere

37. text is fixed
             defined
             closed

                              mind is open
                                      generative
                                      unpredictable

38. text is programmed
             structured

                              mind is programming
                                      structuring

39. Mind

    =   free, creative engineering
            manipulating activities and signs.

            Aggressive, conquering power
            imposing itself on bodies
            dominating time and space

40. =   not reacting only to present stimuli
        much more motivated by the not-yet or no-longer existing.

41. =   always smelling realities behind the visible present,
        like a pig hunting truffles ...

42. -   (what other animal attacks an absent enemy?)

    -     (how many times do your children and wife ask you "why?"
        how many times do your pets ask you "why?")

43. Mind

    =   prefers fishing in the canals of the possible...

        the possessed existing is no more than an exciting
        hors d'oeuvre

44. Metaphors: mind as — operational system
              — orchestra conductor

   mind operation system seems to stand on four pilasters
   = parameters = features = forces = keys

   <u>Capacity of</u> locating and grasping a
   "<u>descriptive-formula</u>"
   (idea, concept, definition, essence,
   pattern, shape, species ....)
   of all the other.

   = capacity of building up theories and
   sciences .... schools and books ...
   writing treatises not only on cars but
   also on speed, on acceleration, on limits, on measures ...

45. itself subdued only by evident truth,
    is a <u>criticising and judging power:</u>
    arguing, contradicting, opposing ...
    (to be gregariously repetitious when "everyone else says or does
    or dresses that way" is considered a comfortable weakness)

46.   is the <u>power of inventing new
       organisations:</u>
       fine arts, technology, business, reli-
       gions, crimes ...

       this <u>idea of that whole</u> was never
       expressed before ... must be patented...
       it is not the mere sum of its compo-
       nents...

       this active organising power
       is a module directed towards
       embracing everything in one coup d'oeil.

       it is a wave rolling towards all the
       shores of the universe = of being.

47. in all the above mind shows a "maniac lust" for logic - proportion - harmony - beauty ...

   it seems constitutionally attuned to it:

   the oxygen vital to mind seems to consist of 3 atoms

                                   the true
                                   the whole
                                   the beautiful ...

   joy and love go with it
   always dreamed of, though so rarely found ...

48. B. What is the hardware of such an operational system?

   the indisputable answer for many will be: the brain ...

   I do not say "no": by examining it linguistically it comes out that:

   the hardware of the mind operating system is first of all the "I".

49. all thinking is implanted in the personality of each individual:

   - ideas are richness and values of each person
   - they shape it = each one is qualified by his mind

50. it is a fact that even all verbs of thinking and talking have a subject, at least a personal pronoun

   that is the undeniable fact
   even though it is difficult to explain

51. among the words, the most amazing of all are the deictic ones, and among those the personal pronouns "I, you, he, we, they ..."

   no one knows how his own "I" entered the world nor does he know the recipe of the cocktail which is his personality: nevertheless it is his most solid reality, his basic concern, his center of everything else.

52. obviously the problem remains: is the brain <u>an</u> organ or the organ of <u>the</u> personality?
   or is the brain the person?

53. <u>C. Hints on linguistic facts, which are "symptoms" of the above</u>

54. they can be summarised as the complexity of the talking phenomenon;
   it is tremendously heterogeneous:
   many various elements structures gadgets

55. information carried by talking is homogeneous only when formalised by the conventional abstraction of the mathematical information theory:

   such a theory packs everything into equal containers i.e. numerical symbols, and sums them up into a global weight, ignoring the specific variations of the content: gold or soya beans, gases or fabrics, carrots or snakes .....

56. - identical repetitions of types: see vocabularies

   - morphostructures: see grammars

   - syntax structures: see all branches of philology

   - from the "semanticity types" some basic universal categories...

   - a spelaeological investigation into the depths of the "parts of speech":
     noun, pronoun, article, adjective; verb, adverb; preposition;
     conjunction;
     interjection:
     all: irreductible - moveable - integrated

- multi-modal inflecting and composing

- taxonomies/of more general/more specific words

- the synonimity

- the metaphor

- in each word layers of meaning — morphological
  — syntactical

- sentence being the elementary unit

- circularity between the value of the sentence as a set and the values of each of its components

- the laborious process of writing: trying and deleting

- the process of translating: contrastive concordances and cross-referenced lexica

- etc.

57. Appendix: have we an immortal soul?

this thinking-talking operating human force:
- has been labelled as mind, soul, spirit;
- has been qualified as invisible - solid - unbreakable power "superior" to (i.e. commanding to and imposing on) bodily matter

58. our theme does not require to discuss it now, though allergies to it are facts requiring scientific investigation also

59. anyway the only experimental encounter with something deserving to be called spiritual, outside of "mystical" experience, is whenever anyone says "I"
(though for anyone else this "I" will remain something "behind").

60. But a scientific inquiry about "spirit" as the programmer of "matter", cannot have any other sound method than

    A. that of excavating the inner roots of all human behaviour: call it "metaphysics" if you wish

61. B. that of investigating and looking for the source-programmes and programmers (team and management) of all the programmes fed, loaded and embodied in all the realms of nature, human included:
call it metaphysics antonomastically.

62. Chapter three: Mind as Interfacing: Intelligence - Reasoning

63. § 1. <u>in all discourse there are</u> ⟨ a content = message / a linguistic vehicle

= ⟨ <u>what</u> I want to communicate / that <u>with which</u> I communicate it

language is a belt conveying messages

64. § 2. <u>language as conveyor belt is already a set of information</u>

65. I describe such information as the first logic of being,

    intuitive
    generative
    critical

66. such a logic generates the language as carrier, enabling to grasp critically and express all other information, like tongs - pliers - pincers con-prehend-ing everything.

67. § 3. <u>that bunch of first certitudes</u>

    is personal   not communicated from other people
    non memorised
    but permanent
    in - native - state

	innerly ignited by any friction with any other thing
is immediately intuitive indubitable	not reasoned, nor deducted
is "natural"	not deliberately acquired
	pre-cultural, i.e. like a pilot-light igniting all first startings of moving to know

68. it is personalising "ontology"

generating	all acting all searching all thinking all talking.

69. § 4. <u>the common - the specific in all vocabulary</u>

in all vocabulary many words express specific aspects of specific subjects	some words express general aspects, universally common to any subject
	e.g. I, you, he, one-many, whole-part, active-passive, starting-arriving .....
	i.e. all which is true of an object not because it is such or such, but just because it is something, i.e. a thing, a reality

70. § 5. <u>these universal values</u> cannot be separated from any of the
specific values but underlay basically each one of them:

              a car is something
              its speed is something
              its acceleration and
              direction are something ....

71. § 6. <u>these first certitudes</u>

        seal the personality of each man
        are the spark-plug of all his <u>conscious</u> vitality

72. they are the (superior) intelligence intuitive pushing to go logically:

73.

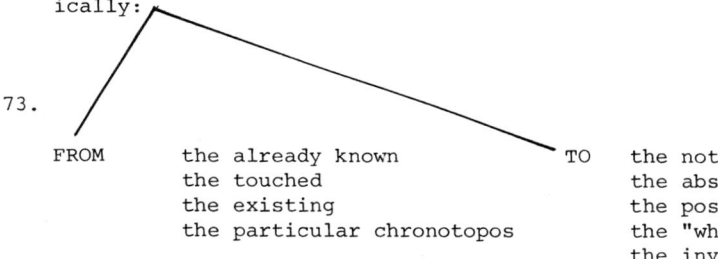

    FROM   the already known   TO   the not yet known
              the touched   the absent, hidden
              the existing   the possible
              the particular chronotopos   the "whole always"
                                                  the invisible

74. in all discourse they are the semantic carrier = the conveyor
belt of all other information

75. § 7.   this first intelligent logic is the "starter":

- there must be a "first": is it possible to think about a train without a first wagon?

- each "I" is a first and unique in the universe: unmultipliable .....

- only by using it, can someone deny the existence of intelligence as when someone would say "I cannot talk", "I do not exist" ....

76. § 8. the other face of that logic is when applied to everything specific

I summarise it into two levels:

- life
- culture

life is:

- personal performance in society
- all family affections
- inner conscience =====> prayer

in culture there is or there should be, a large use of that "reasoning" which I call the other face of intellingence:

- it develops all   calculating
                     deductive
                     cartesian logic

- those logics implied: in all mathematical and physical sciences

- in all operational, personal or social, activities: technology, business, politics etc.

77. § 9. in this way we have completed the round <u>trip between physics and metaphysics</u> of cognitive processes

we started classifying language signs as physical entities

we concluded that such physical entities have been produced by human inner power:

we called it intelligence:
we may call it the metaphysic force also

therefore metaphysics has been "discovered" from physics, but the physics of cognitive processes was originated by that metaphysic force: a cognitive process knowing itself.

78. Chapter four - appendices

79. § 1. "intelligence" cannot be delegated to an instrument:
- "to be the first" cannot be delegated ...
"reasoning", can:
all human manufacturings are "artificial reasoning":

80. § 2.

- therefore by talking all people exercise a "philosophy of being"

consequently in all philosophers there are two philosophies:

the one they intend to assert

the one they exercise in asserting the former

81. in principle no one could be said to be wrong just because contradicting the assertions of other people, not even those of God

    in principle someone is wrong only when (really) contradicting his own inner personal logic

    - which sure enough is identical in him, in others, in God

    - nothing can be found outside of being, different from it, or unconsistent with it ....

82. § 3

    Is so much effort really required to realise that sometimes some people manifest some signs of intelligence?

    I mean original intelligence, not that of gregariously repeating memorised strings of perhaps elegant words ....

83. § 4. <u>let me derive some epistemology from Dante</u>

    E sì come secondo raggio sole uscir del primo e risalire in suso, pur come pellegrin che tornar vole .... (Par. I 49-50)

take an electric torch in the dark of the cellar it will light up objects previously unseen =	point the torch at 90° to a mirror: light will u-turn and through itself point to its source
= reasoning to discover the unknown = sciences	= an emigrant returning home

84. when you scientifically (i.e. reading and writing words socially
    in a frame of public culture) apply your cultured reasoning to
    describe the first light of your mental life and its source,

    you are renewing those disciplines which
    were called rational psychology,
                 ontology,
                 natural theology.

85. when you just think and reflect personally upon yourself, about
    the roots and source of your thinking, you are light gazing into
    itself and questioning itself = certitude realising itself and
    asking itself "why?" ....

    that = meditation = prayer = dialogue with the Father ...

    =   eventually the goal of metaphysics.

86. § 5. Nietzsche said almost the same
        using less poetic and more proper words
        "Ich fürchte, wir werden Gott nicht los,
        weil wir noch an die Grammatik glauben ...."
        Götzen-Dämmerung Ed. Kröner 1964 p. 98

PERCEPTION AS A DYNAMIC INSTABILITY:
a Qualitative Model

Giuseppe Caglioti

Istituto di Ingegneria Nucleare, CESNEF-Politecnico di Milano, Via Ponzio 34/3, 20133 Milano, Italy

ABSTRACT

The process of pattern recognition during perception parallels the process of pattern formation in selforganizing open systems removed far from thermodynamic equilibrium [1]. Both these processes are cooperative, and exhibit dynamic instabilities entailing symmetry breaking.
A qualitative model is presented here according to which the perception of an ambiguous pattern occurs in two steps. First a dynamic instability sets in : the visual thinking falls into one of the two conflicting attractors built in the ambiguous pattern, thus emerging as an order parameter from the heap of disordered sensory stimuli suggested by the pattern to the mind, and collected in memory. Secondly a succession of alternating perspective inversion develops, analogous to that occurring in the charge transfer spectrum during the spectroscopic observation of binary molecules.

1. INTRODUCTION

The aim of this paper is to present an analogy between the perception of ambiguous figures on one hand and a sequence of two processes on the other hand: the synergetic formation of patterns of non-equilibrium systems in nature [1], [2], [3], followed by the spectroscopic observation of binary molecular structures in physics [4].
To this end we focus on the process of the perception of the ambiguous pattern at the bottom of fig.1. This pattern is obtained by a "graphical condensation" of two cubic moduli. Metaphorically, fig.1 can be interpreted as representative of the formation of eg the "hydrogen molecule" starting from two "hydrogen atoms" (the inner white square representing the "proton" and the outer black walls the delocalized "electron"). An introspective observation of this binary pattern allows to propose a qualitative model for perception, physically constructed on the sequence of the synergetic and spectroscopic processes introduced above [5],[6].

2. THE MODEL

Like all dynamic instabilities of physical and chemical systems, perception can be envisaged as a sudden transformation promoted by the action of an external control parameter. Perception occurs right at the moment when, starting from a disordered heap of sensorial stimuli proposed by the object to the mind of the observer, the object itself is recognized: this is the critical point of a dynamic instability: an idea of the object rises, that subsequently develops and takes the form of ordered "visual thinking". Most likely the critical moment of the nucleation of the idea leading to the visual thinking marks the formation of a singularity in the topological organization of neuronal and synaptic signals gathered in the visual cortical system. In this perspective the rising of the "visual" idea would occur at a bifurcation point where the disordered assembly of sensorial stimuli selforganizes under the control action exerted by the mind on the stimuli themselves: also in order to avoid that the stimuli escape from the memory, the mind assembles and correlates them in coherent schemes, and these schemes, suddendly, coalesce as a visual idea subsequently developing as ordered visual thinking.

In order to proceed in the analysis of the process leading to the dynamical perception of the binary structures in fig.1, four ingredients are needed [6],[7].

1. The sensory stimuli. They represent cooperating subsystems selforganizing during the process of perception.

2. The control parameter, c. It measures the control that the observer exerts on the visual stimuli suggested by the interiorized figure, and collected and correlated by the mind.

3. The attention potential, V. This is the analog of the free energy of a second order equilibrium phase transformation: it is a function depending on the order parameter q (see 4.below) and, parametrically, on the control parameter. V is a dynamic field characteristic of the interiorized figure. It is determined both by the morphological interactions impressed by the designer among the signs and symbols, and by the actual value of the control parameter c. c usually increases as the process of perception proceeds. In fig.2 the potential V corresponding to the lowest of the binary structures in fig.1 is plotted vs q for increasing levels of c. Note that as c overtakes a critical level c^* (fig.3), the stable attractor (the minimum at q=0) becomes an unstable focus, and two attractors arise. The visual thinking q precipitates in one of these two attractors (see 4. below)

4. The order parameter, q. This is the R.Arnheim's visual thinking. Again with reference to the lowest binary structure in fig.1, q initially vanishes at low values of the control parameter; afterwards, q selects the right hand (R) or the left hand (L) attractor once $c\star$ exceds c . Subsequently q will resonate with a certain regularity between these two minima of the potential field, producing the illusion of an endless sequence of perspective inversions.

It should be stressed that at present it is not possible to provide an operative definition of c, V and q and to indicate a procedure for measuring these quantities. From the physical point of view the difficulty met in assessing the operative definition of the above quantities represents a limitation of the model proposed here.

Coming back to fig.1, each of the six binary structures in it is composed of a pair of adjacent cubic moduli. Each of these structures, by construction, owns a center of symmetry: by inverting any element of the structure with respect to it, a structure identical with the original one is obtained: after all a transformation is a symmetry transformation if the structure remains invariant against that transformation: symmetry transformation is characterized by the indiscernibility of that transformation. We shall see that, though not explicitly marked, the center of symmetry is the most important point in these structures.

As anticipated in the introduction, two features of the process of perception or observation of the "hydrogen molecule" in fig.1 will be analyzed below:
1. The analogy between the process occurring in our mind, leading dynamically from disordered sensory stimuli to the coherent rise of the third dimension in a 2- dimensional drawing on the one hand, and the dynamic instabilities occurring in the dissipative structures of irreversible thermodynamics distanced enough from equilibrium so as to undergo a spatial, temporal and/or functional selforganization on the other hand [8].
2. The analogy between the endless sequence of perspective inversion from left to right to left, etc., and the charge transfer spectrum occurring in spectroscopic experiments on binary molecular systems that are submitted to a resonant electromagnetic radiation.

Let us first explore the mechanism of the transition from two to three dimensions. During the observation of the "hydrogen molecule" in fig.1, the mind collects in the memory and interiorizes the sensory stimuli coming from the figure. While controlling the initially scattered and disordered stimuli, the mind processes them and correlates them: the aim is to build synergetically a collective and ordered pattern around the center of symmetry of the structure.

Eventually there are **two** possible coherent patterns the stimuli can form: these patterns are mutually in conflict, and yet each one separately is fully compatible and consistent with the logic informing the operative system of our mind. Both patterns assign the central wall 50% to the lhs and 50% to the rhs cube. Both patterns are symmetric and correspond to stationary, time independent states of the system of the stimuli. But while for one of the two patterns, say the g (gerade or even) pattern, the center of symmetry belongs to both the lhs and the rhs cubes, for the other one, say the u (ungerade or odd) pattern, the center of symmetry does not belong either to the lhs or to the rhs cube (fig.3). Being stationary and symmetrical, the g and u patterns are not per se perceptible: they can be of help for describing this picture, they form the basis for its perception, but they cannot enter separately the act of perception itself.

In fact stationarity -i.e invariance with respect to time- is incompatible with the irreversible evolution in time of the process of perception: -**before** perceiving I didn't know, **afterwards** I know-. On the other hand, symmetry -i.e., invariance with respect to a transformation or, equivalently, indiscernibility of a transformation- can be discerned or perceived only while it is being broken. What happens then while the control exerted by the mind on the stimuli offered by the "hydrogen molecule" (the bottom pattern in fig.1) increases ?

This is a difficult question. In order to attempt to answer it, we could postulate that in any figure each sign contributes to build the potential field of our attention, or attention field, V. Once perception will have occurred, the visual thinking, q, tends to drop, tracing the valleys or local depressions of the attention field, formed by the signs, like a ball would do when constrained in a gravimetric profile. For instance, the attention field originated by a black circle in the middle of a piece of a white paper could be envisaged as a circular hole where the visual thinking is bound to drop. The black circle, as any sign, acts with regard to the visual thinking as an attractor of the attention.

Nevertheless the attention field proposed to the observer by an ambiguous structure like that of the "hydrogen molecule" in fig.1 is not immediately defined once for ever, but depends on the amount of the control c exerted by the mind on the ensemble of the signs building the structure itself. As proposed before, during observation the mind collects and controls the stimuli suggested by the signs, trying to organize them in coherent schemes or patterns, analogous to the quantum states g and u. Only when (fig.2), thanks to the increasing level of the control parameter, coherent though mutually contradictory patterns (e.g. g and u) are simultaneously formed in the subconscious, competing attractors (left, L, and right, R) develop where the rising visual thinking q tends to drop in the fully developed potential field of attention.

In other words, the field of attention, $V(q|c)$ (fig.2)

whose dependence on what will emerge as the order parameter or visual thinking q is characterized for c < c* by a single minimum at q = 0 , gradually takes the shape characterized by two symmetric minima. Correspondingly, at the beginning visual thinking disappears (q = 0). But, as the control parameter c exceeds a certain threshold c*, the two patterns g and u counterbalance each other in the mind. As this critical stage finally is reached, one could state:**by construction, the central wall belongs 50% to the left and 50% to the right cube.** Furthermore, (g) the center of symmetry does belong 50% to the left and 50% to the right cube **and, simultaneously,**(u) the center of symmetry **does not** belong either to the left or to the right cube.

The mind is thus trapped in an ambiguous state where two conflicting, unconciliatory patterns occur at the same time. This critical state of the mind is uneasy, the mind is anxious to escape through any possible way out. A small fluctuation, and it's catastrophy! A dynamic instability in the pattern of the stimuli occurs, similar to the bifurcations occurring, according to irreversible thermodynamics, in the pattern of the moduli of dissipative structures when far from equilibrium: the center of symmetry disappears, and, as symmetry breaks, visual thinking seems to lean towards the lhs or the rhs attractor. But visual thinking cannot remain fixed in any of these two positions. In fact as soon as coherent visual thinking of a 3-dimensional structure arises from the disordered stimuli offered by the 2-dimensional drawing of fig.1 , a dynamics of endless alternations from left to right to left prospects sets in.

We could explore the mechanism of these perspective inversions by referring to the spectroscopic observation of the hydrogen molecul. Actually, an analogy could be proposed between the dynamics of ambiguity in visual thinking and the resonant electromagnetic field in spectroscopic observation of this diatomic molecule: visual thinking concentrates on the central wall and drives it from the left to the right and then again to the left cube, much in the same way as the electromagnetic field drives the electron in the hydrogen molecule from left to right and again to left in a charge transfer spectrum experiment [4]. To the question: "where is the covalent electron actually located ? On the left hand or on the right hand proton ?", the official answer from spectroscopy is: one finds the electron where the probe -namely the electromagnetic field- carries it while searching for it.

The above analogies, like all analogies, are not perfect, and should be taken with a grain of salt.

To describe qualitatively a possible mechanism for the formation and development of visual thinking as an order parameter of the system of sensory stimuli is not the same thing as to have written the relevant kinetic equations and to have proved the reliability of these equations by a comparison of their predictions with the actual behaviour of the human brain. Furthermore, the time behaviour of the order parameter is affected in practice by conspicuous noise, so that the time allotted to the left and to the right attractors is not constant, but exhibits an average value with a lot of dispersion. The length of the residence times of the prospects depends not only on the pattern, but also on the individuals perceiving it. As such, the exchange frequency can only be defined as an average value over a distribution of reversal times affected by a remarkable dispersion [9],[10],[11].

3. CONCLUSIONS

This study has been stimulated by the increasing importance of images as communication tools between individuals and between cultures.

At school we learn reading, writing and arithmetic: but no consideration is given to the fact that, at least in western countries, nowadays most of the people spend most of their free time watching television. Overwhelmed by images, we hardly find any time to explore the structures characteristic of the images and the mechanisms through which the images shake our subconsciuous and control our mind.

The process of pattern perception and recognition is a cognitive process. Physics seems to offer appropriate tools to investigate it. In this contribution a qualitative model is outlined, where perception is interpreted in terms of a dynamic instability occurring in our mind. According to the model, immediately after the instability, during the dynamic evolution of the process of perception of ambiguous figures, conflicting prospects (such as convex and concave, left and right attractors etc) alternate following quantum logic rather than classical logic. Nevertheless, an intriguing but important feature of these alternations -not explained by the model- is the distribution function of the residence times of the alternating prospects: as it usually happens when dealing with complex problems, questions rise more numerous than the answer we pretend to give.

REFERENCES

[1] Haken,H. (edr.), Pattern Formation by Dynamic Systems and Pattern Recognition (Springer Series in Synergetics, Springer Verlag, Berlin, 1979)

[2] Teuber,M.L., Sources of Ambiguity in the Prints of Maurits C. Escher, Scientific American, 231 (July 1974) 90-104

[3] Nicolis,G.and Prigogine,I., Self Organization in Non Equilibrium Systems (Wiley, New York, 1977)

[4] Herzberg,G., Molecular Spectra and Molecular Structure. I - Spectra of Diatomic Molecules (Van Nostrand Reinhold Co., New York, 1950)

[5] Caglioti,G., Simmetrie infrante nella scienza e nell'arte (CLUP, Milano, 1983)

[6] Caglioti,G., The world of Escher and Physics, in Proceed. of the M.Escher Congress (H.S.M.Coxeter, R.Penrose, M.Teuber, M.Emmer, Edtrs.) Roma, March 26-28 1985 (North Holland Publ.Co. Amsterdam, (to be published))

[7] Caglioti,G., La catastrofe percettiva, in La Teoria delle Catastrofi (P.Bisogno, A.Forti, Edtrs.), (Franco Angeli, Milano, 1983)

[8] Caglioti,G., On Instability in Perception, in: Proceed. of the Theoretical Physics Meeting, (A.Giovannini, M.Marinaro, F.Mancini, A.Rimini, Edtrs.) Amalfi, May 6-7 1983 (Edizioni Scientifiche Italiane, Napoli, 1984)

[9] Borsellino,A., De Marco,A., Rinesi,S.,and Bartolini,B., Reversal Time Distribution in the Perception of Visual Ambiguous Stimuli, Kybernetics, 10 (1972) 139

[10] Borsellino,A., Carlini,F., De Marco,A., Penengo,P., Trabucco,A Riani,M.,and Tuccio,M.T., Stochastic Model and Fluctuations in Reversal Time of Ambiguous Patterns, Perception, 6 (1977) 645

[11] Caglioti,G., and Caianiello E.R., A Model for Non resolvable Ambiguities, Biol.Cybernetics 31 (1978) 205-208

CAPTION OF FIGURES

fig.1 F.Grignani, Graphical condensation

fig.2 The ensemble of the graphic signs is supposed to constitute a kind of potential for the attention field of the observer. In such a field the attention tends to drop, like a ball constrained along a gravimetric profile. In this figure the potential of the attention field V is plotted vs visual thinking q. In turn, visual thinking is interpreted as the order parameter of perception, and perception is conceived as a dynamic instability controlled by the mind. While the mind controls the sensory stimuli and correlates them, as the control parameter c increases, a selforganization of the potential field $V(q)$ arises, and beyond the instability at $c=c*$ (left and right hand) attractors arise. The rise of attractors is typical of open physical systems undergoing dynamic instabilities promoted by external flows of matter, energy or information. Once the attractors are formed, the length of time of visual thinking devoted to each of them seems to follow a flip-flop kinetics similar to that of a particle in a double potential well, described by quantum mechanics.

fig.3 The symmetric and stationary g and u patterns are schematically represented in terms of amplitude of probability of presence of the common wall (the "electron") on the cube (the "proton") at the left hand side or at the right hand side of the center of symmetry of the "hydrogen molecule" of fig.1. . In the u pattern the geometrical center of the common wall does not belong either to the left or to the right cube. In the g pattern the geometrical center of the common wall belongs both to the left and to the right cube. The g and u patterns are not per se perceptible: they can be of help for describing this picture, they form the basis for its perception, but they cannot enter separately the act of perception itself.

Fig. 1

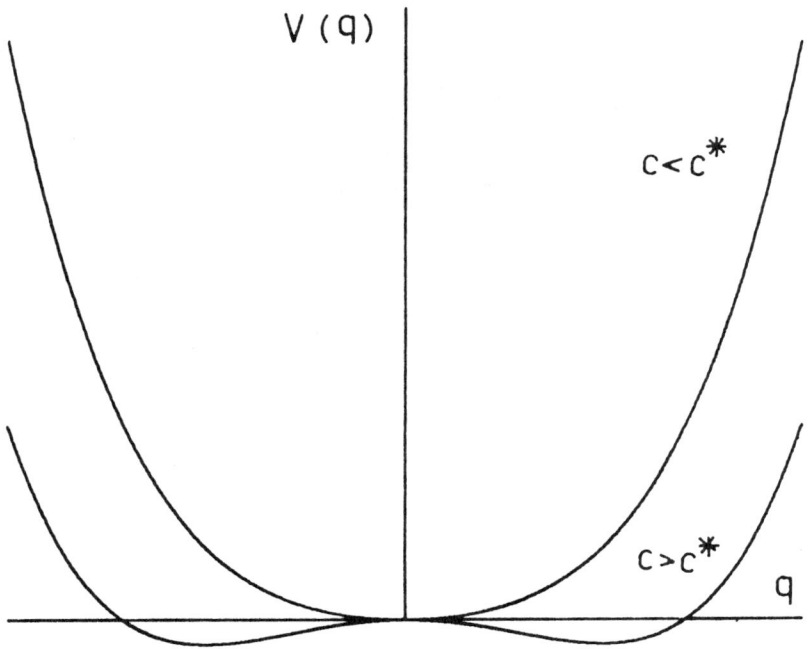

Fig. 2

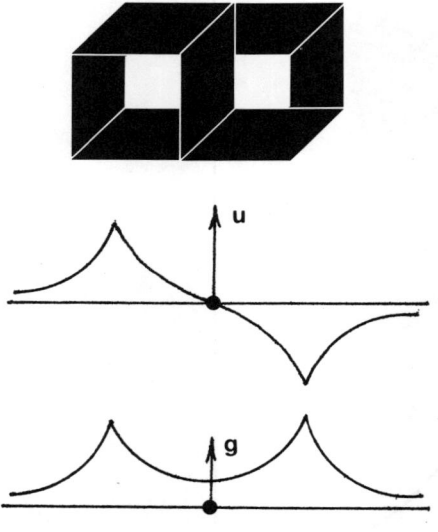

Fig. 3

AFTER THOUGHT

Eduardo R. Caianiello

Dipartimento di Fisica Teorica, Università di Salerno

Salerno, Italy

As the organizer and host of this Meeting, it would be hardly appropriate for me to add an uninvited paper - my own - to those here collected. Equally improper, perhaps, to play AWOL after half a life spent wondering about such issues. I shall compromise by entrusting to a streamlined bibliography the account (not the defense) of my enduring involvement in this fascinating field.

- FROM W. MCCULLOCH'S LOGICAL NETS TO ALGEBRAIC EQUATIONS: "THOUGHT" AS A DYNAMICAL COLLECTIVE PROCESS.

    "Outline of a Theory of Thought-Processes and Thinking Machines", J. Theor. Biol. $\underline{2}$, 204 (1961).

    "Encyclopaedic Dict. of Mathematics", MIT Press (1974)

    "Mathematical and Physical Problems in the study of Brain Models" Symposium on Neural Theory and Modelling, Ojai Valley, December 1962.

    "Non-linear Problems Posed by Decision Equations", Proceedings of International Summer School, Ravello 1965, Academic Press, N.York.

    "Decision Equation for Binary Systems - Application to Neural Behaviour" (in coll. with A. De Luca), Kybernetik 3 Band, 1 Heft, 33-40 (1966)

    "Decision Equation and Reverberations", Kybernetik 3 Band, 3 Heft, 33 (1966)

    "A study of Neural Networks and Reverberations", Wright Patterson Air Force Base Concerence, Dayton, Ohio (1966)

    "Reverberations and Control of Neural Networks" (in coll. with L.M. Ricciardi, and A. De Luca) Kybernetik 4, 10 (1967)

"Reverberations, Constant of Motion, General Behaviour" (in coll. with A. De Luca and L.M. Ricciardi) Proceedings Neural Networks, Ravello, Springer Verlag 1968)

"Synthesis of Reverberating Neural Networks" (in coll. with A. Aiello, and E. Burattini) Kybernetik Band 7, Heft $\underline{5}$, 191 (1970)

"Neural Nets: A Brief Survey" Proceedings NATO School on New Concepts and Technologies in Parallel Information Processing, Capri. Noordhoff Publishing, Series E9, 265 (1975)

"Methods of Analysis of Neural Nets" (in coll. with E. Grimson) Biol. Cybernetics 21, 1-6 (1976)

"Normal Modes in Neural Nets" (in coll. with G. Simoncelli), in Cybernetics Systems..." Ed. E.R. Caianiello, G. Musso, Research Studies Press, Letchworth, 1984.

"From Neuronic Equations to Spin Systems" - 'Progress in Quantum Field Theory' - Eds. H. Ezawa, S. Kamefuchi - North Holland, Amsterdam. 1986.

"Neuronic Equations Revisited and Completely Solved" in "Brain Theory" Eds. G. Palm, A. Aertsen, Springer, Berlin 1986.

- ON BOOLEAN FUNCTIONS (L. S. AND NOT)

"Some Remarks on the Tensorial Linearization of General and Linearly separable Boolean Functions", Kybernetik 12, 90-93 (1973)

"Tensorial Linearization of Threshold Functions" - Proc. NATO School on New Concepts and Technologies in Parallel Information Processing, Capri Noordhoff Publishing Series E9, 313 (1975)

"On Boolean Functions and Nets" - Atti Accademia Lincei, Tomo II, 501-507 (1976)

"Synthesis of Boolean Nets and Time-behaviour of a General Mathematical Neuron" (in coll. with E. Grimson), Biol. Cybernetics 18, 111-117 (1975)

"Polygonal Inequalities as a Key to Neuronic Equations" (in coll. with G. Simoncelli) Biol. Cybernetics 41, 203-209 (1981)

- ON NATURAL LANGUAGES, STRUCTURE AND ORGANIZATION

    "On the Analysis of Natural Languages" Proceedings of the 3rd All Union Conference on Cybernetics, Odessa, SSR, September 1965.

    "The Procrustes Program for the Analysis of Natural Languages" Proceedings Int. School on Automatic Interpretation and Classification of Images, Tirrenia - Academic Press, 1969.

    "On Form and Language: the Procrustes' algorithm for Feature Extraction" (in coll. with R. Capocelli) Kybernetik 8, 223-233 (1971)

    "Neural Nets and Natural Languages" Proceedings, Tokyo JITA Conference, March 1972.

    "Some Remarks on Organization and Structure" Biol. Cybernetics $\underline{26}$, 3 (1977)

    "C-calculus: an Elementary Approach to Some Problems in Pattern Recognition" (in coll. with A. Apostolico, E. Fischetti, S. Vitulano) Pattern Recognition (1978)

    "A Systematic Study of Monetary Systems" (in coll. with G. Scarpetta, G. Simoncelli) J. General Systems $\underline{8}$, 81-92 (1982)

    "Sulla legge di distribuzione delle monete" (in coll. with G. Scarpetta, G. Simoncelli) Rassegna Economica 44, 771-794 (1980)

- ON C-CALCULUS: ARITHMETICS OF (COMPOSITE) STRINGS OF (SIMPLE) SETS.

    "A Calculus of Hierarchical Systems" Proceedings 1st Int. Congress on Pattern Processing IEEE, 73CHO 821-9C ppl e 2 (1973)

    "A New Approach to Some Problems of Pattern Analysis" (in coll. with A. Apostolico and S. Vitulano) Proceedings of Informatics 76, p.104 (1976) Bled, Iugoslavia.

    "Structural Analysis of Hierarchical Systems" (in coll. with R. Capocelli) 3rd J.P.R. Conference, Coronado, California, Nov. 1976.

    "A New Approach to Some Problems of Cell Motion Analysis Based on C-calculus" (in coll. with L. Huimin), in print.

"A Model for C-calculus" (in coll. with A.G.S. Ventre), Preprint 1984.

"On Some Analytic Aspects of C-calculus" (in coll. with A.G.S. Ventre (in print)

- OTHER.

"An Algorithm for Feature Classification and Structural Memorization" (in coll. with A. Aiello, E. Burattini, A. Massarotti), Kybernetik 12, 145-153 (1973)

"Comparison of Two Unsupervised Learning Algorithms" (in coll. with L. Bobrowski), Biol. Cybernetics 37, 1-7 (1980)

"Energetics Versus Communication in the Nervous System" (in coll. with E. Di Giulio), Cybernetics and Systems 13, 187-196 (1982)

"A Geometrical View of Quantum and Information Theories" - Amalfi Conference (1983) in print.

"Entropy, Information and Quantum Geometry" - Proc. of Santa Fé Int. Conference on Non-Linear Phenomena, July 1984. Lecture delivered at the NATO A.S.I. "Non Equilibrium Quantum Statistical Physics", Santa Fé, New Mexico, June, 1984.

"Systems and Uncertainty: a Geometrical Approach" - Academy of Sciences USSR (1984).

A MODEL FOR COMMUNICATION OVER NOISY CHANNELS

R.M. Capocelli, L. Gargano and U. Vaccaro

Dipartimento di Informatica ed Applicazioni
Università di Salerno
84100 Salerno, Italy

ABSTRACT

In this paper the problem of transmission in presence of noise is considered. Encoding systems, multivalued encodings, are studied that associating to each source symbol a set of codewords can be considered as a general description of noisy channels. Indeed, the set of codewords associated to a given source symbol can be considered as the set of codewords in which the original encoding of that symbol can be changed because of the noise. The fundamental properties of multivalued encodings are investigated and algorithms for their testing are furnished. Finally, methods for the construction of sequential decoders, self-synchronizing decoders and invariant decoders for such encodings are proposed.

1. INTRODUCTION

The main goal of information transmission is that of reproducing the output of a given source at some given destination. Usually, the source output, after beeing encoded, is transmitted through a channel which accepts sequences of symbols from the channel input alphabet and produces sequences of symbols from the channel output alphabet. The channel input consists of sequences of words chosen in a subset of all sequences constructed on the input alphabet. Such a subset is referred to as the codeword set or, simply, the code. In order to correctly reconstruct the transmitted information, it is necessary to design the code in such a way one can recover the channel input message from the received sequence. The reconstruction is possible if and only if the code has the property of unique decipherability. Moreover, from a practical point of view, it would be desirable to have a code such that it is possible to start the decoding of a code message before the end of the transmission. This can be done if and only if the code is decipherable with finite delay. The delay of the code corresponds to the maximum number of codewords one has to receive to start the decoding. An interesting particular case is represented by the class of prefix codes. As it is well known their decoding delay is zero in that it is possible to decode each codeword as soon as it has been received. Finally, it would also be desirable to recover the transmitted message both if there are some errors in the received sequence and, roughly speaking, if part of the message has been lost. This is possible if and only if the code has the property of sinchronizability with finite delay. The delay of the code corresponds to the maximum number of codewords one has to receive before knowing an exact separation between codewords of the transmitted message and starting to decode. The three properties of unique decipherability, decipherability with finite delay and synchronizability are central to Coding Theory and have been extensively described and studied. They are related to the possibility of decoding a finite message, an infinite message with known starting point and an infinite message without knowing its beginning and end. Under the assumption of a noiseless channel, Sardinas and Patterson [5] and Leveshtein [6],[7] have provided necessary and sufficient conditions for a variable length code to satisfy these properties. A unified treatment of their results may be found in [8]. We remark that if a code is decipherable with finite delay it is possible to construct a finite memory sequential decoder, whereas, if the code is also synchronizable, the decoder can be constructed in such a way to be able to bound decoding errors due either to misfunctions or errors in the input sequence.

The noiseless channel assumption is rather unrealistic. In practical situations the presence of noise in the channel and, consequently, the occurrence of errors during the trasmission is to be expected. As a consequence, when a sequence of symbols is transmitted over a channel, the output is not uniquely determined but can be any of a set of sequences depending on the transmitted sequence and which pattern of errors has occurred. It is then necessary to know how the channel changes the input sequences, that is, a model of the channel behaviour is needed. A model of the channel behaviour is specified by stipulating the kinds and number of allowable errors; in brief, one describes the permissible error patterns for the channel. To tell it in other words, it is necessary to describe, for each channel input sequence the set of sequences that can occur at the output. First of all, one needs to know which kinds of errors may occur during the transmission. Errors can be classified in three kinds: the change of one symbol into another one (substitution error), the cancellation of one symbol of the transmitted sequence (deletion error) and the insertion of a new symbol in the transmitted sequence (insertion error). Note that if the channel allows not only substitution errors, but also deletion and insertions errors, the output sequences associated to an input sequence may have different lengths. The most general way to describe the behaviour of a channel that suffers of insertion, deletion and substitution errors is by means of multivalued encodings. Indeed, a multivalued encoding is a variable length code system in which not a single word but a set of codewords may correspond to each source symbol. Since each codeword associated to a source symbol can be considered as a noisy version of the original encoding of that symbol, it is clear the importance of multivalued encodings in modeling transmission over noisy channels. This approach, however, can be useful only if the set of sequences associated to each source symbol is not too large. Although, from a theoretical point of view, the set of codewords in which a single transmitted codeword can be changed because of the noise might be very large, in practical situations, however, by ignoring the sequences of errors having very small probability, this set can be considered of limited size.

There are other problems that, even rising in a different context, can be described as transmission over a noisy channel. An interesting example is constitued by the information-theoretic approach to problems such as automatic continuous speech recognition and character recognition processes. It has been pointed out by Bahl and Jelinek [1], Jelinek, Bahl and Mercer [2], [3] that such processes can be modeled as transmission of information through a noisy channel that allows substitution, insertion and deletion errors. In fact, in case of speech recognition, most processes include a segmentation stage followed by a classification stage. In the first stage an acoustic waveform is segmented into units corresponding to a single sound. If the segmentation is not perfect, some sounds can be missed (deletion error), or extra segment boundaries can be inserted (insertion error). In the classification stage when each segment is classified as the sound it is closest to, misclassifications (substution errors) are possible. Thus, the process can be modeled by a not very noisy channel in which only some error patterns, depending on the whole input sentence, are allowed. Studies in this field have shown that in most cases of pratical interest it is reasonable to assume that the output sequences associated to each input sequence constitute a finite set. In this context, they proposed a probabilistic model of the behaviour of channels using finite Markov chains and applied sequential decoding procedures [1-4]. More precisely, they associated to each input a_k a Markov chain $M(a_k)$, such that the set of words $M(a_k)$ can produce when starting from the initial state and stopping in the final state, represents all the outputs words the channel can originate when a_k is the input.

Going back to our approach of modeling the behaviour of noisy channels by means of multivalued encodings, a foundamental problem is the study of the properties such encodings have to satisfy in order to allow the reconstruction of the source sequences. More precisely, it is interesting to give an answer to the following questions: Which properties the sets of codewords defining the multivalued encoding should have in order to recover the source sequence from the received sequence of codewords? Which properties these sets should have to allow the deciphering of code messages sequentially, from left to right, with some finite delay? Which properties these sets should have to allow the deciphering of a code message whose initial part has been lost and whose end is not yet known? The alert reader has certainly realized that we are concerned with the extension of the properties of unique decipherability, decipherability with finite delay and synchronizability to multivalued encodings. In this paper we afford the problem of characterizing the above properties for multivalued encodings and we furnish finite procedures for their testing. Moreover, algorithms for the construction of decoders exhibiting several useful properties for multivalued encodings are given.

In the past, some attempts have been made to deal with properties of unique decipherability, decipherability with finite delay and synchronizability of variable length codes in presence of noise. Studies in this field are due expecially to the Coding Group at Parke Mathematical Laboratories [12], whose most interesting result has been to provide tests for the above properties in case only substitution errors are allowed, [13]. Recently, some work has been done on the construction of block codes having some capability of correcting substitution, insertion and deletion errors [9-11].

The Coding Group at University of Salerno has carried out, by means of multivalued encodings, the characterization of the most general situation of transmission over noisy channels when each kind of errors is allowed. Unfortunately, since for multivalued encodings unique decipherability is not equivalent to unique decomposability (i.e., a code message might be parsed in two different ways both giving the same deciphering in terms of source symbols), the extention of fundamental properties for ordinary codes is not straightforward, neither it does appear possible to use existing methods to test whether a multivalued encoding posses such properties. The aim of the present paper is to give a survey of the studies on multivalued encodings, conducted especially by the Coding Group at University of Salerno, by giving an organized exposition of both new results and already appeared in literature. After the introductory part, the paper is divided in five sections in which the appropiate definitions of the foundamental properties of multivalued encodings are given and the properties themselves are explored. In Section 2 the basic notations which will be used throughout the paper are given. In Section 3 we consider the property of unique decipherability which is central in Coding Theory beeing related to the possibility of recovering the transmitted source sequence from the received code message. We report a procedure, first devised by Sato [14], to test whether a multivalued encoding is uniquely decipherable. Section 4, mainly based on papers by Capocelli [15], and Capocelli and Vaccaro [17], is devoted to the investigation of the property of decipherability with finite delay, which is related to the possibility of beginning the decoding of a code message before its end is known. Indeed, as it is shown, the class of decipherable with finite delay multivalued encodings coincide with the class of multivalued encodings that admit of finite state sequential decoders, that is, decoders that having a code message as input produce as output the source sequence that has originated the message, exception made for a suffix of bounded length. We then afford the problem of actual decoding of such codes and provide an algorithm for the construction of sequential decoders with minimum decipherability delay for multivalued encodings. Finally, a finite procedure to test whether a multivalued encoding is decipherable with finite delay is given. In Section 5, we consider the problem of synchronizability. Such a property is related to the possibility of deciphering a message even though neither its beginning nor its end are known, in such a way that only a prefix and a suffix having finite lengths of the code message might be incorrectly deciphered. We report an algorithm by Capocelli et al. [19] to test whether a multivalued encoding is synchronizable. As a new result, we present an algorithm to construct self-synchronizing decoders for multivalued encodings. Such decoders exhibit the following interesting property: though they start from a wrong state, they are able, within a finite number of steps, to decode the coming part of the input message correctly. We show that the class of synchronizable multivalued encodings coincides with the class of encodings that admit of self-synchronizing decoders and provide an algorithm for the construction of such decoders for synchronizable multivalued encodings. Finally, the problem of fault-tolerant decoding is considered with regard to the problem of the existence of decoders with initial state invariance, that is, decoders that decode correctly no matter which the initial state is. The importance of such decoders rests on the fact that in case of errors like a decoder misfunctioning, they are able to decode correctly the following part of the input message. The difference of such decoders with the self-synchronizing ones is that the former completely eliminate the decoding errors, whereas the latter limit the length of incorrectly deciphered part of the message.

2. NOTATIONS AND DEFINITIONS

Let X be a code alphabet. A finite sequence $w = a_1 \ldots a_m$ of code symbols is called a *word* over X of length $l(w) = m$. Let X^+ represent the set of all finite words over X, $X^+ = \cup_{n=1}^{\infty} X^n$ where X^n is the set of all words of length n, and λ denote the empty word, $\lambda \notin X^+$. Given $w \in X^* = X^+ \cup \{\lambda\}$ and $p, r, s \in X^*$, if $prs = w$ then p is a prefix of w, r is an infix of w and s is a suffix of w. If $p \neq w$ then we call p a proper prefix of w. If p is a prefix of w we write $w \geq p$. For any $Z \subset X^+$ define

$$Prefix(Z) = \{x \in X^* | \exists w \in Z \; \exists r \in X^* \; [w = xr]\}$$
$$Suffix(Z) = \{x \in X^* | \exists w \in Z \; \exists p \in X^* \; [w = px]\}$$
$$Infix(Z) = \{x \in X^* | \exists w \in Z \; \exists p, r \in X^* \; [w = pxr]\}$$
$$B(Z) = \{x \in X^* | \exists w \in Z \; \exists r \in X^+ \; [w = xr]\}$$
$$T(Z) = \{x \in X^+ | \exists w \in Z \; \exists p \in X^+ \; [w = px]\}$$

In words, Prefix(Z), Suffix(Z) and Infix(Z) are the sets of all prefixes, suffixes and infixes of elements of Z, respectively. Moreover, it results $B(Z) \cup Z = Prefix(Z)$ and $T(Z) \cup Z \cup \{\lambda\} = Suffix(Z)$. Let $x \in X$ and $Z \subset X^*$, we denote with Zx and $x^{-1}Z$ the sets $Zx = \{y \in X^+ | y = zx \text{ and } z \in Z\}$ and $x^{-1}Z = \{x^{-1}z | z \in Z\} = \{y \in X^* | xy \in Z\}$, respectively.

Given a finite set A of source symbols, a multivalued encoding is any mapping $F : A \to 2^{X^+}$ from the source alphabet A into the set of subsets of X^+, denoted

by 2^{X^+}. We assume that for each $a \in A$ the set $F(a)$ is finite. In order to define the encoding of source sequences, we expand the domain of F from A to A^* in the following way:
  i) $F(\lambda) = \{\lambda\}$;
  ii) for each $x \in A^*$ and for each $y \in A$

$$F(xy) = F(x)F(y) = \{\alpha\beta | \alpha \in F(x) \quad and \quad \beta \in F(y)\}.$$

For each string of source symbols $x \in A^*$, $F(x)$ denotes the set of all possible encodings of the string x. It is obvious that the above definition, when for each $x \in A$ the set $F(x)$ is a singleton, reduces to the definition of ordinary encodings. Finally, denote by C the set of all codewords. i.e., $C = \cup F(x)$, where the union is taken over all $x \in A$.

3. UNIQUE DECIPHERABILITY

The concept of unique decipherability is central in Coding Theory beeing related to the possibility of recovering the transmitted source sequence from the received code message. Such a request immediately implies that different source sequences must not be encoded by the same code message. Since in case of ordinary encodings the correspondence between source symbols and codewords is one-to-one, it is possible to recover the original source sequence from the received code message only if the latter can be uniquely parsed in terms of codewords. This property does not hold for multivalued encodings where a code message can be parsed in different ways and still allow the exact reconstruction of the source sequence. In order to better explain above concepts, let us consider the following examples.

Example 1. Given the source alphabet $A = \{0,1\}$ and the code alphabet $X = \{a,b\}$, consider the multivalued encoding $F(0) = \{ab, ba, bb\}$, $F(1) = \{ababb, ba\}$. As it is easily seen, the code message $ababab$ can be deciphered in two different ways, namely 01 and 100. Therefore, one cannot recover the transmitted source sequence from the received code message and the encoding is not uniquely decipherable. □

Example 2. Given $A = \{0,1\}$ and $X = \{a,b\}$, consider the multivalued encoding $F(0) = \{ab, aba\}$, $F(1) = \{abb, bb\}$. In this case, the code is uniquely decipherable, even though the code message $ababb$ can be parsed in two different ways, namely $ab|abb$ and $aba|bb$. In fact, since $ab, aba \in F(0)$ and $abb, bb \in F(1)$, both parsings give the same deciphering 01. □

It follows that the most natural generalization of the definition of unique decipherability for multivalued encodings is the following.

Definition 1. A multivalued encoding $F : A \to 2^{X^+}$ is uniquely decipherable if and only if for any $x, y \in A^*$ $(x \neq y) \Rightarrow (F(x) \cap F(y) = \emptyset)$, i.e., there do not exist two sequences x and y having a common encoding. □

A problem which immediately arises is to find an algorithm to test whether a multivalued encoding is uniquely decipherable. Unfortunately, existing tools of classical coding theory are not useful since they only test whether there exists a code message that can be parsed in two different ways. As above seen this does not give any relevant information on the property of unique decipherability of multivalued encodings. In the following part of this section we will expose an algorithm, first devised by Sato [14], for the unique decipherability of multivalued encodings.

A multivalued encoding F can be represented as a transduction by a finite state nondeterministic machine and formalized as a relation between input sequences and output sequences by using a parameter set P^+

$$F = \{(\sigma(p), \tau(p)) | p \in P^+\}$$

$$\sigma(P) = A, \quad \tau(P) = F(A) = C, \quad \sigma(P^+) = A^+, \quad \tau(P^+) = F(A^+),$$

where, $\sigma : T^+ \to A^*$ and $\tau : T^+ \to X^+$ are mappings onto the input (a shortening homomorphism) and onto the output (a length-preserving homomorphism), respectively; T represents the set of the transitions in the state diagram of the finite state machine that realizes F; P represents the set of all elementary successfull transition sequences (i.e., of all transition sequences that produce a codeword), and P^+ represents the set of all possible transition sequences.

Example 3. Let $A = \{0, 1\}$ and $X = \{a, b, c\}$. Consider the multivalued encoding $F(0) = \{aa, bb\}$, $F(1) = \{abb, cab, bca\}$ and denote by T the set $T = \{t_1, t_2, t_3, t_4, t_5, t_6, t_7, t_8, t_9, t_{10}, t_{11}, t_{12}, t_{13}\}$. The two homomorphisms σ and τ are defined by

$$\sigma(t_1) = \sigma(t_3) = 0, \quad \sigma(t_5) = \sigma(t_8) = \sigma(t_{11}) = 1, \quad \sigma(t_i) = \lambda \quad i \neq 1, 3, 5, 8, 11,$$

$$\tau(t_1) = \tau(t_2) = \tau(t_5) = \tau(t_9) = \tau(t_{13}) = a, \quad \tau(t_8) = \tau(t_{12}) = c.$$

$$\tau(t_3) = \tau(t_4) = \tau(t_6) = \tau(t_7) = \tau(t_{10}) = \tau(t_{11}) = b.$$

Finally, the set P is given by $P = \{t_1t_2, t_3t_4, t_5t_6t_7, t_8t_9t_{10}, t_{11}t_{12}t_{13}\}$. □

Let $\Gamma = (T \times T)$. Define the sets
$$R = \{x \in \Gamma^+ | \alpha_1(x) \in P^+ \land \alpha_2(x) \in P^+ \land \tau\alpha_1(x) = \tau\alpha_2(x)\}$$
and
$$S = \{x \in \Gamma^+ | \sigma\alpha_1(x) = \sigma\alpha_2(x)\}$$

where α_1 and α_2 are two (projection) homomorphisms defined by $\alpha_1(t_i, t_j) = t_i$ and $\alpha_2(t_i, t_j) = t_j$, $(t_i, t_j) \in \Gamma$.

R is a regular set and S is a context-sensitive set, and it is easily seen that a multivalued encoding is uniquely decipherable if and only if $R \subset S$. Therefore the problem of unique decipherability can be reduced to the problem of testing if the regular set R is contained in the context-sensitive set S.

In order to obtain a finite procedure to solve this new problem, it is useful to restate the problem in terms of properties of the automaton that accepts the regular set R. Consider the minimal-state finite automaton $A_R = \{K_R \cup \{Q_\emptyset\}, \Gamma, \delta_R, Q_0, F_R\}$ that accepts the regular set R, where K_R is the internal state set, $Q_\emptyset$ is a $\emptyset$-state (a state not in K_R such that any $x \in \Gamma$ $\delta_R(Q_\emptyset, x) = Q_\emptyset$, and $\{Q_\emptyset\} = \emptyset$ if such a state does not exist), and Γ, δ_R, Q_0 and F_R are the input alphabet, the transition function, the initial state and the final state set, respectively. Finally, let us denote by $A(Q_i)$ the set of all input strings that bring A_R from its initial state Q_0 to the state Q_i and by $A_k(Q_i)$ the set of strings in $A(Q_i)$ of length not greater than k. For any $x \in \Gamma^+$, let us define the remainder of x, denoted by $Rem(x)$, as follows

$$Rem(x) = \begin{cases} (y, \lambda) & \text{if } \sigma_1(x) = \sigma_2(x)y \\ (\lambda, y) & \text{if } \sigma_1(x)y = \sigma_2(x), \\ (\lambda, \lambda) & \text{if } \sigma_1(x) = \sigma_2(x), \\ \emptyset & \text{otherwise} \end{cases}$$

where with σ_1 we denote the composition of the two homomorphisms σ and α_1, i.e., $\sigma_1 = \sigma\alpha_1$, and $\sigma_2 = \sigma\alpha_2$.

Example 4. Given $A = \{0, 1\}$ and $X = \{a, b, c, d\}$ consider the multivalued encoding $F(0) = \{ab, abc\}$, $F(1) = \{abd, cd, dab\}$. The sets T and P, and the homomorphism σ are given by:

$$T = \{t_1, t_2, t_3, t_4, t_5, t_6, t_7, t_8, t_9, t_{10}, t_{11}, t_{12}, t_{13}\},$$

$$P = \{t_1 t_2, t_3 t_4 t_5, t_6 t_7 t_8, t_9 t_{10}, t_{11} t_{12} t_{13}\}$$

$\sigma(t_1) = \sigma(t_3) = 0, \quad \sigma(t_6) = \sigma(t_9) = \sigma(t_{11}) = 1, \quad \sigma(t_i) = \lambda, \quad i \neq 1, 3, 6, 9, 11.$

Let $(t_1 t_2 t_9, t_3 t_4 t_5), (t_1 t_2 t_9 t_{10} t_1 t_2, t_3 t_4 t_5 t_{11} t_{12} t_{13}), (t_6 t_7 t_8, t_1 t_2 t_{11}) \in \Gamma^+$. It results

$$\sigma_1(t_1 t_2 t_9, t_3 t_4 t_5) = \sigma(t_1 t_2 t_9) = 01, \quad \sigma_2(t_1 t_2 t_9, t_3 t_4 t_5) = \sigma(t_3 t_4 t_5) = 0,$$

$$\sigma_1(t_1 t_2 t_9 t_{10} t_1 t_2, t_3 t_4 t_5 t_{11} t_{12} t_{13}) = \sigma(t_1 t_2 t_9 t_{10} t_1 t_2) = 010,$$

$$\sigma_2(t_1 t_2 t_9 t_{10} t_1 t_2, t_3 t_4 t_5 t_{11} t_{12} t_{13}) = \sigma(t_3 t_4 t_5 t_{11} t_{12} t_{13}) = 01$$

$\sigma_1(t_6t_7t_8, t_1t_2t_{11}) = \sigma(t_6t_7t_8) = 1 \quad \sigma_2(t_6t_7t_8, t_1t_2t_{11}) = \sigma(t_1t_2t_{11}) = 01$
and then $Rem(t_1t_2t_6, t_3t_4t_5) = (1, \lambda)$, $Rem(t_1t_2t_9t_{10}t_1t_2, t_3t_4t_5t_{11}t_{12}t_{13}) = (0, \lambda)$ and $Rem(t_6t_7t_8, t_1t_2t_{11}) = \emptyset$. □

It is easy to show the following properties of the function Rem
 a) for each $x, y \in \Gamma^+(Rem(x) = Rem(y)) \Rightarrow$ for each $z \in \Gamma^+(Rem(xz) = Rem(yz))$
 b) for each $x \in \Gamma^+(Rem(x) = \emptyset) \Rightarrow$ for each $z \in \Gamma^+(Rem(xz) = \emptyset)$
 c) for each $x, y, z \in \Gamma^+(Rem(xz) = Rem(yz) \neq \emptyset) \Rightarrow (Rem(x) = Rem(y) \neq \emptyset)$

The claim that a multivalued encoding is uniquely decipherable if and only if $R \subset S$ can be now stated as follows.

Theorem 1 [14]. A necessary and sufficient condition for a finite multivalued encoding to be uniquely decipherable is that the remainders of the elements in $\cup A(Q_i)$ are all (λ, λ), where the union is taken over all Q_i in the final state set of A_R. □

The above condition does not give a finite procedure to test the unique decipherability of a multivalued encoding since it requires to test the remainders of all elements of the infinite set $\cup A(Q_i)$. In order to obtain a finite procedure the following technical lemma is needed.

Lemma 1 [14]. Let n be the number of the internal states of the automaton A_R. If for all $Q_i \in K_R$, the remainders of all the elements of $A_n(Q_i)$ take the same value, then the remainders of all the elements of $A(Q_i)$ take this same value. □

By using Lemma 1, it is finally possible to show the following theorem that provides an algorithm for deciding whether a multivalued encoding is uniquely decipherable.

Theorem 2 [14]. A necessary and sufficient condition for a multivalued encoding to be uniquely decipherable is that for all $Q_i \in K_R$ the remainders of all elements in $A_n(Q_i)$ take the same value and the remainders of the elements in $\cup A_n(Q_i)$ are all (λ, λ), where the union is taken over the final state set of A_R.

Proof. If the code is uniquely decipherable, from Theorem 1 one gets that the remainders of the elements of $\cup A(Q_i)$, where the union is taken over all states $Q_i \in F_R$, are all (λ, λ). In particular this holds for the elements of $\cup A_n(Q_i)$, where the union is taken over all states in F_R. Furthermore, suppose that there exist $Q_i \in K_R$ and $x, y \in A_n(Q_i)$ such that $Rem(x) \neq Rem(y)$. Then, for any $z \in \Gamma^+$ such that $\delta_R(Q_i, z) \in F_R$, either $Rem(xz)$ or $Rem(yz)$ is different from (λ, λ) even though xz and yz are both elements of R (i.e., both bring A_R from Q_0 to $Q_i \in F_R$). The reverse condition follows immediately from Lemma 1 and Theorem 1.

Q.E.D.

Example 5. Given $A = \{0,1\}$, $X = \{a,b,c,d\}$, consider the multivalued encoding $F(0) = \{ab, abc\}$, $F(1) = \{cd, dab\}$. The sets T and P and the homomorphism σ are given by

$$T = \{t_1, t_2, t_3, t_4, t_5, t_6, t_7, t_8, t_9, t_{10}\},$$
$$P = \{t_1 t_2, t_3 t_4 t_5, t_6 t_7, t_8 t_9 t_{10}\},$$
$$\sigma(t_1) = \sigma(t_3) = 0, \quad \sigma(t_6) = \sigma(t_8) = 1 \quad \sigma(t_i) = \lambda \quad i \neq 1,3,6,8.$$

For the sequence $x = (t_3 t_4 t_5 t_8 t_9 t_{10}, t_1 t_2 t_6 t_7 t_1 t_2) \in R$ it holds

$$Rem(x) = Rem(t_3 t_4 t_5 t_8 t_9 t_{10}, t_1 t_2 t_6 t_7 t_1 t_2) = (\lambda, 0) \neq (\lambda, \lambda).$$

Hence the code is not uniquely decipherable. □

4. DECIPHERABILITY WITH FINITE DELAY

The property of decipherability with finite delay intuitively means that it is possible to decipher code messages sequentially, from left to right, with some finite delay. This implies that it is possible to start the decoding before the transmission ends. An immediate and quite informal definition of this property is the following: a multivalued encoding F is decipherable with delay P if and only if the individuation in a message of $P + n$ initial consecutive codewords suffices to determine the first n symbols of the source sequence that generated the message. A more formal definition, in terms of generalized sequential machines, is the following.

Definition 2. A multivalued encoding F is decipherable with finite delay P if and only if P is the smallest integer for which the nondeterministic generalized sequential machine (gsm) M that implements F has an inverse machine M^{-1} such that the serial connection $M^{-1}M$ of M in its initial state and M^{-1} in its initial state amounts to a delay machine with maximum delay P. A non deterministic generalized sequential machine D is called a delay machine with maximum delay P if and only if for any arbitrary input $x \in A^+$, $l(x) > P$, any associated output y is a prefix of x, with $l(x) - l(y) \leq P$. □

Example 6. Given $A = \{0,1\}$ and $X = \{a, b, c\}$, consider the multivalued encoding $F(0) = \{ab, bbc\}$, $F(1) = \{abb, bb\}$, and the infinite message $ab(bb)^n \ldots$. Since it results $ab(bb)^n \ldots = abb(bb)^{n-1}b\ldots$, $ab(bb)^n \in F(01^n)$ and $abb(bb)^{n-1} \in F(1^n)$, the prefix of code message $ab(bb)^n \ldots$ has two different decipherings that do not have a common prefix. It follows that one cannot start decoding the message without knowing its end, that is, the multivalued encoding is not decipherable with finite delay. □

A problem that naturally arises is the study of the properties and the structure of the inverse machine M^{-1}. Such an inverse machine is, essentially, a decoder. In order to make our reasoning more precise, let us give the following definitions.

Let $D = (S, s_0, X, A, f, \phi)$ be a finite deterministic, incompletely defined gsm, where S is the internal state set, s_0 is the initial state, X and A are respectively the input and output alphabet, $f : S \times X \to S$ is the transition function and $\phi : S \times X \to A^*$ is the output function.

Definition 3. We say that a finite, deterministic, incompletely defined gsm D is a decoder for a multivalued encoding $F : A \to 2^{X^+}$ if and only if there exists $t \geq 0$ such that for each $a_1 \ldots a_k \in A^+$, $w_1 \ldots w_k \in F(a_1 \ldots a_k)$, $\beta \in C^t$

$$a_1 \ldots a_k \text{ is a prefix of } \phi(s_0, w_1 \ldots w_k \beta) \tag{1}$$

The smallest integer t for which (1) holds is called the delay of the decoder D and will be denoted by τ. □

The following result, confirming the intuition, shows formally the equivalence between decoders and inverse machines.

Theorem 3. Let $M = (Q, q_0, A, X, \lambda, \delta)$ be the nondeterministic gsm that implements F and $M^{-1} = (S, s_0, X, A, f, \phi)$ be a deterministic finite gsm. $M^{-1}M$ is a delay machine with delay P if and only if M^{-1} is a decoder for F with delay P.

Proof. Let $z = a_1 \ldots a_k a_{k+1} \ldots a_{k+P} \in A^+$ and $\delta(q_0, z) = w_1 \ldots w_k \ldots w_{k+P} \in F(z)$. Since M^{-1} is a decoder with delay P, one has $\phi(s_0, w_1 \ldots w_{k+P}) \geq a_1 \ldots a_k$. That is, for any input $z \in A^+$, $l(z) > P$, the output of the machine $M^{-1}M$ with M and M^{-1} in their initial states is a prefix y of z with $l(z) - l(y) \leq P$. Conversely, let us suppose that $M^{-1}M$ is a delay machine with delay P. By definition, any output y associated to an arbitrary input $z = a_1 \ldots a_k a \in A^+$, $l(z) > P$, $a \in A^P$, has $a_1 \ldots a_k$ as prefix. Hence, one gets that $a_1 \ldots a_k$ is a prefix of the output of M^{-1} associate to any encoding $w_1 \ldots w_k \beta$ of z, $w_1 \ldots w_k \in F(a_1 \ldots a_k)$ $\beta \in F(a)$, that enters M^{-1}, that is

$$\phi(s_0, w_1 \ldots w_k \beta) \geq a_1 \ldots a_k.$$

Q.E.D.

The following corollary is immediate.

Corollary 1. A multivalued encoding F has finite decipherability delay P if and only if P is the smallest integer for which there exists a decoder for F with delay P. □

We are left with the problem of the actual construction of decoders for multivalued encodings. To this aim let us introduce the following definitions. Let be given a multivalued encoding $F : A \to 2^{X^+}$. Define initial decomposition of

$\beta \in Prefix(C^+)$ a sequence $\delta(\beta) = w_1 \ldots w_h \beta'$, $(h \geq 0)$, such that $\beta = \delta(\beta)$, $w_i \in C$, $\beta' \in B(C)$. For any initial decomposition of $\beta \in Prefix(C^+)$, say $\delta_i(\beta) = w_{i_1} \ldots w_{i_h} \beta'_i$, define initial deciphering of β the corresponding source sequence $D_i(\beta) = a_{i_1} \ldots a_{i_h} x$ with $a_{i_j} \in A$, $w_{i_1} \ldots w_{i_h} \in F(a_{i_1} \ldots a_{i_h})$ and $x \in A \cup \{\lambda\}$ ($x \in A$ if all codewords beginning with β'_i belong to F(x), $x = \lambda$ otherwise), and partial deciphering of β the source sequence $\Delta_i(\beta) = a_{i_1} \ldots a_{i_h}$. Define now the mappings $G : Prefix(C^+) \rightarrow A^*$ and $H : Prefix(C^+) \rightarrow 2^{Prefix(C^+)} \times 2^{T(C)}$ that will be used to handle the output and the transition function of the decoders in the following way:

$$G(\beta) = a_1 \ldots a_s \in A^*,$$

where $a_1 \ldots a_s$ ($s \geq 0$), is the longest prefix of all initial decipherings of β;

$$H(\beta) = (R_1(\beta), R_2(\beta)),$$

where

$$R_1(\beta) = \{x \in Prefix(C^+) | \exists w_1 \ldots w_s \in F(G(\beta)), \exists \delta_i(\beta)$$
$$[\delta_i(\beta) = w_1 \ldots w_s x]\}$$

$$R_2(\beta) = \{x \in T(C) | \exists w_1 \ldots w_s \in F(G(\beta)), \exists \delta_i(\beta)$$
$$[\delta_i(\beta) x = w_1 \ldots w_s]\}$$

Further, let us extend the domain of G and H. Let $Y \subset Prefix(C^+)$, $Z \subset T(C) \cup \{\lambda\}$, $Y \neq \emptyset \neq Z$. Define G(Y,Z) as the longest common prefix, if any, of all sequences $G(\beta_i)$, $\beta_i \in Y$ if $Z = \emptyset$; λ if $Z \neq \emptyset$; and define

$$H(Y,Z) = \begin{cases} (R_1(Y), R_2(Y)) & \text{if } Z = \emptyset \\ (Y \cup \{\lambda\}, Z - \{\lambda\}) & \text{if } \lambda \in Z \\ (Y, Z) & otherwise \end{cases}$$

where

$$R_1(Y) = \{x \in Prefix(C^+) | \exists w_1 \ldots w_s \in F(G(Y, \emptyset)), \exists \beta_j \in Y, \exists \delta_i(\beta_j)$$
$$[\delta_i(\beta_j) = w_1 \ldots w_s x]\}$$

$$R_2(Y) = \{x \in T(C) | \exists w_1 \ldots w_s \in F(G(Y, \emptyset)), \exists \beta_j \in Y, \exists \delta_i(\beta_j)$$
$$[\delta_i(\beta_j) x = w_1 \ldots w_s]\}$$

Finally, for $x \in X$ such that $(Yx \cap Prefix(C^+), x^{-1}Z) \in 2^{Prefix(C^+)} \times 2^{T(C) \cup \{\lambda\}} - (\emptyset, \emptyset)$ define $G(Y, Z, x) = G(Yx \cap Prefix(C^+), x^{-1}Z)$, $H(Y, Z, x) = H(Yx \cap Prefix(C^+), x^{-1}Z)$.
The following result holds.

Lemma 2 [18]. For each $\beta b \in Prefix(C^+)$, $b \in X$

$$G(\beta b) = G(\beta)G(H(\beta), b) \qquad (2)$$

$$H(\beta b) = H(H(\beta), b) \qquad (3)$$

□

The following theorem (the sufficient part) will provide an algorithm for designing sequential decoders for multivalued encodings having minimum delay. We remark that the condition expressed by Theorem 4 is necessary and sufficient for a multivalued encoding to have finite decipherability delay.

Theorem 4. A necessary and sufficient condition for a decoder to exist for a multivalued encoding F is that there exists $t \geq 0$ such that for each $a_1 \ldots a_k \in A^+$, $w_1 \ldots w_k \in F(a_1 \ldots a_k)$, $\beta \in C^t$

$$G(w_1 \ldots w_k \beta) \geq a_1 \ldots a_k \qquad (4)$$

The smallest integer t, τ, for which (4) holds is equal to the decipherability delay P of F.

Proof. Necessity. Let $D = (S, s_0, X, A, f, \phi)$ be a decoder for F with delay t. One has that for each $\beta \in Prefix(C^+)$

$$G(\beta) \geq \phi(s_0, \beta) \qquad (5)$$

Indeed, suppose that (5) is not true; that is, suppose that there exists $\beta \in Prefix(C^+)$ such that $\phi(s_0, \beta) = a_1 \ldots a_h \ldots a_k$, $G(\beta) = a_1 \ldots a_h$, $0 \leq h < k$. It is then possible to distinguish the following three cases

Case i). There exists an initial decomposition of β, $\delta(\beta) = w_1 \ldots w_h w_{h+1} \ldots w_n \beta'$, $n > h$, $\beta' \in B(C)$ such that $w_1 \ldots w_h \in F(a_1 \ldots a_h)$, $w_{h+j} \in F(a'_{h+j})$, $1 \leq j \leq n-h$, $a'_{h+j} \in A$ and $a'_{h+1} \neq a_{h+1}$. Hence, for each γ such that $\beta'\gamma \in C^t$, it results

$$\phi(s_0, w_1 \ldots w_h \ldots w_n \beta'\gamma) \geq a_1 \ldots a_h a'_{h+1} \ldots a'_n$$

and

$$\phi(s_0, w_1 \ldots w_h \ldots w_n \beta'\gamma) \geq \phi(s_0, \beta) \geq a_1 \ldots a_h a_{h+1} \ldots a_k$$

that contradicts the assumption $a_{h+1} \neq a'_{h+1}$.

Case ii). There exists an initial decomposition of β, $\delta(\beta) = w_1 \ldots w_h \beta'$ with $w_1 \ldots w_h \in F(a_1 \ldots a_h)$ and β' such that $\beta' x = w \in F(a)$, $a \neq a_{h+1}$, for some $x \in X^+$. If $\gamma \in C^t$ we have

$$\phi(s_0, w_1 \ldots w_h \beta'\gamma) \geq a_1 \ldots a_h a$$

and
$$\phi(s_0, w_1 \ldots w_h \ldots w_n \beta' \gamma) \geq \phi(s_0, \beta) \geq a_1 \ldots a_h a_{h+1} \ldots a_k$$
that contradicts the assumption $a \neq a_{h+1}$.

Case iii). There exists an initial decomposition of β, $\delta(\beta) = w_1 \ldots w_{h-1} \beta'$, $w_1 \ldots w_{h-1} \in F(a_1 \ldots a_{h-1})$ such that all codewords beginning with β' belong to $F(a_h)$. For $w \in F(a)$, $a \neq a_{h+1}$, and $\gamma \in C^t$, it then results

$$\phi(s_0, w_1 \ldots w_{h-1} w_h w \gamma) \geq a_1 \ldots a_{h-1} a_h a$$

$$\phi(s_0, w_1 \ldots w_{h-1} w_h w \gamma) \geq \phi(s_0, \beta) \geq a_1 \ldots a_h a_{h+1} \ldots a_k$$

that contradicts the assumption $a \neq a_{h+1}$.

Finally, from (5) and by the definition of a decoder one gets (4).

Sufficiency. We shall provide an algorithm for constructing a decoder D for F with minimum decoding delay τ that, by Corollary 1, coincides with the decipherability delay P of F. Let us consider the decoder $D = (S, s_0, X, A, f, \phi)$ defined as follows:
$S = \{s_{(Y,Z)} | (Y, Z) \in M\}$, where $Y \subset Prefix(C^+), Z \subset T(C)$ and M is constructed according to the following rules

i) $(\{\lambda\}, \emptyset) \in M$

ii) for each $b \in X$, $(Y, Z) \in M$ if $b^{-1}Z$ and $Yb \cap Prefix(C^+)$ are not both empty then $H(Y, Z, b) \in M$.

It is easy to see that M, and therefore S, is finite [18].

The initial state is $s_0 = s_{(\{\lambda\}, \emptyset)}$. The transition function f and the output function ϕ, defined on the pairs $(s_{(Y,Z)}, b) \in S \times X$ such that $Yb \cap Prefix(C^+)$ and $b^{-1}Z$ are not both empty, are defined in the following way

$$\phi(s_{(Y,Z)}, b) = G(Y, Z, b) \tag{6}$$

$$f(s_{(Y,Z)}, b) = s_{H(Y,Z,b)} \tag{7}$$

Notice that the above definitions are consistent. Indeed, if $(s_{(Y,Z)}, b) \in S \times X$ then (6) gives $\phi(s_{(Y,Z)}, b) \in A^*$. On the other hand, because of ii), $H(Y, Z, b) \in M$, i.e., $s_{H(Y,Z,b)} \in S$. We shall show that D is a decoder and has decoding delay τ. For that, because of (4), it is sufficient to show that for each $\beta \in Prefix(C^+)$

$$\phi(s_0, \beta) = G(\beta) \tag{8}$$

We shall prove (8) simultaneously with

$$f(s_0, \beta) = s_{H(\beta)} \tag{9}$$

by induction on the length of β. Let $l(\beta) = 0$ (i.e. $\beta = \lambda$). By definition, one gets $\phi(s_0, \lambda) = \lambda = G(\lambda)$, $f(s_0, \lambda) = s_0 = s_{(\{\lambda\}, \emptyset)} = s_{H(\lambda)}$. Let us now suppose (8) and (9) true for each $\beta \in Prefix(C^+)$ with $l(\beta) = n$. We shall prove that they take place also for each $\beta' = \beta b \in Prefix(C^+)$, $l(\beta) = n$ and $b \in X$, $l(\beta') = n+1$. By Lemma 2, one has $f(s_0, \beta') = f(s_0, \beta b) = f(f(s_0, \beta), b) = f(s_{H(\beta)}, b) = s_{H(H(\beta), b)} = s_{H(\beta b)}$ and then (9). Moreover, $\phi(s_0, \beta') = \phi(s_0, \beta b) = \phi(s_0, \beta)\phi(f(s_0, \beta), b) = G(\beta)\phi(s_{H(\beta)}, b) = G(\beta)G(H(\beta), b) = G(\beta b)$. Finally, from (5) and (8), one gets that the decoding delay τ of the decoder built according to the above described method is P, the decoding delay of the encoding F.

Q.E.D.

We remark that it is possible to modify the algorithm for the construction of decoders in such a way that the output, at each transition, is either a letter of the alphabet or the empty word. Moreover, as showed by Capocelli and Vaccaro [18], one can obtain decoders having the following property: if x is the output of the decoder corresponding to a message β and y is the longest deciphering of β, then $l(y) - l(x)$ is constant and equal to the deciphering delay P of the code. It should be remarked that in case of single valued codes, the above decoders behave essentially as those usually considered for decoding ordinary codes (see for example Levenshtein [7]).

Example 7. Given $A = \{0, 1\}$ and $X = \{a, b, c, d\}$, the multivalued encoding $F(0) = \{aa, aab, bb\}$, $F(1) = \{bbc, cd\}$ has decipherability delay 1. The decoder constructed according to the method described in Theorem 4 is shown in Fig.1. □

Since the condition expressed in the above theorem does not give a finite procedure, we are interested in finding an effective algorithm for testing whether a multivalued encoding is decipherable with finite delay. To this aim we reconsider the formalism introduced in the previous section to transform the problem in an equivalent one expressed in terms of automata.
Let us now consider the minimal-state finite automaton

$$A_P = \{K_P \cup \{q_\emptyset\}, T, \delta_P, q_0, F_P\}$$

that accepts the regular set P^+, where $q_\emptyset$ is a $\emptyset$ state, δ_P is the transition function, and the final state set consists in fact of only the initial state q_0. Let us then construct the incompletely defined automaton

$$A = \{K_P \times K_P, \Gamma, \delta, (q_0, q_0), F_P \times F_P\},$$

where $\delta((q_i, q_j), (t_h, t_k)) = (\delta_P(q_i, t_h), \delta_P(q_j, t_k))$ if $\delta_P(q_i, t_h) \neq \emptyset$, $\delta_P(q_j, t_k) \neq \emptyset$ and $\tau(t_h) = \tau(t_k)$, and is not defined otherwise. Here $(q_i, q_j) \in K_P \times K_P$, $(t_h, t_k) \in \Gamma$.

It is easily seen that A accepts the above defined set R. Note that A is not generally a minimal-state automaton, and many states can exist that either cannot be reached from the initial state or from which the final state cannot be reached. Let s be the number of states that can be reached from the initial state $(q_0, q_0) \equiv Q_0$ and let $A_k(Q_i)$ be the set of strings of length not greater than k that bring A from Q_0 to the state Q_i. For each state $Q_i \in K_P \times K_P$, let us define $Rem(Q_i)$ as follows: if the remainders of all elements of $A_s(Q_i)$ take the same value then $Rem(Q_i)$ take this value; otherwise $Rem(Q_i) = \emptyset$.

Definition 4. A loop of states x,xy with $x, xy \in A_s(Q_i)$ has remainder $\emptyset$ if either $Rem(x) \neq Rem(y)$ or $Rem(x) = Rem(y) = \emptyset$. □

We remark that for any state from which the final state can be reached the remainder can be uniquely assigned and, if the encoding is uniquely decipherable, it is different from $\emptyset$, whereas it cannot generally be uniquely assigned for states from which the final state cannot be reached. These states play an important role in the determination of the finite delay property.

Lemma 3 [18]. If for all states Q_i that can be reached from Q_0 and for each loop $\alpha, \alpha\beta \in A_s(Q_i)$ it results $Rem(\alpha) = Rem(\alpha\beta) \neq \emptyset$ then for each $x \in A_s(Q_i)$, $xy \in A(Q_i)$ it holds $Rem(x) = Rem(xy) \neq \emptyset$. □

Let now $w \in \Gamma^*$ such that $\delta(Q_0, w)$ is definite. It can be shown that it is possible to write

$$w = x_0^{h_0} x_1 x_2^{h_1} x_3 \ldots x_{2n-1} x_{2n}^{h_n} x_{2n+1}$$

$n \geq 0$, $h_j \geq 0$, $0 \leq j \leq n$, $x_0, x_{2n+1} \in \Gamma^*$, $x_i \in \Gamma^+$, $1 \leq i \leq n$, in such a way that
1) $l(x_1 x_3 \ldots x_{2n+1}) \leq s - 1$
2) $\delta(Q_0, x_0) = Q_0$
   $\delta(Q_0, x_1) = Q_1 = \delta(Q_1, x_2)$
   ......
   $\delta(Q_0, x_1 x_3 \ldots x_{2n-1}) = Q_n = \delta(Q_n, x_{2n})$
with $Q_i \neq Q_j$ if $i \neq j$.
Further, if in A there are no loops of states with remainder $\emptyset$ and F is uniquely decipherable, one has that

$$Rem(x_1) = Rem(x_1 x_2) \neq \emptyset$$
$$\ldots\ldots \tag{10}$$
$$Rem(x_1 x_3 \ldots x_{2n-1}) = Rem(x_1 x_3 \ldots x_{2n-1} x_{2n}) \neq \emptyset$$

Indeed, since $x_1 x_3 \ldots x_{2i-1} \in A_s(Q_i)$ $i = 1, 2, \ldots, n$

$$Rem(x_1 x_3 \ldots x_{2i-1}) = Rem(x_1 x_3 \ldots x_{2i-1} x_{2i}) \neq \emptyset$$

if $x_1x_3\ldots x_{2i} \in A_s(Q_i)$. If $x_1x_3\ldots x_{2i} \notin A_s(Q_i)$ one has that $x_1x_3\ldots x_{2i} = xyz$, $x, xy \in A_s(Q_j)$ and $z \in \Gamma^+$ such that $\delta(Q_j, z) = Q_i$. Then, by hypothesis, $Rem(x) = Rem(xy) \neq \emptyset$ and, by Lemma 3, $Rem(x_1x_3\ldots x_{2i-1}) = Rem(x_1x_3\ldots x_{2i-1}x_{2i}) \neq \emptyset$. One has then the following theorem.

Theorem 5. A necessary and sufficient condition for a uniquely decipherable multivalued encoding F to be decipherable with finite delay is that no loop of states exists with remainder $\emptyset$ in the state diagram of A.

Proof. Sufficiency. Let us consider $w = x_0^{h_0} x_1 x_2^{h_1} x_3 \ldots x_{2n-1} x_{2n}^{h_n} x_{2n+1} \in \Gamma^+$ with $\delta(Q_0, w) = Q_n \in K_P \times K_P$. One has that

$$Rem(x_1) = Rem(x_1 x_2) = Rem(x_1 x_2^{h_1}) \neq \emptyset$$

that because of the properties of Rem gives

$$Rem(x_1 x_3) = Rem(x_1 x_2^{h_1} x_3)$$

that in turns, from (10), gives

$$Rem(x_1 x_3) = Rem(x_1 x_2^{h_1} x_3 x_4^{h_2}) \neq \emptyset$$

an so on up to get

$$Rem(x_1 x_3 \ldots x_{2n-1}) = Rem(x_1 x_2^{h_1} x_3 \ldots x_{2n-1} x_{2n}^{h_n}) \neq \emptyset \tag{11}$$

Since, from the unique decipherability of F one has $Rem(x_0) = (\lambda, \lambda)$, from (11) one obtains that

$$Rem(x_1 x_3 \ldots x_{2n-1}) = Rem(x_0^{h_0} x_1 x_2^{h_1} x_3 \ldots x_{2n-1} x_{2n}^{h_n}) \neq \emptyset$$

By the definition of Rem one has that $\sigma_1(w)$ and $\sigma_2(w)$ either coincide or they have a common prefix. The same holds for $\sigma_1(x_1x_3\ldots x_{2n-1}x_{2n+1})$ and $\sigma_2(x_1x_3\ldots x_{2n-1}x_{2n+1})$. In case that $\sigma_1(w)$ and $\sigma_2(w)$ do not coincide, the suffixes for which they differ are the same for which $\sigma_1(x_1x_3\ldots x_{2n-1}x_{2n+1})$ and $\sigma_2(x_1x_3\ldots x_{2n-1}x_{2n+1})$ differ. Since $l(x_1x_3\ldots x_{2n-1}x_{2n+1}) \leq s - 1$, the lengths of these suffixes are upperbounded by $s - 1$. Therefore, for any $w \in \Gamma^+$ for which $\delta(Q_0, w) \in K_P \times K_P$, and then for any code message $x = \tau\alpha_1(w) = \tau\alpha_2(w)$ it is possible to determine a prefix of the source sequence that generated such a message with resulting (undeciphered) suffix of bounded length.

Necessity. Let us suppose that a loop of states exists with remainder $\emptyset$. That is, there exists $Q_i \in K_P \times K_P$ and $x, xy \in \Gamma^+$, $x, xy \in A_s(Q_i)$, such that either $Rem(x) = Rem(xy) = \emptyset$ or $Rem(x) \neq Rem(xy)$. Then it is easily seen that there exists an arbitrary long input sequence $xy^k \in \Gamma^+$ that leads A through the path

$Q_0 \ldots Q_i (\ldots Q_i)^k$ and such that $Rem(xy^k)$ is either $\emptyset$ or (z, λ) [Resp. (λ, z)] with $l(z)$ linearly increasing with k. This is trivial if $Rem(x)$ and/or $Rem(xy)$ is equal $\emptyset$. If, instead, $Rem(x) \neq Rem(xy) \neq \emptyset$, by definition of Rem, one gets $Rem(xy) \neq Rem(xy^2)$ and, in turn, if $Rem(xy^2) \neq \emptyset$ gives $Rem(xy^2) \neq Rem(xy^3)$ and so on. Hence, one easily gets that $Rem(xy^k) = \emptyset$ if for some $i \leq k$ $Rem(xy^i) = \emptyset$, whereas $Rem(xy^k) = (z, \lambda)$ [Resp. (λ, z)] if for all i $Rem(xy^i) \neq \emptyset$. In the latter case it will clearly result that $l(\sigma_1(y)) \neq l(\sigma_2(y))$ and $l(z) = |l(\sigma_1(y)) - l(\sigma_2(y))|k \pm h$ where $h = |l(\sigma_1(x)) - l(\sigma_2(x))|$. The sign will be plus if $l(\sigma_1(x))$ and $l(\sigma_2(x))$ are both larger or smaller, respectively, than $l(\sigma_1(y))$ and $l(\sigma_2(y))$; it will be minus otherwise. That is, there are arbitrarily long messages that do not permit deciphering an arbitrarily large suffix of the source sequences that generate them. Therefore the multivalued encoding does not have finite decipherability delay.

Q.E.D.

For any $x \in A_s(Q_i)$ consider $\sigma_1(x)$ and $\sigma_2(x)$. Denote by L the length of the longest of the sequences $\sigma_1(x)$ and $\sigma_2(x)$ and by l the length of the longest common prefix of $\sigma_1(x)$ and $\sigma_2(x)$.

Corollary 2. The delay of a finite multivalued encoding F decipherable with finite delay is $P = max \mu \leq max L$, where $\mu = L - l$. The maximization is over all $Q_i \in K_P \times K_P$ of A. □

Example 8. Given $A = \{0, 1\}$, $X = \{a, b, c, d, e, f, g, h\}$, consider the multi-valued encoding $F(0) = \{bc, cd, fg\}$, $F(1) = \{a, ab, de, hc, efgh\}$. One has that $\sigma(t_1) = \sigma(t_3) = \sigma(t_5) = 0$, $\sigma(t_7) = \sigma(t_8) = \sigma(t_{10}) = \sigma(t_{12}) = \sigma(t_{14}) = 1$ and $\sigma(t_i) = \lambda$ $i \neq 1, 3, 5, 7, 8, 10, 12, 14$, therefore

$$Rem(x) = Rem(t_7 t_1 t_2 t_{10} t_{11} t_5 t_6 t_{12} t_{13}, t_8 t_9 t_3 t_4 t_{14} t_{15} t_{16} t_{17} t_3) = (\lambda, 1)$$

and

$$Rem(xy) = Rem(t_7 t_1 t_2 t_{10} t_{11} t_5 t_6 t_{12} t_{13} t_{10} t_{11} t_5 t_6 t_{12} t_{13},$$
$$t_8 t_9 t_3 t_4 t_{14} t_{15} t_{16} t_{17} t_3 t_4 t_{14} t_{15} t_{16} t_{17} t_3) = (\lambda, 101)$$

That is, the inputs x and xy bring **A** to the same state. Such a state belongs to a loop and has remainder $\emptyset$. Hence te code is not decipherable with finite delay. □

5. SYNCHRONIZABILITY

The property of synchronizability intuitively means that it is possible to decipher, with some finite delay, a message even though neither its beginning nor its end are known. An informal definition of this property is the following: a multi-valued encoding is synchronizable with delay Q if and only if the individuation in a message of $Q + n$ consecutive codewords suffices to determine an infix of length greater or equal to n of the source sequence that generated it; in particular if the $Q + n$ codewords are the beginning of a message, one can determine a prefix of length not less than Q of the source sequence that generated it. In order to make clear the above concept, let us consider the following example.

Example 9. Given $A = \{0,1\}$ and $X = \{a,b,c\}$, the multivalued encoding $F(0) = \{aa, abc\}$, $F(1) = \{aba, bba, bc\}$ is decipherable with finite delay $P = 0$, but it is not synchronizable. Indeed, it results $\ldots(aa)^n abc \ldots = \ldots a(aa)^n bc \ldots$, with $(aa)^n abc \in F(0^{n+1})$ and $a(aa)^n bc \in F(0^n 1)$. It follows that there exists an arbitrary long code message having $(aa)^n abc$ as infix, which has two decipherings that differ in prefixes and suffixes of unbounded length. Therefore the multivalued encoding is not synchronizable. □

We give now the formal definition of synchronizable multivalued encodings.

Definition 5. A uniquely decipherable multivalued encoding F is synchronizable with delay Q if and only if Q is the smallest integer for which the nondeterministic gsm M that implements F has a self-synchronizing inverse machine M^{-1} such that the serial connection $M^{-1}M$ of M in its initial state and M^{-1} in any state is a self-synchronizing delay machine with maximum delay Q; moreover the the serial connection $M^{-1}M$ of M in its initial state and M^{-1} in its initial state amounts to a delay machine with maximum delay Q. A non deterministic gsm D is called a self-synchronizing delay machine with maximum delay Q if and only if for any initial state, for any input $x \in A^+$, $l(x) > Q$, any associated output y has a suffix z such that $x = x_1 z x_2$ and $l(x_1) + l(x_2) \leq Q$. □

It follows from Definition 5 that if a multivalued encoding $F : A \to 2^{X^+}$ is synchronizable then the encoding and its reversal, i.e., the encoding obtained by reversing in each set $F(a)$, $a \in A$, the order of letters in each codeword, are both decipherable with finite delay. This condition is not sufficient, in fact a multivalued encoding can be not synchronizable even though it and its reversal are both decipherable with finite delay.

We are now interested in the study of the structure and of the properties of self-synchronizing inverse machines. To this aim let us introduce the definition of self-synchronizing decoders.

Definition 6. We say that a finite, deterministic, completely defined gsm D is a self-synchronizing decoder (SSD) for a multivalued encoding $F : A \to 2^{X^+}$ if and only if there exists $r \geq 0$ such that
  1. D is a decoder for F with delay not greater than r;
  2. for each $a_1 \ldots a_k \in A^+$, $w_1 \ldots w_k \in F(a_1 \ldots a_k)$, $s \in S$, $\phi(s_0, w_1 \ldots w_k \beta) = xz$, $x, z \in A^*$, and z is an infix of $a_1 \ldots a_k$ with $k - l(z) \leq r$.

The smallest integer r for which the above conditions hold is called the delay of the SSD D and will be denoted by ρ. □

By using arguments similar to the ones of Theorem 3, it is possible to prove the following result which state the equivalence between self-synchronizing machines and SSD's.

Theorem 6. Let $M = (Q, q_0, A, X, \lambda, \delta)$ be the nondeterministic gsm that implements F and $M^{-1} = (S, s_0, X, A, f, \phi)$ be a deterministic, finite gsm. M^{-1} is a SSD for F if and only if M^{-1} is a self-synchronizing inverse machine for M with delay Q. □

Corollary 3. A multivalued encoding F is synchronizable with delay Q if and only if Q is the smallest integer for which there exists a SSD for F with delay Q. □

In the sequel we will afford the problem of actual construction of self-synchronizing decoders for multivalued encodings.

To this aim, let us give some additional definitions.

For each $\beta \in X^*$ define $Ends(\beta) = Suffix(\{\beta\}) \cap Prefix(C^+)$, that is, the set of all suffixes of β that are prefixes of code messages.

Let $\beta \in Infix(C^+)$, i.e., let β be an infix of code message. Define decomposition of β a sequence $\pi(\beta) = \xi_1 w_1 \ldots w_k \xi_2$, $k \geq 0$, such that $\xi_1 \in T(C) \cup \{\lambda\}$, $\xi_2 \in B(C)$, $w_i \in C$. It is obvious that any $\beta \in Infix(C^+)$ has at least a decomposition, except in case that $\beta \in Infix(C) - (C \cup B(C) \cup T(C))$. For any decomposition of β, say $\pi_i(\beta) = \xi_{i_1} w_{i_1} \ldots w_{i_k} \xi_{i_2}$, consider the corresponding source sequence $d_i(\beta) = a_{i_1} \ldots a_{i_k}$, $w_{i_j} \in F(a_{i_j})$, $1 \leq j \leq k$. Call $d_i(\beta)$ a deciphering of β. The following result holds.

Theorem 7. Given a multivalued encoding $F : A \to 2^{X^+}$. A necessary condition for F to admit of a SSD is that there exist $t_1, t_2 \geq 0$ such that for each $a_1 \ldots a_k \in A^+$, $w_1 \ldots w_k \in F(a_1 \ldots a_k)$, $\beta_1 \in C^{t_1}$, $\beta_2 \in C^{t_2}$

$$a_1 \ldots a_k \text{ is infix of any deciphering of } \beta_1 w_1 \ldots w_k \beta_2. \tag{12}$$

Proof. The proof is by contradiction. Let us suppose that $D = (S, s_0, X, A, f, \phi)$ is a SSD with delay t for F, and that for each $t_1, t_2 \geq 0$ there exist $a_1 \ldots a_k \in A^+$, $w_1 \ldots w_k \in F(a_1 \ldots a_k)$, $\beta_1 \in C^{t_1}$, $\beta_2 \in C^{t_2}$ such that there exists a deciphering of $\beta_1 w_1 \ldots w_k \beta_2$, say $d(\beta_1 w_1 \ldots w_k \beta_2) = a_{j_1} \ldots a_{j_n}$, that has not $a_1 \ldots a_k$ has infix. Then, one can deduce that there exist $a_1 \ldots a_k \in A^+$, $w_1 \ldots w_k \in F(a_1 \ldots a_k)$, $\beta_1 \in C^{t+1}$, $\beta_2 \in C^{t+1}$ such that there exists a deciphering of $\beta_1 w_1 \ldots w_k \beta_2$, say $d(\beta_1 w_1 \ldots w_k \beta_2) = a_{j_1} \ldots a_{j_n}$, that has not $a_1 \ldots a_k$ has infix. Let $\pi(\beta_1 w_1 \ldots w_k \beta_2) = \xi_1 w_{j_1} \ldots w_{j_n} \xi_2$ be the decomposition of $\beta_1 w_1 \ldots w_k \beta_2$ corresponding to the deciphering $d(\beta_1 w_1 \ldots w_k \beta_2) = a_{j_1} \ldots a_{j_n}$. Moreover, let $v_1 = \alpha_1 \xi_1 \in F(b_1)$ and $v_2 = \xi_2 \alpha_2 \in F(b_2)$, $b_1, b_2 \in A$. By the definition of SSD one has that for some $c_1, c_2 \in A$,

$$\phi(s_0, \alpha_1 \xi_1 w_{j_1} \ldots w_{j_n} \xi_2 \alpha_2) = \phi(s_0, v_1 w_{j_1} \ldots w_{j_n} v_2)$$

is a prefix of $b_1 a_{j_1} \ldots a_{j_n} b_2$. Moreover, since t is the delay of the SSD, one has that

$$\phi(s_0, \alpha_1 \xi_1 w_{j_1} \ldots w_{j_n} \xi_2 \alpha_2) = \phi(s_0, \alpha_1) \phi(f(s_0, \alpha_1), \xi_1 w_{j_1} \ldots w_{j_n} \xi_2 \alpha_2)$$
$$= \phi(s_0, \alpha_1) \phi(f(s_0, \alpha_1), \beta_1 w_1 \ldots w_k \beta_2 \alpha_2)$$

has $c_1 a_1 \ldots a_k c_2$ as infix, that contradicts the fact that $a_1 \ldots a_k$ is not infix of $a_{j_1} \ldots a_{j_n}$.

Q.E.D.

We will now prove that condition (12) is also sufficient for the existence of self-synchronizing decoders. To this aim, observe that a multivalued encoding satisfies condition (12) if and only if it satisfies the following condition: there exist $n_1, n_2 \geq 0$ such that for each $\beta \in Infix(C^+)$ there exist $z \in A^*$ for which any deciphering of β can be written as

$$d(\beta) = x_1 z x_2 \qquad l(x_1) \leq n_1, \quad l(x_2) \leq n_2. \tag{13}$$

Roughly speaking, if a multivalued encoding admits of a SSD then any code message has the property that all its decipherings have a common infix and differ in prefixes and suffixes of bounded length.

Given a multivalued encoding F, let us denote by q_1 and q_2 a pair of integers satisfying (13) and such that their sum is minimum; the minimization is over all pairs that satisfies (13).

In the appendix it is shown a method for constructing a SSD whose output function ϕ satisfies the following lemma.

Lemma 4. For each $\beta \in X^*$, $\gamma \in Ends(\beta)$ it holds $\phi(s_0, \beta) = xz$, with z such that any deciphering of γ, $d_i(\gamma)$, can be written as $d_i(\gamma) = x_{1_i} z x_{2_i}$, with

$$l(x_{1_i}) \leq \lceil \frac{3(q_1 + q_2 + 1)L - 1}{l} \rceil + \lceil \frac{(q_2 + P)L - 1}{l} \rceil + 2q_1$$

$$l(x_{2_i}) \leq q_1 + 2q_2 + P.$$

where P is the deciphering delay of the encoding, L and l denote the maximum and the minimum length of a codeword, respectively. □

From the above results, one has then the following theorem.

Theorem 8. Given a multivalued encoding $F : A \to 2^{X^+}$, condition (12) is necessary and sufficient for F to admit of a self-synchronizing decoder. □

Example 10. Given $A = \{0, 1\}$ and $X = \{a, b, c\}$ the multivalued encoding $F(0) = \{aa, b\}$, $F(1) = \{c\}$ is decipherable with finite delay $P = 0$. Moreover, it is synchronizable and condition (13) is satisfied for $q_2 = 0 = Q_2$ and $q_1 = 1 = Q$. The SSD for F constructed according to the method described in the Appendix and parameters q_1 and q_2, is showed in Fig. 2. □

We are left with the problem of finding a finite condition that leads to an algorithm to test whether a multivalued encoding is synchronizable, that is, if it satisfies condition (13). To this aim we again consider the formalism introduced in Section 3.

For each $x \in \Gamma^+$ define the remainder of x with respect to the pair $(a,b) \in A^* \times A^*$ as follows

$$Rem(x;(a,b)) = \begin{cases} (y,\lambda) & \text{if } a^{-1}\sigma_1(x) = b^{-1}\sigma_2(x)y, \\ (\lambda,y) & \text{if } a^{-1}\sigma_1(x)y = b^{-1}\sigma_2(x), \\ (\lambda,\lambda) & \text{if } a^{-1}\sigma_1(x) = b^{-1}\sigma_2(x), \\ \emptyset & \text{otherwise} \end{cases}$$

Recall that if α, β are two strings on the same alphabet, $\alpha^{-1}\beta$ is equal to γ if $\beta = \alpha\gamma$, is not defined if α is not a prefix of β. It is easy to see that properties a), b) and c) of the function Rem holds also for this extended definition.

Let now m be the number of internal states of the automaton $A = \{K_P \times K_P, \Gamma, \delta, (q_0, q_0), F_P \times F_P\}$ defined in Section 4. Denote by $A(Q_i, Q_j)$ the set of all strings that bring A from the state Q_i to the state Q_j and by $A_k(Q_i, Q_j)$ the set of all strings in $A(Q_i, Q_j)$ of length not greater than k.

Definition 7. We say that $x, xy \in \Gamma^+$ ($xy \neq \lambda$) is a loop of states in the state diagram of A if and only if $x, xy \in A(Q_i, Q_j)$, for some Q_i, Q_j states in **A**. We say that the loop x, xy has length n if $l(xy) = n$. □

Definition 8. Let $x, xy \in A(Q_i, Q_j)$ and $\alpha, \alpha\beta \in A(Q_i, Q_r)$ be loops of states in A. We say that x, xy and $\alpha, \alpha\beta$ are linked if following conditions hold:
  i) $\alpha = x_1 x_2$, $x = x_1 x_3$, $l(x_1) \leq m$, $(x_1, x_2, x_3 \in \Gamma^*)$;
  ii) $x_1 \in A_m(Q_i, Q_k) \Rightarrow A_m(Q_k, Q_r) \neq \emptyset$. □

Definition 9. Let D be the set $D = \{(a,b) \in (A^* \times A^*) | \exists Q_i, Q_j \text{ states of } \mathbf{A}, \exists x, xy \in A_m(Q_i, Q_j) \text{ such that } \sigma_1(xy) = az \text{ and } \sigma_2(xy) = bzc \text{ or } \sigma_1(xy) = azc \text{ and } \sigma_2(xy) = bz, z, c \in A^*\}$. □

Definition 10. We say that $x, xy \in A(Q_i, Q_j)$ is a loop of states $n-favorable$ if there exists $(a,b) \in D$ such that $Rem(x;(a,b)) = Rem(xy;(a,b)) \neq \emptyset$, and for each loop of states $\alpha, \alpha\beta$ ($l(\alpha\beta) \leq n$) linked to the loop of states x, xy, $Rem(\alpha;(a,b)) = Rem(\alpha\beta;(a,b)) \neq \emptyset$. We say that a loop of states $x, xy \in A(Q_i, Q_j)$ is favorable if it is n-favorable whatever n. □

In order to obtain the desired test for the synchronizability of multivalued encodings, we need the following technical lemma.

Lemma 5 [20]. For each Q_i, Q_j, if each loop of states $x, xy \in A_m(Q_i, Q_j)$ is m-favorable then for each $\alpha \in A_m(Q_i, Q_j)$, $\alpha\beta \in A(Q_i, Q_j)$ the loop of states $\alpha, \alpha\beta$ is favorable. □

Let now $w \in \Gamma^*$, $\delta(Q, w) = Q'$ for some Q, Q' states of A. It can be shown that it is possible to write $w = x_0^{h_0} x_1 x_2^{h_1} x_3 \ldots x_{2n-1} x_{2n}^{h_n} x_{2n+1}$ $n \geq 0$, $h_j \geq 0$, $0 \leq j \leq n$, $x_0, x_{2n+1} \in \Gamma^*$, $x_i \in \Gamma^+$, $1 \leq i \leq n$, in such a way that $l(x_1 x_3 \ldots x_{2n+1}) \leq m - 1$ and

$$\delta(Q, x_0) = Q,$$
$$\delta(Q, x_1) = Q_1 = \delta(Q_1, x_2),$$
$$\ldots, \ldots$$
$$\delta(Q, x_1 x_3 \ldots x_{2n-1}) = Q_n = \delta(Q_n, x_{2n})$$

with $Q_i \neq Q_j$ if $i \neq j$. Further, if for any Q_i, Q_j, for any $x, xy \in A_m(Q_i, Q_j)$ the loop of states x, xy is favorable, from Lemma 5 one has that there exists $(a, b) \in D$ such that

$$\begin{aligned}
Rem(\lambda; (a, b)) &= Rem(x_0; (a, b)) \neq \emptyset \quad \text{(if } x_0 \neq \lambda\text{)} \\
Rem(x_1; (a, b)) &= Rem(x_1 x_2; (a, b)) \neq \emptyset \\
&\ldots \\
Rem(x_1 x_3 \ldots x_{2n-1}; (a, b)) &= Rem(x_1 x_3 \ldots x_{2n-1} x_{2n}; (a, b)) \neq \emptyset
\end{aligned} \quad (14)$$

We are now able to prove the following theorem that provides an algorithm for testing whether a multivalued encodings is synchronizable.

Theorem 9. A necessary and sufficient condition for a multivalued encoding $F: A \to 2^{X^+}$ to be synchronizable is that any loop of states in the state diagram of A of length not greater than m is m-favorable, where m is the number of states of A.

Proof. Sufficiency. Let us consider $w = x_0^{h_0} x_1 x_2^{h_1} x_3 \ldots x_{2n-1} x_{2n}^{h_n} x_{2n+1} \in \Gamma^+$, with $\delta(Q, w) = Q'$, $Q, Q' \in K_P \times K_P$. One has that there exists $(a, b) \in D$ such that $Rem(\lambda; (a, b)) = Rem(x_0; (a, b)) \neq \emptyset$, and, from the properties of Rem, one gets $Rem(\lambda; (a, b)) = Rem(x_0^{h_0}; (a, b)) \neq \emptyset$. From (14) and from the properties of Rem, one has then $Rem(x_1; (a, b)) = Rem(x_0^{h_0} x_1; (a, b)) \neq \emptyset$, $Rem(x_1; (a, b)) = Rem(x_0^{h_0} x_1 x_2^{h_1}; (a, b)) \neq \emptyset$ and so on up to get $Rem(x_1 x_2 \ldots x_{2n-1}; (a, b)) = Rem(x_0^{h_0} x_1 x_2^{h_1} x_3 \ldots x_{2n-1} x_{2n}^{h_n}; (a, b)) \neq \emptyset$. By definition of Rem one has that $\sigma_1(w)$ and $\sigma_2(w)$ either coincide or have a common infix. The same holds for $\sigma_1(x_1 x_3 \ldots x_{2n-1})$ and $\sigma_2(x_1 x_3 \ldots x_{2n-1})$. In case $\sigma_1(w)$ and $\sigma_2(w)$ do not coincide, the prefixes and the suffixes in which they differ are the same in which $\sigma_1(x_1 x_3 \ldots x_{2n-1})$ and $\sigma_2(x_1 x_3 \ldots x_{2n-1})$ differ. Since $l(x_1 x_3 \ldots x_{2n-1}) \leq m - 1$, the length of these prefixes and suffixes are themselves not greater than $m - 1$. Then one has that for each $w \in \Gamma^+$ such that $\delta(Q, w)$ is defined, for some state Q of A, and then for each infix of code message $\tau \alpha_1(w) = \tau \alpha_2(w)$, it is possible to determine an infix of the source sequence that generated the message with resulting (undeciphered) prefix and suffix having bounded length.

Necessity. Let us suppose that there exists a loop of states $x, xy \in A_m(Q_i, Q_j)$ that is not m-favorable. It is possible to distinguish the following two cases.

Case a). For each (a, b) $Rem(x; (a, b)) = Rem(xy; (a, b)) = \emptyset$ or $Rem(x; (a, b)) \neq Rem(xy; (a, b))$. We will show that for each $R > 0$ there exists k such that $\sigma_1(xy^k)$ and $\sigma_2(xy^k)$ either do not have a common infix or differ in prefixes and suffixes of length greater than R. Indeed, if this is not true, one has $\sigma_1(y) = cd$ and $\sigma_2(y) = dc$, $c, d \in A^*$, from which it follows $Rem(x; (\sigma_1(x), \sigma_2(x)d)) = Rem(xy; (\sigma_1(x), \sigma_2(x)d)) = (d, \lambda) \neq \emptyset$, that contradicts the hypothesis, because $(\sigma_1(x), \sigma_2(x)d) \in D$. Then, there exists a code message such that either its decodings do not have a common infix or they differ in prefixes and suffixes of unbounded length, and the encoding is not synchronizable.

Case b). There exists $\alpha, \alpha\beta \in A_m(Q_i, Q_k)$ loop of states linked to x,xy such that there does not exist $(a, b) \in D$ for which $Rem(x; (a, b)) = Rem(xy; (a, b)) \neq \emptyset$ and $Rem(\alpha; (a, b)) = Rem(\alpha\beta; (a, b)) \neq \emptyset$. Since the loop of states $\alpha, \alpha\beta$ is linked to the loop of states x, xy, one has $x = x_1 x_2$, $\alpha = x_1 x_3$ and $x_1 \in A_m(Q_i, Q_r)$ $x_4 \in A_m(Q_k, Q_r)$. It is possible to show that for each $R > 0$ there exist h_1, h_2, h_3 such that $\sigma_1(x_1(x_3\beta^{h_1}x_4)^{h_2}x_2y^{h_3})$ and $\sigma_2(x_1(x_3\beta^{h_1}x_4)^{h_2}x_2y^{h_3})$ differ in prefixes and suffixes of length greater than R. Indeed, if this were not true, one would have that for each $h \geq 0$ $\sigma_1(x_3\beta^h x_4) = a(h)b(h)$ and $\sigma_2(x_3\beta^h x_4) = b(h)a(h)$, $a(h), b(h) \in A^*$, i.e., the two decodings of the cycle $x_3\beta^h x_4$ must be one a cyclic permutation of the other. In addition $\sigma_1(\beta) = cd$ and $\sigma_2(\beta) = dc$ $c, d \in A^*$. Hence, it follows that $\sigma_1(x_3\beta^h x_4) = ab(h)$ and $\sigma_2(x_3\beta^h x_4) = b(h)a$ [Resp. $\sigma_1(x_3\beta^h x_4) = a(h)b$ and $\sigma_2(x_3\beta^h x_4) = b(h)b$]. This implies that for some $z \in A^+$, either $c^{-1}\sigma_1(x_3)z = \sigma_2(x_3)$ and $c^{-1}\sigma_1(x_3\beta)z = \sigma_2(x_3\beta)$, or $c^{-1}\sigma_1(x_3) = \sigma_2(x_3)z$ and $c^{-1}\sigma_1(x_3\beta) = \sigma_2(x_3\beta)z$. Moreover, if $\sigma_1(x_1(x_3\beta^{h_1}x_4)^{h_2}x_2y^{h_3})$ and $\sigma_2(x_1(x_3\beta^{h_1}x_4)^{h_2}x_2y^{h_3})$ differ in prefixes and suffixes of bounded length, it must hold that for each h_1, h_2, h_3 either $c^{-1}\sigma_1(x_2y)s = \sigma_2(x_2y)$ and $c^{-1}\sigma_1(x_2y^2)s = \sigma_2(x_2y^2)$, or $c^{-1}\sigma_1(x_2y) = \sigma_2(x_2y)s$ and $c^{-1}\sigma_1(x_2y^2) = \sigma_2(x_2y^2)s$. It follows that $Rem(x_1x_3; (\sigma_1(x_1)c, \sigma_2(x_1))) = Rem(x_1x_3\beta; (\sigma_1(x_1)c, \sigma_2(x_1))) \neq \emptyset$ and $Rem(x_1x_2; (\sigma_1(x_1)c, \sigma_2(x_1))) = Rem(x_1x_2y\beta; (\sigma_1(x_1)c, \sigma_2(x_1))) \neq \emptyset$ contradicting the hypothesis. Then one gets that there exists an arbitrary long infix of a sequence of codewords $\tau_1(x_1(x_3\beta^{h_1}x_4)^{h_2}x_2y^{h_3})$ such that its decodings either do not have a common infix or differ in prefixes and suffixes of unbounded length.
Q.E.D.

6. DECODERS WITH INITIAL STATE INVARIANCE

In the previous section we have studied multivalued encodings that admit of sequential decoders able to re-synchronize themselves within a finite number of steps. Generally speaking, the above decoders may output wrong simbols during the re-synchronization time giving rise to decoding errors. Although the re-synchronization time of a SSD is bounded, in some situations it would be desiderable to have decoders able to eliminate completely the effects due to a loss of synchronization. A solution to this problem is represented by decoders with initial state invariance, i.e., decoders which decipher correctly the incoming code message whichever their initial time state is. A formal definition is the following.

Definition 11. A decoder $D = (S, s_0, X, A, f, \phi)$ for a multivalued encoding $F : A \to 2^{X^+}$ is called invariant with respect to the initial state (or simply invariant) if and only if for each $s_j \in S$ the gsm $D_j = (S, s_j, X, A, f, \phi)$ is a decoder for F. □

We shall give a necessary and sufficient condition for a multivalued encoding to admit of an invariant decoder.

Let $F : A \to 2^{X^+}$ be a multivalued encoding and $C = \cup_{a \in A} F(a)$. For each $a \in A$, $w \in F(a)$ denote by $p_1(w)$ the smallest prefix of w such that, for some integer p

$$w = p_1(w)\gamma_1 \ldots \gamma_p,$$

where each γ_i ($\gamma_i \neq \lambda$) is a prefix of some codeword in $F(a_i)$, $a_i \neq a$. Denote by $P_1(C)$ the set

$$P_1(C) = \{p_1(w) | w \in C\}.$$

Then for $i > 1$, $a \in A$, $w \in F(a)$ define $p_i(w)$ as the smallest proper prefix of $p_{i-1}(w)$ such that either there exist $u, v \in F(a)$, $\xi_1, \xi_2 \in X^*$ for which $p_i(w)v = p_{i-1}(w)\xi_1 p_{i_1}(u)\xi_2$ or there exist $c \in A$ ($c \neq a$), $v \in F(c)$, $\gamma \in X^*$ for which $p_i(w)v = p_{i-1}(w)\gamma$, if such a proper prefix exists, $p_i(w) = p_{i-1}(w)$ otherwise. Then for each $i > 1$ denote by $P_i(C)$ the set

$$P_i(C) = \{p_i(w) | w \in C\}.$$

Finally, define

$$P(C) = \{p(w) | w \in C\} = P_n(C)$$

where n is the smallest integer such that $P_n(C) = P_{n+1}(C)$. It is easy to see that if C is finite, such an integer exists.

Definition 12. A multivalued encoding F is called fault-tolerant if and only if the following conditions hold
  a) for each $a, b \in A$ ($a \neq b$), $v \in F(a)$, $w \in F(b)$ $w \neq \beta_1 p(v) \beta_2$, for each $\beta_1, \beta_2 \in X^*$
  b) for each $a \in A$, $u, v \in F(a)$, $w \neq \xi_1 p(u) \xi_2 p(v) \xi_3$, for each $\xi_1, \xi_2, \xi_3 \in X^*$. □

Example 11. Given $A = \{0, 1\}$, $X = \{a, b, c\}$, the multivalued encoding $F(0) = \{ac, ab\}$, $F(1) = \{cbb, cabb\}$ is not fault-tolerant. Indeed, $P_1(C) = \{a, ab, cbb, cabb\} = P_2(C) = P(C)$ and $p(ac) = a$, $cabb = cp(ac)bb$ with $ac \in F(0)$ and $cabb \in F(1)$. It follows that condition a) of Definition 12 is not satisfied. □

The following theorem states that a multivalued encoding admits of an invariant decoder if and only if it is fault-tolerant.

Theorem 10. Let $F : A \to 2^{X^+}$ be a multivalued encoding. A necessary and sufficient condition for an invariant decoder for F to exist is that F is fault-tolerant.

Proof. Necessity. Let $D = (S, s_0, X, A, f, \phi)$ be an invariant decoder for F. We shall show that for each $s \in S$, $a \in A$, $w \in F(a)$

$$\phi(s, w) = a. \tag{15}$$

Indeed, if t is the decoding delay of D, by definition of decoder, one has that for each $b \in A$, $v \in F(b)$, $\gamma \in C^t$

$$\phi(s, wv\gamma) \geq ab. \tag{16}$$

On the other hand, since D is invariant, one has

$$\phi(s, wv\gamma) = \phi(s, w)\phi(f(s, w), v\gamma) \geq \phi(s, w)b. \tag{17}$$

From (16) and (17), it follows (15). Reinforcing the above result, it is possible to show that for each $s \in S$, $a \in A$, $w \in F(a)$

$$\phi(s, p(w)) = a. \tag{18}$$

The proof is by induction. We shall prove that for each $s \in S$, $a \in A$, $w \in F(a)$, $i \geq 1$

$$\phi(s, p_i(w)) = a. \tag{19}$$

Let $i = 1$. By definition of $p_1(w)$, one gets $w = p_1(w)\gamma_1 \ldots \gamma_p$, $\gamma_j \in Prefix(F(b_j))$, for some $b_j \in A$, $b_j \neq a$, $1 \leq j \leq p$. Since $\phi(s, w) = \phi(s, p_1(w)\gamma_1 \ldots \gamma_p) = a$, in order to prove (18), it suffices to show that

$$\phi(f(s, p_1(w)\gamma_1 \ldots \gamma_{j-1}), \gamma_j) = \lambda, \qquad 1 \leq j \leq p. \tag{20}$$

Suppose that (20) is not true. From (15), one has

$$\phi(f(s, p_1(w)\gamma_1 \ldots \gamma_{j-1}), \gamma_j) = a \tag{21}$$

for some $1 \leq j \leq p$. From the definition of γ_j, one gets that there exist $b_j \in A$ ($b_j \neq a$) and $w_j \in F(b_j)$ such that $w_j = \gamma_j \gamma$, $\gamma \in X^*$. Moreover, because of (15), one has that $\phi(f(s, p_1(w)\gamma_1 \ldots \gamma_{j-1}), w_j) = b_j$, whereas, from (21), one gets $\phi(f(s, p_1(w)\gamma_1 \ldots \gamma_{j-1}), w_j) \geq \phi(f(s, p_1(w)\gamma_1 \ldots \gamma_{j-1}), \gamma_j) = a$ that contradicts the assumption $a \neq b_j$. Thus (19) holds for $i = 1$. We now prove that if it holds for $i - 1$ then it holds also for i. If $p_i(w) = p_{i-1}(w)$ then (19) is obviously true.

Suppose then $p_i(w) = p_{i-1}(w)\gamma$, $\gamma \neq \lambda$. It is possible to distinguish the following two cases.

Case i). There exist $u, v \in F(a)$ such that $p_i(w)v = p_{i-1}(w)\xi_1 p_{i-1}(u)\xi_2 = p_i(w)\gamma p_{i-1}(u)\xi_2$, $\xi_1, \xi_2 \in X^*$.
Suppose that $\phi(s, p_i(w)) = \lambda$. Because of the inductive hypothesis $\phi(s, p_{i-1}(w)) = a$, it follows $\phi(f(s, p_{i-1}(w)), \gamma\xi_1 p_{i-1}(u)\xi_2) \geq aca$, $c \in A^*$. On the other hand, from (15), one has $\phi(f(s, p_i(w)), v) = a$, that contradicts the above relation.

Case ii). There exist $b \in A$ ($b \neq a$), $v \in F(b)$ such that $p_i(w)v = p_{i-1}(w)\alpha$, $\alpha \in X^*$.
Let $p_i(w)\gamma = p_{i-1}(w)$, $\gamma \in X^+$. From the inductive hypothesis one has $\phi(s, p_{i-1}(w)) = a$. Then, in order to prove (19), it suffices to show that $\phi(f(s, p_i(w)), \gamma) = \lambda$. Suppose, on the contrary, $\phi(f(s, p_i(w)), \gamma) = a$; one gets $\phi(f(s, p_i(w)), v) \geq \phi(f(s, p_i(w)), \gamma) = a$. On the other hand, by definition of invariant decoder, one has $\phi(s, p_i(w)) = b \neq a$. This contradicts the above relation.

Since for each $a \in A$, $w \in F(a)$ there exists n such that $p(w) = p_n(w)$, it follows that (18) is satisfied.

In order to prove that the condition stated in the theorem is necessary, let us suppose that the multivalued encoding F is not fault-tolerant. It is possible to distinguish the following two cases.

Case 1). Condition a) of Definition 12 does not hold.
It follows that there exists $a, b \in A$ ($a \neq b$), $w \in F(a)$, $v \in F(b)$ such that $w = \beta_1 p(v)\beta_2$, $\beta_1, \beta_2 \in X^*$. Let $s \in S$. By using (18), one gets

$$\phi(s, w) = \phi(s, \beta_1 p(v)\beta_2) = \phi(s, \beta_1)\phi(f(s, \beta_1), p(v))\phi(f(s, \beta_1 p(v)), \beta_2) \\ = \phi(s, \beta_1)\phi(f(s, \beta_1 p(v)), \beta_2) \quad (22)$$

On the other hand, from (15) one has that $\phi(s, w) = a$, that contradicts (22).

Case 2). Condition b) of Definition 12 does not hold.
It follows that there exist $a \in A$, $u, v, w \in F(a)$ such that $w = \xi_1 p(u)\xi_2 p(v)\xi_3$, $\xi_1, \xi_2, \xi_3 \in X^*$. Let $s \in S$. By using (18), one gets

$$\phi(s, w) = \phi(s, \xi_1 p(u)\xi_2 p(v)\xi_3) \geq \phi(s, \xi_1 p(u)\xi_2 p(v)) \\ = \phi(s, \xi_1)a\phi(f(s, \xi_1 p(u)), \xi_2)a. \quad (23)$$

On the other hand, from (15) it follows $\phi(s, w) = a$ that contradicts (23).

Sufficiency. Let $F : A \to 2^{X^+}$ be a multivalued encoding. We shall give an algorithm for constructing an invariant decoder for F. Denote by M the set

$$M = \{x \in X^* | \exists y \in P(C), \exists z \in X^+ [y = xz \text{ and for each } \beta_1, \beta_2 \in X^*, \\ \alpha \in P(C) \quad x \neq \beta_1 \alpha \beta_2]\},$$

i.e., M is the set of all proper prefixes of elements in $P(C)$ that have no elements of $P(C)$ as infix. Moreover, for each $\beta \in X^+$ denote by $suf(\beta)$ the longest suffix of β that belongs to M. Define then the gsm $D = (S, s_0, X, A, f, \phi)$ in the following way

a) $S = \{s_y | y \in M\}$, $s_0 = s_\lambda$,
b) for each $(s_y, b) \in S \times X$, the transition function f and the output function ϕ are determined as follows

$$f(s_y, b) = \begin{cases} s_\lambda & \text{if } yb \text{ has a suffix belonging to P(C)} \\ s_{suf(yb)} & \text{otherwise,} \end{cases}$$

$$\phi(s_y, b) = \begin{cases} a \in A & \text{if } yb = \beta p(w),\ w \in F(a),\ \beta \in X^* \\ \lambda & \text{otherwise.} \end{cases}$$

This definition implies that D performs as follows: if it receives a string $\beta \in X^*$ that has no infixes belonging to $P(C)$ then

$$f(s_\lambda, \beta) = s_{suf(\beta)} \quad \text{and} \quad \phi(s_\lambda, \beta) = \lambda \tag{24}$$

Moreover, if $b \in X$ is such that βb has a suffix $p(w) \in P(C)$, $w \in F(a)$, $a \in A$, then

$$f(s_\lambda, \beta b) = f(s_{suf(\beta)}, b) = s_\lambda \quad \text{and} \quad \phi(s_\lambda, \beta b) = \phi(s_{suf(\beta)}, b) = a \tag{25}$$

In order to show that D is an invariant decoder for F, we shall prove that for each $s_y \in S$, $a \in A$, $w \in F(a)$ $\phi(s_y, w) = a$. Let α be the shortest prefix of yw such that there exist $b \in A$ and $v \in F(b)$ for which $p(v)$ is a suffix of α. One can write $yw = \alpha\beta_2 = \beta_1 p(v)\beta_2$, $\beta_1, \beta_2 \in X^*$. By definition of S, it follows that y is a proper prefix of $\beta_1 p(v)$. From (24), one gets $f(s_\lambda, y) = s_{suf(y)}$ and $\phi(s_\lambda, y) = \lambda$. In order to show that $\phi(s_\lambda, w) = a$, let us distinguish the following two cases.

Case i). $p(v)$ is infix of w.

From the definition of fault-tolerant multivalued encoding, it results that there exists $a \in A$ such that $v, w \in F(a)$. Let $w = \beta p(v)\beta_2$. Since no infixes of $y\beta$ belongs to P(C), from (24) one gets

$$f(s_\lambda, y\beta) = s_{suf(y\beta)} \in S \quad \text{and} \quad \phi(s_\lambda, y\beta) = \lambda$$

Whereas, from (25) one has

$$f(s_\lambda, y\beta p(v)) = s_\lambda = f(f(s_\lambda, y), \beta p(v)) = f(s_y, \beta p(v)),$$
$$\phi(s_\lambda, y\beta p(v)) = a = \phi(s_\lambda, y)\phi(f(s_\lambda, y), \beta p(v)) = \phi(s_y, \beta p(v)).$$

In order to prove that $\phi(s_y, w) = a$, it suffices to show that $\phi(s_\lambda, \beta_2) = \lambda$. From the definition of fault-tolerant multivalued encoding it follows that no words $u \in C$

exist for which $w = \xi_1 p(v)\xi_2 p(u)\xi_3$, $\xi_1, \xi_2, \xi_3 \in X^*$. This implies that no elements of $P(C)$ are infix of β_2. From (24) it follows

$$f(s_\lambda, \beta_2) = s_{suf(\beta_2)} \in S \quad \text{and} \quad \phi(s_\lambda, \beta_2) = \lambda.$$

Case ii). $\beta_1 p(v) = y\mu$, μ prefix of w, $\mu \neq \lambda$.
From (24), one has

$$f(s_\lambda, y) = s_{suf(y)} \in S \quad \text{and} \quad \phi(s_\lambda, y) = \lambda.$$

Moreover, it is possible to show that $v \in F(a)$. Indeed, assuming $v \in F(b)$, $b \in A$, $b \neq a$, one has $p(v)\beta_2 = \delta w$, δ proper prefix of $p(v)$, which contradicts the definition of $p(v)$. From (25) one gets

$$f(s_\lambda, \beta_1 p(v)) = s_\lambda \quad \text{and} \quad \phi(s_\lambda, \beta_1 p(v)) = a.$$

Moreover,

$$\phi(s_\lambda, \beta_1 p(v)\beta_2) = \phi(s_\lambda, yw) = \phi(f(s_\lambda, y), w) = \phi(s_y, w),$$
$$\phi(s_\lambda, \beta_1 p(v)\beta_2) = \phi(s_\lambda, \beta_1 p(v))\phi(f(s_\lambda, \beta_1 p(v)), \beta_2) = a\phi(s_\lambda, \beta_2);$$

that is,

$$\phi(s_y, w) = \phi(s_\lambda, \beta_2).$$

Finally, in order to show that $\phi(s_y, w) = a$ it suffices to prove that $\phi(s_\lambda, \beta_2) = \lambda$. Since the encoding is fault-tolerant, is possible to see that there does not exist $u \in F(b)$, $b \in A$, $b \neq a$, with $p(u)$ infix of β_2. Otherwise one would have that $p(u)$ is infix of $w \in F(a)$, which contradicts the fact that the multivalued encoding is fault-tolerant. Suppose that there exists $u \in F(a)$ such that $p(u)$ is infix of β_2. It follows that $p(v)\beta_2 = \delta w = p(v)\xi_1 p(u)\xi_2$, with δ proper prefix of $p(v)$, that contradicts the definition of $p(v)$. Since β_2 has no infixes belonging to $P(C)$, from (24) one gets

$$f(s_\lambda, \beta_2) = s_{suf(\beta_2)} \in S \quad \text{and} \quad \phi(s_\lambda, \beta_2) = \lambda.$$

This completes the proof.

Q.E.D.

Example 12. Given $A = \{0, 1\}$, $X = \{a, b, c\}$, consider the multivalued encoding $F(0) = \{cabcb, ab, cab\}$, $F(1) = \{bbc\}$. One gets $P(C) = \{ca, a, ca, bb\}$ and $M = \{\lambda, c, b\}$. The invariant decoder obtained applying the previous algorithm is shown in Fig. 3. □

APPENDIX

We give some definitions that will be used for the construction, for a synchronizable multivalued encoding, of a SSD satisfying Lemma 4.

For each $Z \subset X^+$, $Z \neq \emptyset$, we denote with $max(Z)$ the longest element of Z. Let n be a positive integer, $Y \subset Prefix(C^+)$, $Y \neq \emptyset$. Define the function $G^n : 2^{Prefix(C^+)} \rightarrow A^*$ in the following way: $G^n(Y) = a_1 \ldots a_s \in A^*$, $(s \geq 0)$, where $a_1 \ldots a_s$ is the longest common prefix of all partial deciphering of the elements of Y, obtained by ignoring the last n symbols of the longest of them.

Let $Z = (Z_1 \ldots Z_m)$ be a m-tupla of $Z_i \subset Prefix(C^+)$, $Z_i \neq \emptyset$. Moreover suppose that each Z_i has the following property, it is possible to partition Z_i into disjoint subsets $X_{i_1}, \ldots, X_{i_k}$ in such a way that for each X_{i_j}, $(1 \leq j \leq k_i)$, there exists $a_{i_j} \in A^+$ such that for each $\alpha \in X_{i_j}$, there exist $\beta \in X_{i_{j+1}}$, $w \in F(a_{i_j})$ for which $\beta = w\alpha$. In such a case we will say that Z_i has the property of k_i-disjunction and refer to the elements a_{i_j}'s as the separation elements between X_{i_j} and $X_{i_{j+1}}$. This property implies that all the elements of $X_{i_{j+1}}$ are obtainable by adding a suitable coding of a same $a_{i_j} \in A^+$ at the beginning of an element of X_{i_j}.

For each $Y \subset Prefix(C^+)$, Z having the property of k-disjunction, Y and Z not both empty, define the function I as follows

$$I(Y, Z) = \begin{cases} G^{Q_2}(Y) & \text{if } l(G^{Q_2}(Y)) > q_1, Y \neq \emptyset \\ G^{Q_2}(X_{1k}) & \text{if } l(G^{Q_2}(X_{1_1})) > q_1, Y = \emptyset \\ \lambda & \text{otherwise} \end{cases}$$

where $Q_2 = q_2 + P$, P equal to the deciphering delay of F.

Moreover, for each $Y \subset Prefix(C^+)$, Z m-tupla of subsets of $Prefix(C^+)$ having the property of k-disjunction and $b \in X$ define

$$I(Y, Z, b) = I(Yb \cap Prefix(C^+), Zb \cap Prefix(C^+))$$

where

$$Zb \cap Prefix(C^+) = (Z_1 b \cap Prefix(C^+), \ldots, Z_m b \cap Prefix(C^+))$$

considering in this m-tupla only the sets $Z_i \cap Prefix(C^+) \neq \emptyset$. Moreover, define

$$(Z_1, \ldots, Z_m) + Z' = \begin{cases} (Z_1, \ldots, Z_m) & \text{if } Z' = \emptyset \text{ or } Z' = Z_i, 1 \leq i \leq m \\ (Z_1 = Z') & \text{if } (Z_1, \ldots, Z_m) = (\emptyset) \\ (Z_1, \ldots, Z_m, Z_{m+1} = Z') & \text{otherwise.} \end{cases}$$

$$(Y, (Z_1 \ldots Z_m) + Z') = \begin{cases} (Y, (Z_1 \ldots Z_m)) & \text{if } Z' = Y \\ (Y, ((Z_1 \ldots Z_m) + Z')) & \text{otherwise.} \end{cases}$$

and
$$(Y,(Z_1,\ldots,Z_m)+(Z_1',\ldots,Z_m'))=(((Y,(Z_1,\ldots,Z_m)+Z_1')+\ldots)+Z_m')$$

We introduce now some functions that will be used for constructing a SSD. Let $Y \subset Prefix(C^+)$ and $Z=(Z_1,\ldots,Z_m)$, $Z_i \subset Prefix(C^+)$ having the property of k_i-disjunction. Denote with

$$R_1(Y,Z) = \{x \in Prefix(C^+) | \exists w_1 \ldots w_k \in F(I(Y,Z)), \exists \beta \in Y,$$
$$\exists \delta(\beta)[\delta(\beta) = w_1 \ldots w_k x]\}$$

the set obtained by removing from any initial decomposition of any element of Y, which corresponding partial deciphering has I(Y,Z) as prefix, the first k codewords. In a similar way define

$$R_2^i(Y,Z) = \{x \in Prefix(C^+) | \exists w_1 \ldots w_k \in F(I(Y,Z)), \exists \beta \in Z_i,$$
$$\exists \delta(\beta)[\delta(\beta) = w_1 \ldots w_k x]\}$$

and

$$R_3^i(Y,Z) = \{x \in Prefix(C^+) | \exists w_1 \ldots w_k \in F(G^{Q_2}(X_{1_j})), 1 \leq j \leq k, \exists \beta \in Z_i,$$
$$\exists \delta(\beta)[\delta(\beta) = w_1 \ldots w_k x \text{ and the length of the longest partial}$$
$$\text{deciphering of x is at most } q_2 + Q_2\}$$

Then set
$$R_2(Y,Z) = (((R_2^1(Y,Z))+\ldots)+R_2^m(Y,Z))$$
$$R_3(Y,Z) = (((R_3^1(Z_1,Z))+\ldots)+R_3^m(Z_1,Z))$$

Given $\beta \in Prefix(C^+)$, define
$$R(\beta) = (((R'(\beta_1))+\ldots)+R'(\beta_k))$$

where
i) $k = |Ends(\beta) - \{\beta\}| = |\{\beta_1,\ldots,\beta_k\}|$,
ii) $R'(\beta_i) = \{x \in Prefix(C^+) | \exists x \in A^+ \text{ such that } I(Y,Z) = yz, G^P(\beta_i) = azb, l(a) \leq q_1, l(b) \leq 2q_2, \exists \delta(\beta_i) \text{ with } \delta(\beta_i) = w_1 \ldots w_r \beta = v_1 v_2 x, v_1 \in F(a), v_2 \in F(z)\}$.

By using the functions just introduced, define

$$S(Y,Z) = \begin{cases} (((R_1(Y,Z),(\emptyset))+R_2(Y,Z))+R(max(Y))) & \text{if } Y \neq \emptyset \\ (Y,Z) & \text{if } Y = \emptyset \text{ and } l(G^{Q_2}(X_{1_1})) \leq q_1 \\ (((R_1(X_{1_1},Z),(\emptyset))+R_3(Z_1,Z))+R(max(X_{1_1}))) & \text{otherwise.} \end{cases}$$

Finally, for each $b \in X$ define

$$S(Y,Z,b) = \begin{cases} (Yb \cap Prefix(C^+), Zb \cap Prefix(C^+)) \\ \quad \text{if } Yb \cap Prefix(C^+) \neq \emptyset \text{ or } l(G^{Q_2}(X_{1_i})) > q_1 \\ (max(Yb), Zb \cap Prefix(C^+)) \\ \quad \text{if } Yb \cap Prefix(C^+) = \emptyset \text{ and } l(G^{Q_2}(X_{1_i})) \leq q_1 \\ S(max(Ends(max(Yb))), (\emptyset)) \quad \text{otherwise} \end{cases}$$

By using the functions I and S we are now able to furnish, in the hypothesis that a multivalued encoding F satisfies (13), an algorithm for constructing a SSD for F.

Let M be a set defined as follows
 i) $(\lambda, (\emptyset)) \in M$,
 ii) for each $(Y, Z) \in M$, $b \in X$ $S(Y, Z, b) \in M$.

The SSD $D = (S, s_0, X, A, f, \phi)$ for F is defined in the following way: $S = \{s_{(Y,Z)} | (Y, Z) \in M\}$ is the state set, $s_0 = s_{(\lambda,(\emptyset))}$ is the initial state, X and A are the input and output alphabet, respectively, the transition function f and the output function ϕ, defined over the pairs $(s_{(Y,Z)}, b) \in S \times X$, are defined in the following way

$$f(s_{(Y,Z)}, b) = s_{S(Y,Z,b)} \in S$$

$$\phi(s_{(Y,Z)}, b) = I(Y, Z, b) \in A^*$$

Finally, denote by $|\beta|$, $\beta \in Prefix(C^+)$, the length of the longest deciphering of β. Observe now that a definition of SSD equivalent to the given one is the following.

Definition A1. Let $F : A \to 2^{X^+}$ be a multivalued encoding. We say that the gsm $D = (S, s_0, X, A, f, \phi)$ is a SSD for F if and only if there exists $r \geq 0$ such that
 i) D is a decoder for F with delay r;
 ii) for each $\beta \in X^*$, $\gamma \in Ends(\beta)$ $\phi(s_0, \beta) = xz$, with z infix of any partial deciphering of γ and $|\gamma| - l(z) \leq r$. □

We are now able to prove Lemma 4 for the SSD constructed according to the above described method.

Lemma 4. For each $\beta \in X^*$, $\gamma \in Ends(\beta)$ it holds $\phi(s_0, \beta) = xz$, with z such that any deciphering of γ, $d_i(\gamma)$, can be written as $d_i(\gamma) = x_{1_i} z x_{2_i}$ with

$$l(x_{1_i}) \leq \lceil \frac{3(q_1 + q_2 + 1)L - 1}{l} \rceil + \lceil \frac{Q_2 L - 1}{l} \rceil + 2q_1$$

$$l(x_{2_i}) \leq Q + q_2. \tag{A.1}$$

Proof. The proof is by induction on the length of β. Let

$$N = \lceil \frac{3(q_1 + q_2 + 1)L - 1}{l} \rceil + \lceil \frac{Q_2 L - 1}{l} \rceil + 2q_1 + Q + q_2$$

One has that for each $\beta \in X^*$, $\gamma \in Ends(\beta)$ with $|\gamma| > N$, $\phi(s_0, \beta) = xz$ with z such that each deciphering of γ is of the form $d_i(\gamma) = x_{1_i} z x_{2_i}$ with

$$l(x_{1_i}) \leq \lceil \frac{3(q_1 + q_2 + 1)L - 1}{l} \rceil + \lceil \frac{Q_2 L - 1}{l} \rceil + 2q_1 \qquad (A.2)$$

$$l(x_{2_i}) \leq Q + q_2.$$

Moreover, if $f(s_0, \beta) = s_{(Y,Z)}$ one of the following condition holds
  i) $Y = \{y \in Prefix(C^+) | \exists d_i(\gamma) = x_1 z x_{2_i}, \exists w \in F(x_{2_i}), \exists v \in B(C)$ such that $\gamma = uwv$ for some $u \in C^+$ and $y = wv\}$
  ii) there exists $Z_t \in Z$ such that Z_t has the property of k_t-disjunction and there exists r such that $X_{t_r} = \{x \in Prefix(C^+) | \exists d_i(\gamma) = x_1 z x_{2_i} \exists w \in F(x_{2_i}) \exists v \in B(C)$ such that $\gamma = uwv$, for some $u \in C^+$ and $x = wv\}$. Moreover, if $\alpha_{t_j} = a_{t_k t-1} \ldots a_{t_j}$ it holds $z\alpha_{t_j} = \mu_{t_j} \tilde{z}$, $\mu_{t_j} \in A^*$.

It is obvious that (A.2), i) and ii) hold for each $\beta \in X^*$ with $l(\beta) = 0$. We will show that if they are true for each $\beta \in X^*$ with $l(\beta) = n$ then they are true for each $\beta b \in X^*$ wichever $b \in X$. Let $f(s_0, \beta) = s_{(Y,Z)}$ and $f(s_{(Y,Z)}, b) = f(s_0, \beta b) = s_{(Y'',Z'')}$. It is obvious that (A.2) is true for each $\gamma \in Ends(\beta b)$ with $|\gamma| \leq N$. In order to show that (A.2) and either i) or ii) hold for each $\gamma \in Ends(\beta b)$ with $|\gamma| \geq N+1$, let us suppose that γ satisfies either i) or ii). Let us distinguish same cases.

Case 1). $Yb \cap Prefix(C^+) \neq \emptyset$

Let $\phi(s_{(Y,Z)}, b) = \lambda$. Let $\gamma b \in Ends(\beta b)$, $\gamma \in Ends(\beta)$ satisfying (A.2) and i). It follows that $\phi(s_0, \beta b) = \phi(s_0, \beta) = xz$ and each deciphering of γb can be written as $d_i(\gamma b) = x_1 z x_{2_i} a_i$, $a_i \in A \cup \{\lambda\}$, $x_1 z x_{2_i}$ is some deciphering of γ. Then, γb satisfies i). Since for each $y \in Yb \cap Prefix(C^+)$ it holds $|y| < Q$ then $l(x_{2_i} a_i) \leq Q$ and (A.2) holds for γb. Let now $\gamma b \in Ends(\beta b)$, $\gamma \in Ends(\beta)$ satisfying (A.2) and ii). It follows that each deciphering of γb can be written as $d_i(\gamma b) = x_1 z x_{2_i} a_i$, $a_i \in A \cup \{\lambda\}$ where $x_1 z x_{2_i}$ is some deciphering of γ. One has to show that $l(x_{2_i} a_i) \leq Q + q_2$. From (13), we know that all decipherings of any suffix of an element of $Ends(\beta b)$ satisfying (A.2) and i), and a suffix of $\gamma b \in Ends(\beta b)$ have a common infix and differ in suffixes of length not greater than q_2. One has then that $l(x_{2_i} a_i) \leq Q + q_2$ and (A.2) holds for γb. It is obvious that γb satisfies ii). Let now $\phi(s_{(Y,Z)}, b) = a_1 \ldots a_k = G^{Q_2}(Yb \cap Prefix(C^+))$. Recall that $f(s_{(Y,Z)}, b) = s_{(Y'',Z'')}$. Let $\gamma b \in Ends(\beta b)$ with $\gamma \in Ends(\beta)$ for which (A.2) and i) hold. One has that for each $y \in Y''$ $|y| \leq Q_2$ and $a_1 \ldots a_k$ is prefix of all deciphering of elements in $Yb \cap Prefix(C^+)$. From this, by using the inductive hypothesis on γ, one gets that $\phi(s_0, \beta b) = xza_1 \ldots a_k$ and each deciphering of

γb has the form $d_i(\gamma b) = x_1 z a_1 \ldots a_k x\prime_{2_i}$ with $l(x\prime_{2_i}) \leq Q_2$. It follows that γb satisfies (A.2) and i). Let now $\gamma b \in Ends(\beta b)$ such that for $\gamma \in Ends(\beta)$ (A.2) and ii) hold for some Z_i in Z. From (13), we know that the decipherings of any suffix of an element $\omega \in Ends(\beta b)$ satisfying (A.2) and i), and of a suffix of $\gamma b \in Ends(\beta b)$ have a common infix and differ in suffixes of length not greater than q_2. Moreover, all decipherings of the suffix of ω have the form $y_1 z y_{2_i}$ with y_{2_i} deciphering of some element of $Yb \cap Prefix(C^+)$, and then having $a_1 \ldots a_k$ as prefix and resulting suffix of length not greater than Q_2. It then follows that each deciphering of γb has the form $d_i(\gamma b) = x_1 z a_1 \ldots a_k x'_{2_i}$, $l(x'_{2_i}) \leq Q + q_2$. Let us now show that γb satisfies ii). By inductive hypothesis, one has that $d_i(\gamma) = x_1 z x_{2_i}$ and there exists r such that $X_{t_r} = \{x \in Prefix(C^+) | \exists d_i(\gamma) = x_1 z x_{2_i} \exists w \in F(x_{2_i}) \exists v \in B(C)$ such that $\gamma = uwv$ for some $u \in C^+$ and $x = uv\}$. It follows that $d_i(\gamma b) = x_1 z a_1 \ldots a_k x'_{2_i}$ and $Z"_s = \cup_{j=1}^{k_\sqcup} X"_{t_j}$, where $X"_{t_j} = \{x \in Prefix(C^+) | \exists y \in X_{t_j} \exists w_1 \ldots w_k \in F(a_1 \ldots a_k) \ Y = w_1 \ldots w_k x\}$. It is obvious that $Z"_s$ has the property of k_s-disjunction ($k_s \geq 0$). Moreover, $X"_{t_r} = \{x \in Prefix(C^+) | \exists d_i(\gamma b) = x_1 z x\prime_{2_i} \exists w \in F(x\prime_{2_i}) \exists v \in B(C)$ such that $\gamma = uwv$ for some $u \in C^+$ and $x = uv\}$. In order to complete the proof that γb satisfies ii), let us make the following observations. If $k_s = 1$ then ii) is true for γb. Let $k_s > 1$, by inductive hypothesis, one has $z\alpha_{i_j} = \mu_{i_j} z$, $(j = 1 \ldots k_s)$. From this, one gets

$$zG^0(X_{tk_t}b \cap Prefix(C^+)) = z\alpha_{t_j} G^0(X_{t_j}b \cap Prefix(C^+)) \\ = \mu_{t_j} z G^0(X_{t_j}b \cap Prefix(C^+)). \quad (A.3)$$

The separation elements $\alpha\prime_{s_j}$ of $Z"_s$ are obtained by removing the prefix $\alpha_{t_j} G^{Q_2}(Yb \cap Prefix(C^+))$ from the beginning of $G^0(X_{tk_t}b \cap Prefix(C^+))$. From (A.3) it follows $zG^{Q_2}(Yb \cap Prefix(C^+))\xi_{t_j} = \mu_{t_j} z G^{Q_2}(Yb \cap Prefix(C^+))\xi'_{t_j}$, that is, putting $\xi_{t_j} = \beta_{t_j} \xi'_{t_j}$, $zG^{Q_2}(Yb \cap Prefix(C^+))\beta_{t_j} = \mu_{t_j} z G^{Q_2}(Yb \cap Prefix(C^+))$. Rearranging the indexes, one has that ii) holds for γb.

Case 2). $Yb \cap prefix(C^+) = \emptyset$ and $Zb \cap Prefix(C^+) \neq \emptyset$.
Let $Zb \cap Prefix(C^+) = Z' = (Z'_1, \ldots, Z'_m)$ and $\gamma b \in Ends(\beta b)$ such that $\gamma \in Ends(\beta)$ satisfies (A.2) and ii) for some Z_i in Z such that $Z_i b \cap Prefix(C^+) = Z'_1$. By definition, one has

$$\phi(s_{(Y,Z)}, b) = \begin{cases} G^{Q_2}(X\prime_{1k_1}) & \text{if } l(G^{Q_2}(X\prime_{1k_1})) \geq q_1 + 1 \\ \lambda & \text{otherwise} \end{cases}$$

By inductive hypothesis, one has that $\phi(s_0, \beta) = xz$ with $z \in A^*$ such that each deciphering of γ satisfies (A.2). Moreover, Z_i has the property of k_i-disjunction ($k_i > 0$), there exists r such that $X_{t_r} = \{x \in Prefix(C^+) | \exists d_i(\gamma) = x_1 z x_{2_i} \exists w \in F(x_{2_i}) \exists v \in B(C)$ such that $\gamma = uwv$, for some $u \in C^+$ and $x = wv\}$ and $z\alpha_{i_j} = \mu_{i_j} z$, $(j = 1 \ldots k_i)$. In case $k_i = 1$, it is evident that by using arguments similar to the one used in Case 1) and considering the set $Z'_1 = Z_i \cap Prefix(C^+)$ instead of $Yb \cap Prefix(C^+)$ one can show that if $\phi(s_{(Y,Z)}, b) \neq \lambda$ then γb satisfies (A.2) and i), and satisfies (A.2) and ii) otherwise. Let then $k_i > 1$. Suppose first $\phi(s_{(Y,Z)}, b) = \lambda$. One has that each deciphering of γb has the form $d_i(\gamma b) = x_1 z x_{2_i} a_i$, $a_i \in A \cup \{\lambda\}$, and for each $y \in X'_{1_i} = X_{i_1} b \cap Prefix(C^+)$ it holds $|y| \leq Q$. It follows that γb satisfies (A.2) and ii). Let now $\phi(s_{(Y,Z)}, b) = G^{Q_2}(X'_{1k_1}) = \alpha_{1_j} G^{Q_2}(X'_{1_j})$. We recall that $\alpha_{1_i} = a_{1k_1-1} \ldots a_{1_i}$, where a_{i_j} is the separation element between X'_{1_j} and $X'_{1_{j+1}}$, $1 \leq j > k_1$. By inductive hypothesis, one has that each deciphering of γb satisfies $d_i(\gamma b) = x_1 z x_{2_i} a_i$, $a_i \in A \cup \{\lambda\}$,

$$l(x_{1_i}) \leq \lceil \frac{3(q_1 + q_2 + 1)L - 1}{l} \rceil + \lceil \frac{Q_2 L - 1}{l} \rceil + 2q_1$$

$$l(x_{2_i}) \leq Q + q_2 + 1$$

and there exists r such that

$$X'_{1_r} = \{x \in Prefix(C^+) | \exists d_i(\gamma b) = x_1 z x_{2_i} a_i \exists w \in F(x_{2_i} a_i) \exists v \in B(C)$$
$$\text{such that } \gamma = uwv, \text{ for some } u \in C^+ \text{ and } x = wv\} \quad (A.4)$$

and $z\alpha_{i_r} = \mu_{i_r} z$. One gets that $\phi(s_0, \beta b) = xza_1 \ldots a_k = xz\alpha_{1_r} G^{Q_2}(X'_{1_r}) = x\mu_{1_r} z G^{Q_2}(X'_{1_r})$ and each deciphering of γb satisfies $d_i(\gamma b) = x_1 z \alpha_{1_r} G^{Q_2}(X'_{1_r}) x'_{2_i}$. Recall that $f(s_0, \beta b) = s_{(Y'', Z'')}$. From the definition of the transition function f, one gets

$$Y'' = R_1(X'_{1_1}, Z')$$
$$= \{x \in Prefix(C^+) | \exists w_1 \ldots w_s \in F(G^{Q_2}(X'_{1_1})) \exists y \in X'_{1_1} \quad y = w_1 \ldots w_s x\}$$
$$= \{x \in Prefix(C^+) | \exists w_1 \ldots w_t \in F(G^{Q_2}(X'_{1_r})) \exists y \in X'_{1_r} \quad y = w_1 \ldots w_t x\}$$

From this and (A.4), one has $Y'' = \{x \in Prefix(C^+) | \exists d_i(\gamma b) = x_1 z G^{Q_2}(X'_{1k_1}) x'_{2_i} \exists w \in F(x'_{2_i}) \exists v \in B(C) \text{ such that } \gamma = uwv \text{ for some } u \in C^+, x = wv\}$ and $l(x'_{2_i}) \leq Q_2 \leq Q + q_2$. Then (A.2) and ii) hold for γb.

Let now $\gamma' b \in Ends(\beta b)$, $\gamma' \in Ends(\beta)$ satisfying (A.2) and ii), for some Z_i in Z such that $Z_i b \cap Prefix(C^+) = Z''_j$, $j \neq 1$. In order to show that (A.2) and ii) hold for $\gamma' b$, observe that, by inductive hypothesis, one has that $\phi(s_0, \beta) = x'z'$ and each deciphering of γ' has the form $d_i(\gamma') = x'_1 z x'_{2_i}$ with

$$l(x'_1) \leq \lceil \frac{3(q_1 + q_2 + 1)L - 1}{l} \rceil + \lceil \frac{Q_2 L - 1}{l} \rceil + 2q_1$$

$$l(x'_{2_i}) \leq Q + q_2.$$

Let us observe that $\phi(s_0,\beta) = xz = x'z'$ with z' suffix of z and z infix of each deciphering of $\gamma \in Ends(\beta)$, $|\gamma| - l(z) \leq N$, and γ such that ii) holds for Z_i in Z for which $Z_i b \cap Prefix(C^+) = Z'_1$.

Let $\phi(s_{(Y,Z)}, b) = \lambda$. Then all deciphering of γ and γ' have a common infix and, from (13) differ in at most q_2 final symbols. Since each deciphering of γb can be written as $x_1 z x_{2_i}$ with $l(x_1) \leq Q$, each deciphering of $\gamma' b$ has the form $x'_1 z' x'_{2_i}$ with $l(x'_{2_i}) \leq Q + q_2$. Then $\gamma' b$ satisfies (A.2); moreover, it is obvious that ii) is true.

Let now $\phi(s_{(Y,Z)}, b) = G^{Q_2}(X'_{1k_1}) = \alpha_{1_t} G^{Q_2}(X'_{1_t})$, $t = 1, \ldots, k_1 - 1$. By inductive hypothesis, one has that γ' satysfies (19) and $z\alpha_{1_t} = \mu_{1_t} z$, $t = 1, \ldots, k_1 - 1$. From this, one gets that for each t, it is possible to write $\phi(s_0, \beta b) = xz\phi(s_{(Y,Z)}, b) = xz\alpha_{1_t} G^{Q_2}(X'_{1_t}) = x_t z G^{Q_2}(X'_{1_t})$, $x_t = x\mu_{1_t}$, where α_{1_t}'s, $(1 \leq t < k_1)$, are the separation elements with respect to Z'_1. Moreover, from the definition of G^{Q_2}, one has that there exists $d(\gamma b)$ such that $d(\gamma b) = x_1 x \alpha_{1_t} G^{Q_2}(X'_{1_t}) x_2$, $x_2 \in A^*$, $l(x_2) = Q_2$. Let us show that there exists r such that for each $d_i(\gamma b)$

$$d_i(\gamma b) = y_1 z' G^{Q_2}(X'_{1_r}) y_{2_i}$$
$$l(y_1) \leq \lceil \frac{3(q_1 + q_2 + 1)L - 1}{l} \rceil + \lceil \frac{Q_2 L - 1}{l} \rceil + 2q_1 \quad l(y_{2_i}) \leq Q + q_2. \tag{A.5}$$

Let us suppose that (A.5) is not true. By inductive hypothesis, one has that each deciphering of γ' has the form $d_i(\gamma') = y'_1 z' y'_{2_i}$; then each deciphering of $\gamma' b$ can be written as $d_i(\gamma' b) = y'_1 z' y'_{2_i} a_i$, $a_i \in A \cup \{\lambda\}$. If (A.5) is not true, it follows that for each r there exists $d_i(\gamma' b)$ such that either $y'_{2_i} a_i$ has $G^{Q_2}(X'_{1_r})$ as prefix with resulting suffix of length greater than $Q_2 + q_2$ or it has not $G^{Q_2}(X'_{1_r})$ as prefix. In both cases, there exists a prefix of code message (the shortest between γb and $\gamma' b$) having two deciphering that do not satisfy (13). One obtains then that $\gamma' b$ satisfies (A.2). Let us now show that it satisfies ii). Let Y_{sr} $(1 \leq s \leq k_j)$ be the set obtained by removing a coding of $G^{Q_2}(X'_{1_r})$ from the elements of X'_{j_s} that have it as prefix,

$$Y_{sr} = R(X'_{1_r}, (X'_{j_s}))$$
$$= \{x \in Prefix(C^+) | \exists w_1 \ldots w_k \in G^{Q_2}(X'_{1_r})) \exists y \in X'_{j_s}$$
$$\exists \delta(\beta) \quad [\delta(\beta) = w_1 \ldots w_k x]\}$$

From definition of R_3 it follows that $Z'_j = \cup Y_{sr}$, where the union is taken over all sets $Y_{sr} \neq \emptyset$. Let $x_1 \in Y_{sr}$ and $x_2 \in Y_{kt}$. By definition one has that there exist $w_1 \in F(G^{Q_2}(X'_{1_r}))$ and $y_1 \in X'_{j_s}$ such that $y_1 = w_1 x_1$, and $w_2 \in F(G^{Q_2}(X'_{1_t}))$ and $y_2 \in X'_{k_j}$ such that $y_2 = w_2 x_2$. Moreover, since Z'_j has the property of disjunction one has that there exist $a \in A^+$ and $w \in F(a)$ such that $y_1 = w y_2$ or $y_2 = w y_1$. Without loss of generality, suppose that $y_1 = w y_2$ and $G^{Q_2}(X'_{1_r}) =$

$\alpha G^{Q_2}(X'_{1_t})$, $\alpha \in A^*$. It follows that $y_1 = w_1 x_1 = w y_2 = w w_2 x_2 = v w_2 x_1$, for some $v \in F(\alpha)$. One can then deduce that there exist $c \in A^*$, $w \in F(c)$ such that either $x_1 = w x_2$ or $x_2 = w x_1$. This implies that it is possible to partition $Z"_j$ into disjoint subsets Y_{j_r} such that for each Y_{j_r} there exists a_{j_r} such that for each $\alpha \in Y_{j_r}$ there exist $w \in F(a_{j_r})$, $\alpha' \in Y_{j_{r+1}}$ such that $\alpha' = w\alpha$. In order to complete the proof that ii) is true for $\gamma' b \in Ends(\beta b)$, let us make the following observations. From above results, we know that $z = z"z'$, $z" \in A^*$, where z' is infix of each deciphering of $\gamma' \in Ends(\beta)$ and z is infix of each deciphering of $\gamma \in Ends(\beta)$ with γ such that ii) holds with respect to Z_i in Z for which $Z_i b \cap Prefix(C^+) = Z'_1$. Moreover, by inductive hypothesis, $z\alpha_{1_j} = \mu_{1_j} z$ $(1 \le j \le k_1)$. Let now $x_1 \in Y_{s_r}$. Since Z'_j has the property of k_j-disjunction, it follows that there exists $x_2 \in Y_{k_t}$ such that there exist $a \in A^*$, $\alpha \in F(a)$ for which either $x_1 = \alpha x_2$ or $x_2 = \alpha x_1$. Moreover, we know that

$$\alpha_{1_k} = \alpha_{1_s} \xi \quad \text{or} \quad \alpha_{1_s} = \alpha_{1_k} \xi, \quad \xi \in A^*$$

For sake of semplicity, suppose $x_1 = \alpha x_2$ and $\alpha_{1_k} = \alpha_{1_s} \xi$ In what follows, we denote with α_{j_t} the separation elements in $Z'_j = Z_i b \cap Prefix(C^+)$. We know that

$$z' \alpha_{j_r} G^0(X'_{1_r}) = z' \alpha_{j_t} G^0(X'_{j_t})$$

Since $x_1 \in Y_{s_r}$, $x_2 \in Y_{k_t}$ and $x_1 = \alpha x_2$, $\alpha \in F(a)$, $a \in A^*$, one has

$$z' \alpha_{j_r} G^0(X'_{1_s}) a = z' \alpha_{j_t} G^0(X'_{1_k})$$

Since $\alpha_{1_k} = \alpha_{1_s} \xi$, one gets $G^{Q_2}(X'_{1_s}) = \xi G^{Q_2}(X'_{1_k})$ and

$$z' \alpha_{j_r} \xi G^0(X'_{1_k}) a = z' \alpha_{j_t} G^{Q_2}(X'_{1_k})$$

$$\mu_{j_r} z' \xi G^0(X'_{1_s}) a = \mu_{j_t} z' G^{Q_2}(X'_{1_k}) \quad (A.6)$$

Let us show now that

$$z' \xi = \mu z' \quad \mu \in A^*. \quad (A.7)$$

From the inductive hypothesis on $\gamma \in Ends(\beta)$, one has that $z\alpha_{1_s} \xi = \mu_{1_k} z$, from which it follows $z\alpha_{1_s} \xi = \mu_{1_k} z$ and $\mu_{1_s} z \xi = \mu_{1_k} z$. Since z' is a suffix of $z = z"z'$, one has $\mu_{1_s} z" z' \xi = \mu_{1_k} z" z'$ and (A.6) holds. Substituting (A.6) in (A.5), one gets

$$\mu_{j_s} \mu z' G^0(X'_{1_k}) a = \mu_{j_t} z' G^0(X'_{1_k})$$

from which

$$[z' G^0(X'_{1_k})] a = \mu[z' G^0(X'_{1_s})]$$

This proves that ii) holds for $\gamma' b \in Ends(\beta b)$.

Case 3). $Yb \cap Prefix(C^+) = \emptyset$ and $Zb \cap Prefix(C^+) = (\emptyset)$.
One has that for each $\gamma b \in Ends(\beta b)$, $\gamma \in Ends(\beta)$ do not satisfy neither i) nor ii). Let now $\gamma b \in Ends(\beta b)$ such that for $\gamma \in Ends(\beta)$ neither i) nor ii) hold. We shall prove that if (A.2) and either i) or ii) do not hold for γb then $|\gamma b| \leq N$. Let us suppose first that $Yb \cap Prefix(C^+) = \emptyset$ and that γb is a suffix of $max(Yb \cap Prefix(C^+))$. It is possible to write $max(Yb \cap Prefix(C^+)) = y_1 y_2$, with $y_1, y_2 \in Prefix(C^+)$, $y_2 = \gamma b$. From the definition of the output function ϕ, one has

$$|max(Yb \cap Prefix(C^+))| \leq Q + 1$$

Then, from (13), it follows

$$|\gamma b| \leq Q + 1 + q_1 + q_2 \leq N \qquad (A.8)$$

Notice that in case $\phi(s_{(Y,Z)}, b) = G^{Q_2}(Yb \cap Prefix(C^+)) \neq \lambda$, if $y_1 \in T(C)$ then it is easily seen from the definition of the transition function f that γ satisfies ii) and each deciphering of γ can be written as $d_i(\gamma) = x_1 a_k x_{2i}$, with $l(x_1) \leq 2q_1$ and $l(x_2) \leq Q + q_2$. Let now $y_1 \notin T(C)$. Recall that $f(s_{(Y,Z)}, b) = s_{(Y'',Z'')}$. Since $l(G^{Q_2}(Yb \cap Prefix(C^+))) = q_1 + 1$ one can write $\gamma = \gamma_1 \gamma_2$ and either

$$|\gamma_1| \leq \lceil \frac{(q_1 + 1)L - 1}{l} \rceil \quad \text{and} \quad \gamma_2 = \xi max(Y'') \quad \xi \in T(C) \qquad (A.9)$$

or γ is a suffix of $max(Y'')$. Suppose now $Yb \cap Prefix(C^+) = \emptyset$ and $Zb \cap Prefix(C^+) \neq (\emptyset)$. Let $Z' = Zb \cap Prefix(C^+)$ and $Z'_i = Z_i b \cap Prefix(C^+)$. Suppose that γ is a suffix of $max(X_{i_1} b \cap Prefix(C^+)) = max(X'_{1_i})$. If $\phi(s_{(Y,Z)}, b) = \lambda$ then from the definition of the function ϕ, one has $|max(X'_{1_i})| \leq Q$. From (13) it follows that

$$|\gamma| \leq Q + q_1 + q_2 \leq N \qquad (A.10)$$

Let now $\phi(s_{(Y,Z)}, b) = G^{Q_2}(X'_{1k_1})$. From the definition of the function ϕ, for each element z in Z and y in Y it holds $|z| \leq |y| + q_2$. One then gets $|max(X_{ij})| \leq Q + q_2$ and

$$|max(X'_{1k_1})| \leq Q + q_2 + 1. \qquad (A.11)$$

It then follows that

$$|\gamma| \leq Q + q_2 + q_1 + q_2 \leq N. \qquad (A.12)$$

Consider now the worst case, that is, $max(X'_{1_i})$ suffix of $max(Yb)$. One can write $\gamma = \gamma_1 \gamma_2$ with $\gamma_2 = max(X'_{1_i})$. Observe that, by definition of f, if we denote with $s_{(Y'',Z'')}$ the state in which the decoder is after it has given an output for the last time, it results that for each $y \in Y''$, it holds $|\gamma| \leq Q_2$, moreover, for each $z \in Z''$ one has that $||z| - |y|| \leq Q_2$. This implies

$$|\gamma_1| \leq \lceil \frac{Q_2 L - 1}{l} \rceil \qquad (A.13)$$

and by (A.12)
$$|\gamma_1| \leq \lceil \frac{Q_2 L - 1}{l} \rceil + Q + q_1 + 2q_2 \qquad (A.14)$$

Notice that if $\phi(s_{(Y,Z)}, b) = G^{Q_2}(X'_{1k_1}) = a_1 \ldots a_k$, $f(s_{(Y,Z)}, b) = s_{(Y'',Z'')}$ and $\gamma = \gamma_1 \gamma_2$ with $\xi \gamma_2 = max(X'_{1_1})$, $\xi \in T(C)$. From definition of N, it follows that each deciphering of γb can be written as $d_i(\gamma b) = x_1 a_{q_1+1} \ldots a_k x_{2_i}$ with

$$l(x_1) = |\gamma| + 2q_1 \leq \lceil \frac{Q_2 L - 1}{l} \rceil + 2q_1$$

$$l(x_{2_i}) \leq Q_2 \leq Q + q_2$$

It is easy to see that ii) holds for γb. Let now $max(X'_{1_1}) = \xi \gamma b$, $\xi \notin T(C)$. From definition of N, it follows that neither i) nor ii) hold for γb. Since $l(G^{Q_2}(X'_{1k_1})) \leq q_1 + q_2 + 1$, one has that either

$$\gamma b = \gamma' \gamma'', \qquad |\gamma'| \leq \lceil \frac{(q_1 + q_2 + 1)L - 1}{l} \rceil, \quad \gamma'' = max(Y'') \qquad (A.15)$$

or γb is a suffix of $max(Y'')$.

Consider now the case $Yb \cap Prefix(C^+) = \emptyset$ and $Zb \cap Prefix(C^+) = (\emptyset)$. Let $\gamma b \in Ends(\beta b)$ and γb suffix of $max(Yb)$. Notice that $max(Ends(max(Yb)))$ is the longest suffix of $max(Yb)$ belonging to $Prefix(C^+)$. Let $\phi(s_{(Y,Z)}, b) = \lambda$. From the definition of ϕ, one has $|max(Ends(max(Yb)))| \leq Q$. Then, because of (13), $|\gamma| \leq Q + q_1 + q_2$. Consider now the case $\phi(s_{(Y,Z)}, b) \neq \lambda$, that is, by definition of ϕ, $\phi(s_{(Y,Z)}, b) = G^{Q_2}(max(Ends(max(Yb)))) = a_1 \ldots a_k$. One can write $max(Ends(max(Yb))) = \xi \gamma b$, $\xi \in X^*$. Suppose that $\xi \in T(C)$. Then, from (13), one gets that each deciphering of γb satisfies $d_i(\gamma b) = x_1 a_{q_1+1} \ldots a_k x_{2_i}$ with $l(x_1) \leq 2q_1$ and $l(x_{2_i}) \leq Q + q_2$. Then (A.2) holds for $\gamma b \in Ends(\beta b)$; moreover, it is easy to see that ii) holds too. Suppose now that $\xi \notin T(C)$. Recall that $f(s_{(Y,Z)}, b) = s_{(Y'',Z'')}$. From the definition of ϕ, it follows that $max(Ends(max(Yb))) = \alpha max(Y'')$, $\alpha \in C^+$. In order to evaluate the length of γb, one can bound the length k of the output $a_1 \ldots a_k = G^{Q_2}(max(Ends(max(Yb))))$. In the worst case, one has that $Z \neq (\emptyset)$, with $G^{Q_2}(X_{1_1}) \leq q_1$ and $max(Z)$ suffix of $max(Y)$. It is possible to write $max(Ends(max(Yb))) = \alpha_1 \alpha_2$, $\alpha_1 \in X^*$, $\alpha_2 = max(Z)$. One gets $|\alpha_2 b| \leq Q + 1 + q_1 + q_2$, and then $|max(Ends(max(Yb)))| \leq Q + 1 + q_1 + q_2 + q_2 + q_1 + q_2 = Q + 2q_1 + 3q_2 + 1$. This implies $k \leq Q + 1 + 2q_1 + 3q_2 - Q_2 = 3q_1 + 3q_2 + 1$. One can write $\gamma b = \gamma_1 \gamma_2$ with $\gamma_2 = \xi max(Y'')$, $\xi \in T(C)$. As seen above, it results

$$|\gamma_1| \leq \lceil \frac{(3q_1 + 3q_2 + 1)L - 1}{l} \rceil$$

$$|\gamma_2| \leq Q_2 + q_1 + q_2 \qquad (A.16)$$

Consider now the case $max(Yb)$ suffix of γb. Note that property (13) and the definition of f imply that, in case $\phi(s_{(Y,Z)}, b) \neq \lambda$, γb satisfies either i) or ii). Then, one has to prove that if $\phi(s_{(Y,Z)}, b) = \lambda$ and neither i) nor ii) hold for γb then $|\gamma b| \leq N$. We shall prove, at the same time, that if $\phi(s_{(Y,Z)}, b) \neq \lambda$ than γb satisfies (19). Let $Yb \cap Prefix(C^+) \neq \emptyset$ and $\gamma = \gamma_1 \gamma_2$ with $\gamma_1 \in X^*$ and $\gamma_2 = \xi max(Yb \cap Prefix(C^+))$, $\xi \in T(C)$. Suppose $\phi(s_{(Y,Z)}, b) = \lambda$. One has that $|max(Yb \cap Prefix(C^+))| \leq Q$. Then, from (13), one gets $|\gamma_2| \leq Q + q_1 + q_2$. Denote by $b_1 \ldots b_r$ the last output of the decoder when it starts from s_0 and has βb as input. As seen, (A.9), (A.13), (A.15), (A.16), one has $r \leq (3q_1 + 3q_2 + 1)$ and

$$|\gamma| \leq \lceil \frac{(3q_1 + 3q_2 + 1)L - 1}{l} \rceil + Q + q_1 + q_2$$

Suppose now $\phi(s_{(Y,Z)}, b) = a_1 \ldots a_{q_1+1} = G^{Q_2}(Yb \cap Prefix(C^+))$. From (13), one has that each deciphering of γb can be written as $d_i(\gamma b) = x_1 a_{q_1+1} x_{2_i}$ with

$$l(x_1) \leq |\gamma_1| + 2q_1 \leq \lceil \frac{(3q_1 + 3q_2)L - 1}{l} \rceil + 2q_1$$
$$l(x_{2_i}) \leq Q + q_2$$

Moreover, it is easy to see that ii) holds. Let now $Yb \cap Prefix(C^+) = \emptyset$ and $Z' = Zb \cap Prefix(C^+) \neq (\emptyset)$, with $max(Yb)$ suffix of $\gamma b \in Ends(\beta b)$. One can write $\gamma = \gamma_1 \gamma_2$ with $\gamma_1 \in X^*$, $\xi \gamma_2 = max(Yb)$, $\xi \in T(C)$. As above, one has

$$|\gamma_1| \leq \lceil \frac{(3(q_1 + q_2)L - 1}{l} \rceil.$$

Let $\phi(s_{(Y,Z)}, b) = \lambda$. Since γ_2 is a suffix of $max(Yb)$, from (32), one has that in the worst case, that is, $\gamma_2 = \gamma_3 max(X'_{1_i})$,

$$|\gamma_2| \leq \lceil \frac{Q_2 L - 1}{l} \rceil + Q + q_1 + 2q_2$$

and then

$$|\gamma| \leq \lceil \frac{3(q_1 + q_2)L - 1}{l} \rceil + \lceil \frac{Q_2 L - 1}{l} \rceil + Q + q_1 + 2q_2.$$

Let now $\phi(s_{(Y,Z)}, b) = a_1 \ldots a_k = G^{Q_2}(X'_{1k_i})$. Suppose that γb has $max(X'_{1_i})$ as suffix. Note that if γb is suffix of $max(Yb)$, it is possible to repeat the above reasoning. Since $max(X'_{1_i})$ is a suffix of γb, from (13), one has that ii) holds for γb. Moreover, writing $\gamma b = \gamma' max(X'_{1_i})$ one gets that each deciphering of γb satisfies $d_i(\gamma b) = x_1 a_{q_1+1} \ldots a_k x_{2_i}$ with $l(x_1) \leq |\gamma'| + 2q_1$ and $l(X_{2_i}) \leq Q + q_2$.

In the worst case $\gamma = \gamma_1 max(Yb) = \gamma_1 \gamma'' max(X'_{1_i}) = \gamma' max(X'_{1_i})$, from (A.13) and (A.15), one gets

$$|\gamma'| \leq \lceil \frac{(3(q_1+q_2)L-1}{l} \rceil + \lceil \frac{Q_2L-1}{l} \rceil$$

and

$$|x_1| \leq \lceil \frac{(3(q_1+q_2)L-1}{l} \rceil + \lceil \frac{Q_2L-1}{l} \rceil + 2q_1.$$

Finally, consider the case $Yb \cap Prefix(C^+) = \emptyset$ and $Zb \cap Prefix(C^+) = (\emptyset)$, with $max(Yb)$ suffix of γb. Let $\gamma b = \gamma_1 \gamma_2$ with $\gamma_1 \in X^*$ and $\gamma_2 = max(Yb)$. Suppose $\phi(s_{(Y,Z)}, b) = \lambda$. One has that $|\gamma_2| \leq Q + q_1 + q_2$. Moreover, from (A.11) and (A.15), one gets

$$|\gamma_1| \leq \lceil \frac{(3(q_1+q_2)L-1}{l} \rceil$$

$$|\gamma_2| \leq N.$$

Let $\phi(s_{(Y,Z)}, b) = a_1 \ldots a_k = G^{Q_2}(max(Ends(max(Yb))))$. From (13), one has that each deciphering of γb can be written as $d_i(\gamma b) = x_1 a_{q_1+1} \ldots a_k x_{2_i}$ with

$$|x_1| \leq |\gamma_1| + 2q_1 \leq \lceil \frac{(3(q_1+q_2)L-1}{l} \rceil + 2q_1$$

$$|x_2| \leq Q + q_2.$$

Moreover, it is obvious, from the definition of f, that either (A.2) and ii) hold or, in case $\gamma b = max(Ends(max(Yb)))$, γ satisfies (A.2) and ii).

Q.E.D.

REFERENCES

[1] Bahl, L.R. and Jelinek, F., "Decoding for Channels with Insertions, Deletions and Substitutions with Applications to Speech Recognition" IEEE Trans. Inform. Theory, $IT-21$, 404-411, (1975).

[2] Jelinek, F., Bahl, L.R. and Mercer, R.L., "Design of a Linguistic Statistical Decoder for the Recognition of Continuous Speech", IEEE Trans. Inform. Theory, $IT-21$, 250-256, (1975).

[3] Jelinek, F., Bahl, L.R. and Mercer, R.L., "A Maximum Likelihood Approach to Continuous Speech Recognition", IEEE Trans. Pattern Analysis and Machine Intelligence, $PAMI-5$, 179-190, (1983).

[4] Zigangirov, K.Sh. and Sorokin, V.N., "Use of Sequential Decoding in recognition of Continuous Speech", Problemi Pederachi Informatsii, 13, 81-88, (1977).

[5] Sardinas, A.A. and Patterson, G.W., "A Necessary and Sufficient Condition for Unique Decomposition of Encoded Messages" IRE Conv. Rec., $pt8$, 104-108, (1953).

[6] Levenshtein, V.I., "Certain Properties of Code Systems", Soviet Phis. Dokladi, 6, 858-860, (1962).

[7] Levenshtein, V.I., "Some Properties of Coding and Self Adjusting Automata for Decoding Messages", Problemi Kibernetiki, 11, 63-121, (1964).

[8] Capocelli, R.M., "A Note on Uniquely Decipherable Codes", IEEE Trans. Inform. Theory, $IT-25$, 90-94, (1979).

[9] Tanaka, E. and Kasai, T., "Synchronization and Substitution Error-Correcting Codes for Levenshtein Metric", IEEE Trans. Inform. Theory, $IT-22$, 159-162, (1976).

[10] Green, E.P., "Burst-Trapping Decoding for Correction of both Additive Noise and Deletion Errors within Cyclic Block Codes", IEEE Trans. Inform. Theory, $IT-23$, 618-620, (1977).

[11] Iizuka, I., Kasahara, M. and Namekawa, T., "Block Codes Capable of Correcting both Additive and Timing Errors", IEEE Trans. Inform. Theory, $IT-26$, 393-400, (1980).

[12] Hartnett, W.E., Editor, "Foundations of Coding Theory", Boston MA: Reidel, (1974).

[13] Hartnett, W.E., "Generalization of Tests for Certain Properties of Variable Length Codes", Inform. Contr., 13, 20-45, (1968).

[14] Sato, K., "A Decision Procedure for the Unique Decipherability of Multivalued Encodings" IEEE Trans. Inform. Theory, $IT-25$, 356-360, (1979).

[15] Capocelli, R.M., "A Decision Procedure for Finite Decipherability and Synchronizability of Multivalued Encodings", IEEE Trans. Inform. Theory, $IT-28$, 307-318, (1982).

[16] Capocelli, R.M. and Vaccaro, U., "Decoding Automata for Multivalued Encodings", ISIT '83, Canada, (1983).

[17] Capocelli, R.M. and Vaccaro, U., "Finite Decipherability of Multivalued Encodings" Twenty-first Annual Allerton Conf. on Comm., Control and Computing, 528-536, (1983).

[18] Capocelli, R.M. and Vaccaro, U., "Structure of Decoders for Multivalued Encodings", submitted, (1984).

[19] Capocelli, R.M., Gargano, L. and Vaccaro, U., "Decoders with Initial State Invariance for Multivalued Encodings", IEEE Int. Symp. on Inform. Theory, Brighton, England, (1985).

[20] Capocelli, R.M., Gargano, L. and Vaccaro, U., "Synchronizability of Multivalued Encodings", Tenth Prague Conference on Information Theory, Statistical Decision Functions and Random Processes, Prague, Czechoslovakia, (1986).

[21] Dobrushin, R.L., "Shannon Theorems for Channels with Synchronization Errors", Problemy Peredachi Informatsii, $Vol.\ 3$, 18-36, (1967).

[22] Levenshtein, V.I., "Decoding Automata with Initial State Invariance", All-Union Conference on Theory of Coding and Its applications, 201-207, Odessa, (1963).

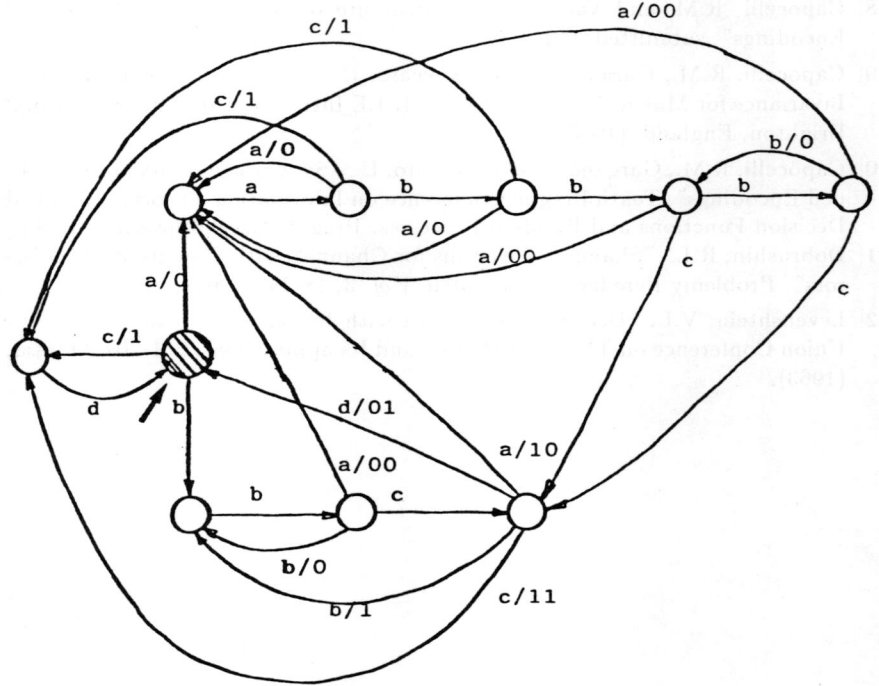

Fig. 1

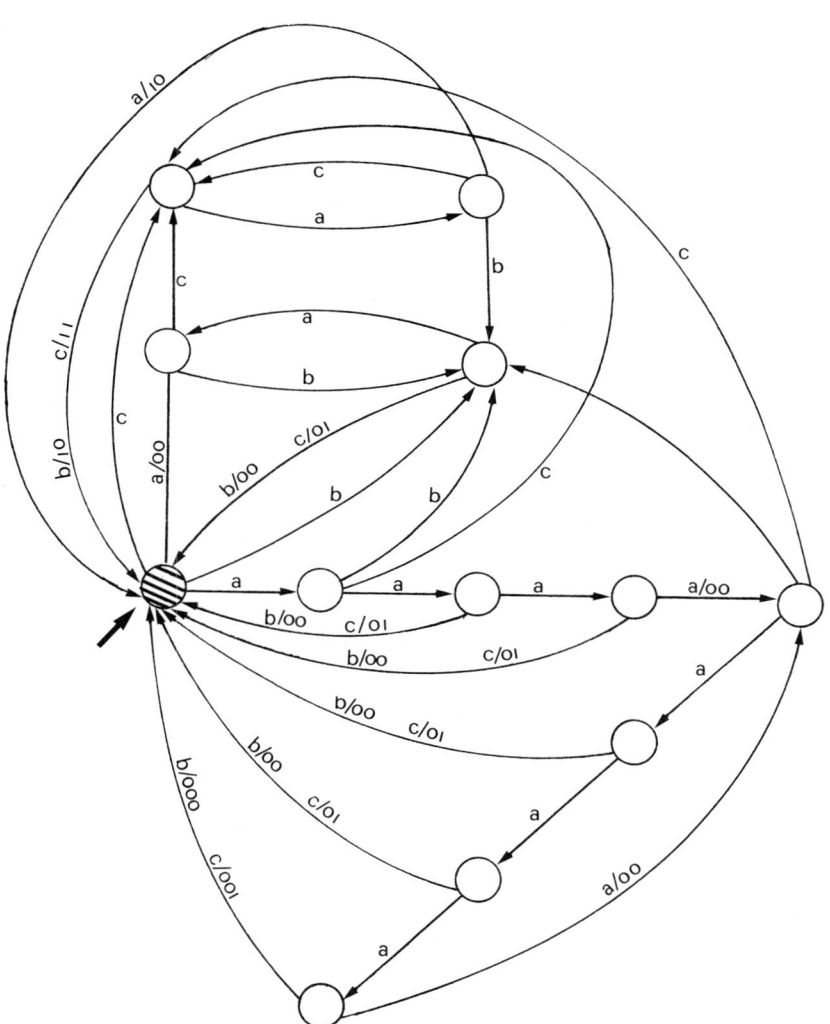

Fig. 2

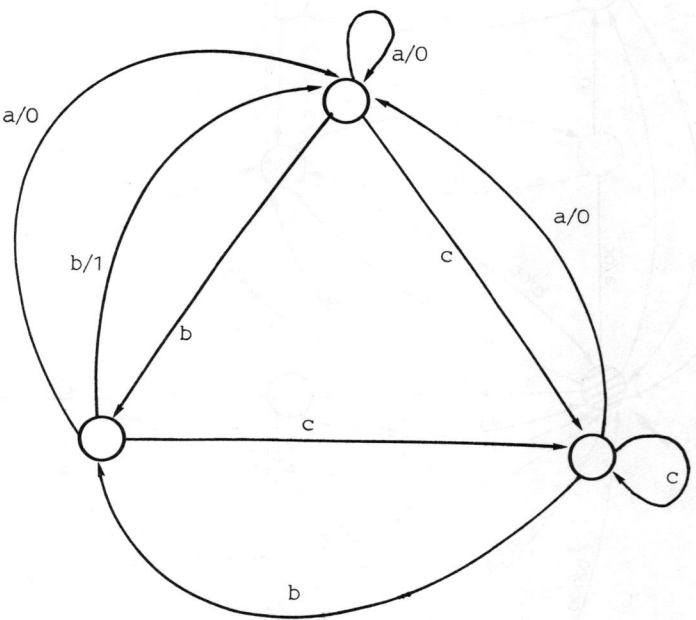

Fig. 3

DISSIPATIVE STRUCTURES EMERGING WITHIN AN ABSTRACT MODEL OF NEURAL NETWORK

Giovanni Maria Guazzo
Instituto Walden
Via D. Scaramella 15bis, 84100 SALERNO
ITALY

ABSTRACT

This preliminary study shows an abstract model of the learning process; it will be identified as the creating process of a dissipative structure.

The dissipative structure will provoke a correlation within the whole system by means of the associated symmetry-breaking process.

1. INTRODUCTION

This work shows an abstract model of the learning process.

The main property of the model is to uniquely define the concept of learning in a given system.

Genarally, the learning activity is considered as a structural type of transaction which is triggered off by means of the appropriate external stimulus and which is associated with the acquisition of new patterns of behavoiur (Caianiello, 1961). The first approaches towards the definition of a system model able to show the learning process development fulfilling such a requirement dates back to Hebb's work (Hebb, 1949). The main characteristics of Hebb's model, as regards our objective, may be synthetized as follows:

a. the behaviour can be identified with the activity patterns of a neural network;
b. the learning process consists in the variation of neuronic interconnection coefficients;
c. the memory is the coefficient status at a given time;
d. the values of such coefficients tend to stable states depending on the network parameters and on the nature of the external stimuli (Guazzo, 1983).

Some works (Feigenbaum, 1983; Grossberg, 1974; Cohen, Grossberg, 1983) show that the attempts to construct an adequate model of the learning processes following the Hebb's hypothesis clash very hard against mathematical problems. A possible solution of the problem can be found by using the continuous mathematical model of morphogenesis within the area of the self-organizing systems (Prigogine, 1973; Haken, 1983). In this way, the learning process will no longer be considered as interconnection coefficients transactions but it will be identified as the creating process of a dissipative structure. The dissipative structure will provoke a correlation within the whole system by means of the associated symmetry-breaking process; this can be considered to be memory (Amari, 1983; Ermentrout, Cowan, 1980a, 1980b; Oosovets, Ginzburg, Gurfinkel', Zenkov, Latash, Malkin, Mel'nichuk, Pasternak,1983; Ricciardi, Umezawa, 1967).

2. THE MODEL

The considerations of the previous section suggest imposing the condition that the model should describe, in some way, a correlation of one of the two variables representing the components "activity" of the choosen system. Such a condition is compatible with the appearance of the stable dissipative structures only within systems of the following type (Belintsev, Livshits, Volkenstein, 1981):

$$\begin{cases} \dfrac{\partial \alpha}{\partial t} = D_1 \dfrac{\partial^2 \alpha}{\partial x^2} + \dfrac{1}{\pi} \beta \alpha + S \\ \dfrac{\partial \beta}{\partial t} = D_2 \dfrac{\partial^2 \beta}{\partial x^2} + r\alpha^2 + r\beta \end{cases} \qquad (1)$$

where D_1 and D_2 are the diffusion coefficients of the variables α and β, r is an opportune parameter and S is a constant.

System (1) describes a self-organization process; therefore a non-omogeneous dissipative structure will appear when some parameters overcome particular "thresholds". The equilibrium points of a such system can be found by using a standard procedure (Auchmuty, Nicolis, 1975):

$$\alpha_0 = -\sqrt[3]{\pi S} \qquad (2.a)$$

$$\beta_0 = \sqrt[3]{(\pi S)^2} \qquad (2.b)$$

In order to analyze the stability of the equilibrium points, we have to linearize system (1) in the neighbours of those points. Therefore let (Pessa, 1983,1985):

$$\alpha = \alpha_0 + \xi \qquad (3.a)$$

$$\beta = \beta_0 + \eta \qquad (3.b)$$

By replacing (3.a) and (3.b) in system (1) and neglecting exponential terms of ξ, η overcoming the first order, we have:

$$\begin{cases} \dfrac{\partial \xi}{\partial t} = D_1 \dfrac{\partial^2 \xi}{\partial x^2} + \dfrac{\beta_0}{\pi} \xi + \dfrac{\alpha_0}{\pi} \eta \\ \dfrac{\partial \eta}{\partial t} = D_2 \dfrac{\partial^2 \eta}{\partial x^2} + 2\alpha_0 r \xi - r\eta \end{cases} \qquad (4)$$

If we assume now that the solution has the following form:

$$\xi = c_1 e^{\lambda t} \operatorname{sen} kx \qquad (5.a)$$

$$\eta = c_2 e^{\lambda t} \operatorname{sen} kx \qquad (5.b)$$

And we replace (5.a) and (5.b) within system (4), we have:

$$\begin{cases} (-K^2 D_1 - \lambda + \frac{\beta_0}{\pi}) C_1 + \frac{\alpha_0}{\pi} C_2 = 0 \\ 2\alpha_0 r C_1 + (-r - K^2 D_2 - \lambda) C_2 = 0 \end{cases} \quad (6)$$

In order to have no null solutions in system (6), it is necessary that the determinant of the system coefficient matrix will be equal to zero.

Then we have the relation:

$$\lambda^2 - \lambda(-K^2 D_1 + \frac{\beta_0}{\pi} - K^2 D_2 - r) + (-K^2 D_1 + \frac{\beta_0}{\pi})(-r - K^2 D_2) + $$

$$- \frac{2\alpha^2 r}{\pi} = 0 \quad (7)$$

The solution of (7) considered as a λ-equation, is stable if both roots of λ have a negative real part. The critical point appears when either the known term is equal to zero or the coefficient of λ-term is equal to zero.

Let us now suppose that one of the value of λ is positive and assume:

$$r - r_0 = \gamma_1 \varepsilon + \gamma_2 \varepsilon^2 + \ldots \quad (8)$$

where r_0 is the critical value of the parameter and ε is a small parameter.

Furthermore, let us assume that:

$$\xi = \varepsilon \xi_0 + \varepsilon^2 \xi_1 + \varepsilon^3 \xi_2 + \ldots \quad (9.a)$$

$$\eta = \varepsilon \eta_0 + \varepsilon^2 \eta_1 + \varepsilon^3 \eta_2 + \ldots \quad (9.b)$$

By replacing (8), (9.a) and (9.b) within system (4) and making equal terms of the same order in ε, we obtain an infinite system of equation of the following type:

at 1-st ε-order: $L_1(\xi, \eta) = Q_1(\xi, \eta)$ \quad (10.a)
at 2-nd ε-order: $L_2(\xi, \eta) = Q_2(\xi, \eta)$ \quad (10.b)

system of equation of the following type:

at 1-st ε-order: $L_1(\xi,\eta) = Q_1(\xi,\eta)$ (10.a)

at 2-nd ε-order: $L_2(\xi,\eta) = Q_2(\xi,\eta)$ (10.b)

....

If the operator $L_1(\xi,\eta)$ is equal to system (4) with linear terms and $Q_1(\xi,\eta) = 0$, then it is possible by applying the Fredholm theorem, to impose the following condition of solvability:

$$\int_0^1 (\xi_0 R_1 + \eta_0 R_2) dx = 0 \qquad (11)$$

where 1 is the network length.

By developing the integral, where ξ_0 and η_0 are known, we have an equation in γ_1.

If $\gamma_1 > 0$ then the dissipative structure is stable.

If $\gamma_1 < 0$ the the dissipative structure is unstable.

Therefore the dissipative structure form is:

$$\xi = \xi_0 + \frac{(r-r_0)}{\gamma_1}\xi_1 \qquad (12.a)$$

$$\eta = \eta_0 + \frac{(r-r_0)}{\gamma_1}\eta_1 \qquad (12.b)$$

where ξ_1 and η_1 are determined by resolving with approximation method the system of ε^2-order.

If the system (1) does not depend on x then it assume the form:

$$\begin{cases} \frac{\partial \alpha}{\partial t} = \frac{1}{\pi}\beta\alpha + S \\ \frac{\partial \beta}{\partial t} = r\alpha^2 + r\beta \end{cases} \qquad (13)$$

Therefore, by deriving the first equation, we have:

$$\frac{\partial^2 \alpha}{\partial t^2} - \frac{\pi}{\alpha^2}\left(\frac{\partial \alpha}{\partial t}\right)^2 + \left(-\frac{\pi S}{\alpha} + r\right)\frac{\partial \alpha}{\partial t} - \frac{r\alpha^3}{\pi} - \frac{\pi S^2}{\alpha^2} - rS + S^2 = 0 \qquad (14)$$

By successive approximation of (14) for new small values of α and S it happens that the dissipative structure is not stable at solution α_0.

3. CONCLUSIONS

The results of section 2 show that, for critical values of the parameter, there is the conceiving of either temporal or spatial dissipative structure.

Therefore, for appropriate values of r, it is impossible to have an homogeneous network and then α and β assume the morphogenetic agents rules. The network connectivity can be automatically determined from the dynamic evolution of α and β which privilege some distribution and remove some others (Grossberg,1980).

Moreover, from this model, it is possible to see that a learning process may exist only if the system is capable of a self-organization process and if the external intensity stimulus overcomes a given threshold and finally if there are two types of internal activity, excitatory and inhibitory, mutually interacting in a non-linear way.

Acknowledgement

The author wishes to tank Prof. E.Pessa for his suggestions and comments on an earlier draft of this paper and Prof. E.R.Caianiello for useful discussions.

4. REFERENCES

Amari, S., "Field theory of self-organizing neural nets", IEEE Transaction on System, Man and Cyberbetic, SMC-13,741-748 (1983).

Auchmuty, J.F.G., Nicolis, G., "Bifurcation analysis of nonlinear reaction-diffusion equations. I: Evolution equations and steady state solutions", Bull. Math. Biol., <u>37</u> , 325-365 (1975).

Belinstev, B.N., Livshits, M.A., Volkenstein, M.V., "Pattern formation in system with non local interactions", Z. Phis., 44B, 345-351 (1981).

Caianiello, E.R., "Outline of a theory of thought processes and thinking machines", Journal of Theoretical Biology, $\underline{1}$, 204-235 (1961).

Cohen, M.A., Grossberg, S., "Absolute stability of global pattern formation and parallel memory storage by competitive neural networks", IEEE Transactions on System, Man and Cybernetics, SMC-13, 815-823 (1983).

Ermentrout, G.B., Cowan, J.D., "Large scale spatially organized activity in neural nets" SIAM J. Appl. Math., $\underline{38}$, 1-21 (1980a).

Ermentrout, G.B., Cowan, J.D., "Secondary bifuraction in neuronal nets" SIAM J. Appl. Math., $\underline{39}$, 323-340 (1980b).

Feigenbaum, M.J., "Universal behaviour in nonlinear systems" Physica, 7D. 16-39 (1983).

Grossberg, S., "Classical and instrumental learning by neuronal networks", Progress in Theoretical Biology, $\underline{3}$, 51-141 (1974).

Grossberg, S., "How a brain build a cognitive code?", Psychological Review, $\underline{58}$, 1-51 (1980).

Guazzo, G.M., "Introduzione allo studio di un modello di rete nervosa soddisfacente le ipotesi di Neisser", N.P.S., $\underline{2}$, 320-352 (1983).

Haken, H., Advanced Synergetics, Springer-Verlag, Berlin-Heidelberg-NewYork (1983).

Hebb, D., The organization of behaviour, Wiley&Sons, NewYork (1949).

Oosovets, S.M., Ginzburg, D.A., GUrfinkel', V.S., Zenkov, L.P., Latash,

L.P., Malkin, V.B., Mel'nichuk, P.V., Pasternak, E.B., "Electrical activity of the brain: mechanisms and interpretation" Soviet Physics Uspekhi, 26, 801-828 (1983).

Pessa, E., "Un nuovo modello matematico del processo di apprendimento di Grossberg", Comunicazioni Scientifiche di psicologia generale, 11 , 105-146 (1983).

Pessa, E., Stabilità e auto-organizzazione, Veschi, Roma (1985).

Prigogine, I., "Time, irreversibility and structure", Merha J, (ed), D. Reidl Publ. Company, Dordrecht-Holland, Boston-USA, 561-593 (1973).

Ricciardi, L.M., Umezawa, H., "Brain and physics of many-body problems", Kibernetik, 4 , 443-469 (1978).

THE RECOGNITION OF PATTERN FORMATION IN SELFORGANIZING SYSTEMS BY MEANS OF INFORMATION ENTROPY

H. Haken

Institute for Theoretical Physics, University of Stuttgart
Pfaffenwaldring 57/IV, 7000 Stuttgart 80, Germany

ABSTRACT

Self-organizing systems are systems which can acquire spatial, temporal or functional structures without specific interference from the outside. In this paper, we focus our attention on systems which acquire spatial or temporal patterns by change of a control parameter. We first give a short outline of the microscopic or mesoscopic theory where at specific critical points these patterns evolve. The dynamics is governed by the so-called order parameters which slave the subsystems. The equations to be treated are either of the Langevin type with nonlinear terms or of the Fokker-Planck type. Examples of these systems are lasers, fluids showing specific pattern formations, chemical reactions leading to macroscopic patterns or specific models of morphogenesis in biology and of muscle coordinations. A phenomenological access to pattern formation is provided by the maximum information entropy principle which is briefly outlined. But in contrast to systems in thermodynamic equilibrium new kinds of constraints, namely special types of moments must be used. We then show, how the distribution function obtained by this principle can be cast into a form entirely analogous to those of the microscopic or mesoscopic theory. We then show how by a proper transformation of the exponent of the distribution function patterns can be recognized. This method is a new approach to the recognition of emergent patterns in systems close to their points of non-equilibrium phase transitions. In this way it becomes possible to recognize patterns by means of a specific algorithm rather than to recognize patterns by means of human inspection of the individual systems. In the concluding section we consider some aspects of semantic information, information compression a.s.o..

1. SELFORGANIZING SYSTEMS. SYNERGETICS.

We define selforganizing systems as those which obtain their spatial, temporal or functional structures without specific interference from the outside. Quite evidently the animate world is abundant of specific examples of the processes of self-organization. But there are a number of examples available also in the inanimate world. Lasers can perform coherent oscillations, fluids when heated from below can acquire a spatial or temporal pattern, chemical reactions can lead to the formation of macroscopic oscillations, stripe structures, spirals or moving concentric waves. In the following we shall first describe what has been done in synergetics[1] to treat these processes of self-organization. At the microscopic level we may treat the individual atoms or molecules. On the other hand it has turned out that in the present context a mesoscopic level is appropriate in which many atoms or molecules are lumped together into a volume element so that the volume element can be treated by its density, by its velocity field, temperature a.s.o.. Such a description is also applicable to biological systems when we think of densities of cells in a tissue of a specific kind, densities of neurons firing at specific rates, etc..

Our following treatment will refer to this mesoscopic level. First of all we introduce a state vector

$$\underline{q}(\underline{x},t) = (q_1, q_2, \ldots q_N) \qquad (1.1)$$

which contains the various components $q_1, \ldots q_N$ describing the system. We shall assume that the temporal evolution of the state vector (1.1) is governed by nonlinear evolution equations of the type

$$\dot{q} = N(q,\alpha) + F(t) \tag{1.2}$$

where N is a nonlinear function which depends on the state vector and on control parameters, α. F is a fluctuating force which may or may not depend on the state vector. In the following we shall assume that the control parameter is fixed from the outside. In biological systems, in many cases, however, we must assume that the control parameters are fixed from a lower or higher hierarchical level than the one treated by (1.1). When we change the control parameter from a value $\alpha_0 \dashrightarrow \alpha$, we may expect that the solution q_0 changes. In the following we shall be interested in qualitative changes which we define by those situations where the linear stability of q_0 gets lost. Therefore we study the stability of the solution q_0 by making the hypothesis

$$q_0 \dashrightarrow q = q_0 + w \tag{1.3}$$

which leads in the linearized case to an equation of the form

$$\dot{w} = Lw. \tag{1.4}$$

It is well known that the solutions can be written in the form

$$w = \exp(\lambda t)v. \tag{1.5}$$

To solve the fully nonlinear equations (1.2) we make the hypothesis

$$q = q_0 + \sum_u \xi_u(t)v_u + \sum_s \xi_s(t)v_s \tag{1.6}$$

where we have introduced the indices u and s referring to unstable and stable modes corresponding to whether λ in (1.5) is positive or negative. Inserting (1.6) into (1.2) we obtain after some trivial transformations the equations

$$\dot{\xi}_u = \lambda_u \xi_u + \hat{N}_u(\underline{\xi}_u, \underline{\xi}_s) + \hat{F}_u \qquad (1.7)$$

and

$$\dot{\xi}_s = \lambda_s \xi_s + \hat{N}_s(\underline{\xi}_u, \underline{\xi}_s) + \hat{F}_s. \qquad (1.8)$$

We now make use of the slaving principle which allows us to express the stable modes governed by (1.8) by the unstable mode amplitudes ξ_u, i.e. we have a relation of the form

$$\xi_s = g_s(\underline{\xi}_u, t) \qquad (1.9)$$

where g is a well-defined function which depends on ξ_u and via the stochastic forces $\underline{F}$ on time t. By this formula we can eliminate ξ_s and obtain instead of equation (1.7) the set of equations

$$\dot{\underline{\xi}}_u = \tilde{\underline{N}}_u(\underline{\xi}_u) + \tilde{\underline{F}}_u. \qquad (1.10)$$

These equations can be transformed into a generalized Fokker-Planck equation. Its solution can be written in the steady state case in the form

$$P(\underline{\xi}_u, \underline{\xi}_s) = \prod_s P_s(\xi_s | \underline{\xi}_u) \, p(\underline{\xi}_u) \qquad (1.11)$$

where on the right hand side we made use of the slaving principle. The left hand side is a joint probability for the vectors $\underline{\xi}_u$ and $\underline{\xi}_s$ whereas the right hand side has the following meaning: p is the probability distribution for the order parameters, while P_s is the conditional probability for the slaved mode amplitude ξ_s under the condition that the vector $\underline{\xi}_u$ is given. In the following we shall call the $\underline{\xi}_u$ order parameters.

2. PATTERN FORMATION CLOSE TO NONEQUILIBRIUM PHASE TRANSITIONS

In the following we shall call those points in the control parameter space where at least one λ changes its sign a non-equilibrium phase transition. We briefly indicate how spatial or temporal patterns are formed close to these points. In such case $\underline{v}_u(\underline{x})$ is a space dependent function. For instance when a single λ becomes positive, the whole sum (1.6) is dominated by a single term containing the function

$$\underline{v}_u(\underline{x}) \tag{2.1}$$

or in the specific example of a sinusoidal wave

$$v_u(x) = \sin k_u x. \tag{2.2}$$

In the case of the onset of a so-called hard mode, where oscillations occur, v has the form

$$v_u(t) = e^{i\omega t} \tag{2.3}$$

or

$$v_u(t) = e^{-i\omega t}. \tag{2.4}$$

Of course, a number of more complicated cases can appear, for instance, when several λ's become positive or when a complex λ acquires a positive real part. Since these kinds of pattern formations have been treated in great detail elsewhere we shall not dwell upon giving here more examples.

3. INFORMATION ENTROPY

We define the information entropy as usual by the formula[2]

$$i = -\Sigma_j p_j \ln p_j \qquad (3.1)$$

where we interpret the p_j as probabilities. In the following we shall write ln instead of $\log_2$ which changes i just by a constant.

$$\ln \longrightarrow \log_2 \qquad (3.2)$$

In the case of the probability distribution in the form (1.11) we may write the information entropy quite generally as[3]

$$i = -\Sigma_{\underline{\xi}_s, \underline{\xi}_u} P(\underline{\xi}_s, \underline{\xi}_u) \ln P(\underline{\xi}_s, \underline{\xi}_u) \qquad (3.3)$$

It is easy to show that because of the form (1.11), (3.3) can be written in the general way[3]

$$i = i_f + \Sigma_{s,\underline{\xi}_u} P(\underline{\xi}_u) i_s(\underline{\xi}_u) \qquad (3.4)$$

where we have used the following definitions:

$$i_f = -\sum_{\underline{\xi}_u} f(\underline{\xi}_u) \ln f(\underline{\xi}_u) \qquad (3.5)$$

$$i_s = -\sum_{s} P_s(\underline{\xi}_s|\underline{\xi}_u) \ln P_s(\underline{\xi}_s|\underline{\xi}_u). \qquad (3.6)$$

In the following we shall be mainly interested in the change of the information entropy close to points of non-equilibrium phase transitions. As a consequence we shall introduce the change

$$\Delta i(\alpha) = i(\alpha_2) - i(\alpha_1) \qquad (3.7)$$

where it can be shown that

$$\Delta i(\alpha) \simeq \Delta i_f(\alpha). \tag{3.8}$$

So far, we have derived the information entropy (3.4) by means of the microscopic or mesoscopic theory where the distribution function was found as the solution of a Fokker-Planck equation. For systems in <u>thermal equilibrium</u> the distribution function can also be derived by the maximum (information) entropy principle[4].

More recently, by the introduction of new constraints it has become possible to derive the distribution function by means of the maximum entropy principle[5]. To this end we start as usual, quite generally, from the definition of the information entropy

$$S = i = -\Sigma p_j \ln p_j. \tag{3.9}$$

We introduce further the constraints of normalization

$$\Sigma_j p_j = 1 \tag{3.10}$$

and of a number of average values by

$$\Sigma_j p_j f_j^{(k)} = f_k. \tag{3.11}$$

When we require

$$S = \text{Max}! \tag{3.12}$$

under the constraints (3.10) and (3.11) we readily obtain as usual the distribution function p_j by means of[4]

$$p_j = \exp(-\lambda - \Sigma_k \lambda_k f_j^{(k)}) \tag{3.13}$$

and the entropy in the form

$$S = \lambda + \sum_k \lambda_k f_k. \qquad (3.14)$$

Now let us turn to the central topic of this talk, namely to pattern recognition.

4. PATTERN RECOGNITION

As we have mentioned above specific patterns can occur or change when control parameters are changed. In the following we wish to describe a method by which the evolving patterns can be discovered by machines or by means of an algorithm and not by human perception. We have mentioned a number of examples for phase transitions above. Those can be the phase transition of a laser where the light emission from normal lamps is replaced by laser light emission, the onset of patterns in fluid dynamics, or in chemical reactions. But pattern formation may occur in biological systems, also.

An example we have been treating in detail more recently is the phase transition in involuntary hand movements of humans which may serve as a paradigm for the change of gaits of horses, cats and other changes of coordination of muscles[6]. In the following we want to devise a general theory for the recognition of these patterns. Our way will be guided by a knowledge of the distribution function of the order parameters determined by the microscopic theory and in addition by the slaving principle.

The distribution function of the order parameters can be written in most cases in the form

$$p(\underline{\xi}_u) = N \exp(\sum_u \Lambda_u \xi_u^2 + \overline{N}(\underline{\xi}_u)). \qquad (4.1)$$

In the following we wish to derive distribution functions of these forms from phenomenologically given data. Phenomenologically, the state vector

$$\underline{q} = (q_1, q_2, \ldots, q_N)$$

can be measured. But we must be aware of the fact that due to fluctuations the state vector with its components $q_1, \ldots, q_n$ is a stochastic quantity, so that we have to average over measurements. The constraints

$$f_k = \langle q_k \rangle, \ \langle q_k q_{k'} \rangle, \ldots, \langle q_k q_{k'} q_{k''} q_{k'''} \rangle \qquad (4.2)$$

occurring in (3.11) are now interpreted as the given moments which are measured and averaged over. By means of the maximum entropy principle we can derive the probability distribution function P in the form

$$P(\underline{q}) = \exp(V(\underline{\lambda}, \underline{q})) \qquad (4.3)$$

where V is a function depending in a linear fashion on the Lagrange multipliers $\underline{\lambda}$ and in a nonlinear fashion on the state vector components q where we have included according to (4.2) only powers up to the 4. order.

We now wish to bring V into a normal form by means of the following steps: We first make the transformation

$$\underline{q} \dashrightarrow \underline{\bar{q}} \qquad (4.4)$$

so that $\underline{\bar{q}}$ is shifted against $\underline{q}$ so that we find the condition

$$\frac{\partial V}{\partial q_k} = 0 \qquad (4.5)$$

fulfilled for $\dot{\underline{q}} = 0$. After the transformation (4.4) we obtain

$$V \longrightarrow \tilde{V}(\underline{\tilde{\lambda}},\underline{\tilde{q}}) = \tilde{\lambda} + \Sigma \tilde{\lambda}_{jk} \tilde{q}_j \tilde{q}_k + \tilde{N}(\underline{\tilde{\lambda}},\underline{\tilde{q}}) \qquad (4.6)$$

where N is a nonlinear function going from the 2. to the 4. order. Now the main transformation consists in diagonalizing the bilinear part of (4.6) by means of the transformation

$$\tilde{q}_k = \sum_j a_{kj} \xi_j . \qquad (4.7)$$

This transforms (4.6) into

$$\tilde{V} \longrightarrow \tilde{\tilde{V}} = \tilde{\lambda} + \Sigma \tilde{\tilde{\lambda}}_j \xi_j^2 + \tilde{\tilde{N}}(\tilde{\lambda},\underline{\xi}) . \qquad (4.8)$$

We now distinguish the eigenvalues $\tilde{\tilde{\lambda}}$ according to the scheme

$$\tilde{\tilde{\lambda}}_j \geq 0 \quad j \longrightarrow u, \ \tilde{\tilde{\lambda}}_u, \ \tilde{\tilde{\xi}}_u$$

$$\tilde{\tilde{\lambda}}_j < 0 \quad j \longrightarrow s, \ \tilde{\tilde{\lambda}}_s, \ \tilde{\tilde{\xi}}_s . \qquad (4.9)$$

The index u refers to "unstable" in the sense described in the beginning of this article, the index s refers to the stable modes. We now introduce this notation (4.9) into (4.8) and rewrite (4.8) in the following form

$$\tilde{\tilde{V}} = \hat{V}_u + \hat{V}_s(\underline{\xi}_s, \underline{\xi}_u) . \qquad (4.10)$$

In it $\hat{V}_u$ contains only the unstable mode amplitude whereas $\hat{V}_s$ contains all the other terms. When we choose the constant term of $\hat{V}_u$ properly, we may write the following form for the joint probability of $\underline{\xi}_u$ and $\underline{\xi}_s$, namely

$$P(\underline{q}) \Longrightarrow P(\underline{\xi}_u,\underline{\xi}_s) = P(\underline{\xi}_s|\underline{\xi}_u) P(\underline{\xi}_u) \qquad (4.11)$$

where

$$P(\underline{\xi}_u) = \exp \hat{V}_u \qquad (4.12)$$

$$P(\underline{\xi}_s | \underline{\xi}_u) = \exp \hat{V}_s$$

are normalized. One may easily show that because of the smallness of the slaved mode amplitudes one may approximate the conditional probabilities by a product in the form (1.11) where P_s is a bilinear function in ξ_s only. A comparison of the result (4.11), (4.12), (4.8), (4.9) with (4.1) reveals that we have found the probability distribution for the mode skeleton. In particular it turns out that the transformation (4.7) provides us with the spatial pattern, provided we interpret the index of q_k as the coordinate of a space point. In such a case the elements a_{kj} of the transformation matrix occurring in (4.7) describe for each order parameter with index u a specific pattern and the total pattern appears as a superposition of a_{ku} with the order parameters as amplitudes.

Of course, we may also include the fine structure by including the slaved modes. If the system is still below its threshold of a non-equilibrium phase transition then we may expect that all $\tilde{\lambda}$ are negative. In this case it is possible to neglect the nonlinear terms higher than the 2. order so that we have to deal with a Gaussian distribution which can be treated by conventional pattern recognition methods.

If, on the other hand, we are well above threshold then again V has new minima around which we can approximate (4.8) by means of a bilinear function and again we may reduce the problem to the recognition of patterns of conventional pattern recognition theories. The particularly new feature of our approach is the emergence of the newly evolving pattern which requires a fully nonlinear treatment.

The comparison with the microscopic theory lets us expect that our new method is a powerful tool to recognize patterns close to non-equilibrium phase transitions by machines, i.e. by evaluating the matrix a_{kj}.

5. INFORMATION COMPRESSION, EMERGENT PATTERNS, SEMANTICS.

In the context of the present meeting it might be desirable to attempt at somewhat more general statements deduced from our above proceedure. First of all we know that close to non-equilibrium phase transitions patterns emerge and our method aims at recognition of these patterns. It is well known that close to these transition points strong fluctuations occur so that we have devised a method to recognize patterns in the presence of strong fluctuations.

The order parameter potential $\hat{V}_u$ defines a set of basins of attraction into which the total system is driven beyond the point of threshold. Because the numbers of order parameters are in general much smaller than the components of the vector $\underline{q}$, we obtain a pronounced compression of information. In a way the perceived pattern can be described both by the transformation matrix a_{kj} and the specific minima of V acquired by the order parameters. In view of human pattern recognition, it seems interesting to discuss the question whether this information compression has something to do with semantics.

As I have mentioned at various occasions[7], I believe that semantic information or meaning can be defined only with respect to a receiver. In general, it is assumed that the receiver possesses a number of templates which it checks against the incoming information and if a sufficiently good match is achieved, the pattern is recognized by identifying it with a template and now a meaning can be attributed to the information. I personally believe that nature proceeds in a more clever way, namely that it can organize templates by means of few data and by alternating these templates.

Indeed, in synergetics we see that upon the change of one or a few control parameters specific patterns can be formed. Therefore, it is tempting to study the formation of templates by means of self-organization under a number of given control parameters which in a way store the patterns. Incoming information will first evoke a hyothesis in the receiver upon which specific templates selforganize which then can be brought closer to the incoming pattern by symmetry breaking caused by various control parameters.

Of course, in nature pattern recognition will be much more involved, so one may think of a hierarchy of hypothesis under which order parameters are formed which govern the self-organization of various patterns of templates under the impact of specific control parameters. In addition, we may expect feed back loops between order parameters and the individual elements of the template or of the incoming information so that specific patterns are stabilized. A more detailed exposition of these ideas and its relationship to the results reported above will be published elsewhere.

The self-organized formation of macroscopic patterns described by order parameters may also serve as a metaphor for the creation of meaning. I wish to thank Professor Shimizu for the stimulating discussion on section 5.

6. REFERENCES

1) Haken, H., "Synergetics. An Introduction.", 3rd edit., Springer (1983),
   also available in Chinese, German, Hungarian, Italian, Japanese and Russian.

   Haken, H., "Advanced Synergetics", Springer (1983)
   also available in Japanese and Russian.

   Haken, H., "Synergetics: The Science of Structure",
   Van Nostrand Reinhold (1983)
   also available in German, Italian and Spanish,
   Japanese edition in prep..

2) Shannon, C.E., Bell System Techn.J.$\underline{27}$, 370 (1948).

3) Haken, H., Z.Phys.B, $\underline{61}$, 329-334 (1985)

4) Jaynes, E.T., Phys.Rev.$\underline{106}$, 4, 620 (1957)
   Phys.Rev.$\underline{108}$, 171 (1957).

5) Haken, H., Z.Phys.B, $\underline{61}$, 335-338 (1985)

6) Haken, H., Kelso, J.A.S. and Bunz, H., Biol.Cybern.$\underline{51}$, 347-356 (1985)
   Schoener, G., Haken, H. and Kelso, J.A.S., Biol.Cybern.$\underline{53}$ 247-257 (1986)

7) Haken, H., "Some Basic Ideas on a Dynamic Information Theory" in "Stochastic Phenomena and Chaotic Behavior in Complex Systems", ed.by P.Schuster, Springer Publ.Co. (1984)

THE INFORMATICS OF HIGH AND LOW CONTEXT SYSTEMS

Edward T. Hall
Emeritus Professor of Anthropology
Northwestern University

When discussing the context theory of culture, I am not concerned so much with the explicit meaning of a statement as I am with what kinds of meaning a particular language (or culture) generates and how the structure of a language guides and restricts the possible meanings. A language of course can be anything that generates, communicates, or organizes information. Furthermore, a new language changes meaning. This principle was stated in more specific terms by David Sharp in a conversation with Gian-Carlo Rota: ". . . if you want to apply mathematics, you have to live the life of differential equations. When you live this life, you can then go back to your molecular biology with a new set of eyes that will see things you couldn't otherwise see."[1]

The objective of this paper is to describe a basic but not generally recognized level of culture, informatics,[2] in which information is the primary locus of interest. Because the subject is vast and has been previously described,[3] I am restricting the main thrust to a single idea in which the ratio of stored information

---

[1] Rota and Sharp, 1985.
[2] "Informatics" is a term suggested by MIT's Professor Rota to designate an interdisciplinary and intercultural approach to the study of interpersonal transactions as well as of perceptual transactions between individuals and their environment. The term was needed because nonverbal communication has been popularly read as "body language," which omits such concerns as cultural differences in the organization and meaning of time, space, and matter (cf. Hall 1959).
[3] See Hall 1959, 1966, 1974, 1976, 1983.

to transmitted information varies in natural settings with widely differing results.[4]) A highly contexted communication is one in which most of the meaning has already been stored in memory, so that in order to release a response it is only necessary to transmit a

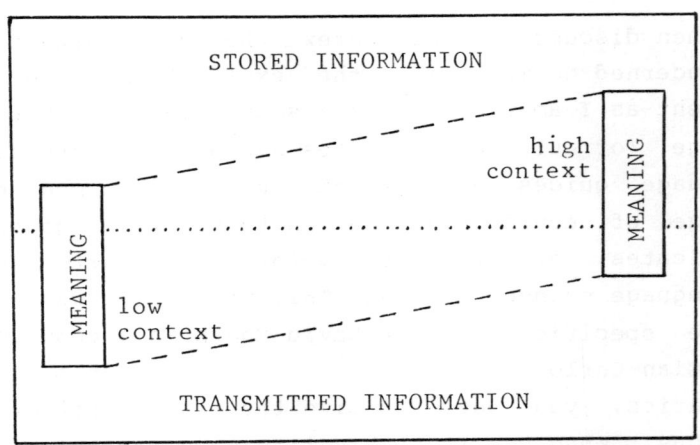

FIG. 1

4) Before proceeding further it should be made clear that my theoretical position regarding information transmission, storage, and flow between human beings is at variance with both the Shannon and Weaver (1949) and the behaviorist stimulus-response Pavlovian models. The informatic approach is more closely linked with the transactional school of psychology in which perception is viewed as a transaction with the environment. That is, the state of any human system is never neutral and must be considered in all transactions whenever and wherever the perceptual systems are involved. My remarks will be addressed to the topic of context, a subject that has confounded philosophers, linguists, and psychologists for years. As will become clear, my approach has proved to be at variance with that of the majority of my anthropological colleagues, who with few exceptions work with the manifest levels of culture which produce consciously-treated images, rather than working with emergent underlying structures.

minimum of information. Twin brothers reared together share so much that at times even words are unnecessary. A low context communication (L/C) is the opposite -- the ratio of transmitted information to stored information is relatively large. Proceedings in an American court of law and the act of programming a computer are low context communications.

A related and well-known phenomena is the effect of context on the meaning of a word, utterance, or statement. The word "man," for example, can be a reference to gender or maturity, a sexual object in the eyes of an admiring woman, tribal affiliation in some languages, a sexist reference when combined with "chair" ("chairman") and "kind" ("mankind"), and the entirety of white society (when used by blacks).

All meaning is dependent upon two functionally interrelated sets: information and context, which translates to transmitted information, and stored information, both of which are required to produce the third element, meaning.[5] The issue of context was recognized by Stanislaw Ulam before his death as relevant to AI (artifical intelligence) and also by Rota and

___

5) When George Trager and I were developing the culture matrix in the early 1950s, the inherent design of the matrix called for the two different types of mentalities it was possible to identify at the time in literary texts, which we called "point" and "line" integrations (Hall and Trager 1953). Point integrators jumped from point to point without filling in the intervening spaces. Line integrators were more linear and "felt uncomfortable until they had put all the little clothespins on that line." Halprin's (1969) open and closed scoring bears some resemblance to our subject. Albert Szent-Gyorgyi (1972) classifies the two types according to the Greek tradition: Apollonian (low context) and Dionysian (high context). If one looks to the brain as the source of high and low context, the logical place to begin would be with the right and left hemispheres, although MacLean's (1980) triune model would place high context in the neo-cortex and low context in

Stein.6) Stored information increases in complexity with use (experience). As I hope to demonstrate, the two ends of the context continuum produce vastly different results. Also, any <u>shifts in context</u> in the course of a transaction is a meta-communication, a subtle but apparently universal feedback mechanism used to provide ongoing information concerning the state of the relationship of the interlocutors -- increase in context signals warming of the relationship whereas lowering of context signals the opposite.

Another factor in my analysis has to do with the mind. I have, over the years, been deeply involved in the observable realities of how people's minds work in different cultures. But first, a few contexting words concerning the term <u>culture</u>.

There are two popular meanings attached to this word. It was for a long time associated with, and symbolic of, the education and social standing of individuals, e.g., "She is a very cultured woman." For the past hundred years, however, this same term has been used to describe the field of study of New World cultural anthropologists -- the beliefs, customs, ceremonies, myths, social organization, religion, etc., of exotic peoples.7) Culture in this second sense has been used and

----

the brain stem. An after-the-fact association can also be found in the analog and digital formats for watches and computers, the analog providing more contexted information.

6) Context is of course central to solving the multiple problems of artificial intelligence (see Rota and Sharp 1985, Rota and Stein 1985, Rota and Ulam 1982).

7) In Europe it was for some time assumed that what Europeans had was "civilization," an extension of the concept of the "cultured" individual. "Natives" were thought less than civilized, but when anthropologists began studying them seriously, one of the first discoveries was that all societies have their own "culture."

is understood to apply to those features of social and material life that the people could <u>talk about and describe</u> either to themselves or to the anthropologist.8) Yet when the visible, manifest aspects of culture disappear, something else remains; the people still identify with each other as long as that something else is present. My research has been with that something else -- a third level of culture which in the briefest possible sense can be said to be "the underlying principles which control our lives."

This level -- the <u>informatic</u> level -- operates almost totally out of awareness, is roughly synonomous with the informal level of culture described by me in the 1950s,9) and is also reminiscent of Polanyi's tacit dimension.10) In other words, it is that set of regularities that govern behavior which are so automatic that people do not realize that there are even any rules.

As with the unconscious in the various schools of psychoanalysis, there has been great resistance among the educated to accepting the idea that there is anything about themselves that is not fully within their control. The same reactions were observed when descriptive linguists were developing grammars and writing systems for languages not previously reduced to writing.11)

    8) A recent article by Louis Sass in <u>Harper's</u>, "Anthropology's Native Problems," summarizes the current state of the field. Regardless of approach, the preoccupation of anthropologists has been with the visible aspects of culture. Anthropology at this level has altered our view of the world. I only mean to stress that the informatic level should not be confused with the visible and more conventional, tangible level.
    9) Hall 1959.
    10) Polanyi 1966.
    11) The linguist Charles Ferguson, while engaged in a study of colloquial Arabic in the eastern Mediterranean area, was constantly faced with the conviction on the part of his subjects that he was engaged in a fruitless chase, that colloquial Arabic could not possibly have a

The hard scientists, of course, have accustomed themselves to the fact that there are laws of nature yet to be described. Human beings have yet to arrive at that stage of understanding of their own dimension. Since the estimates of the ratio of verbal to nonverbal communication run from 90 percent or more on the nonverbal side, it would appear that those who are involved in informatic studies have our work cut out.

At this point it is appropriate to address the matter of how it was that the informatic level was identified at all. The following examples illustrate how I was faced with evidence that simply did not fit the more conventional paradigms in vogue at the times the events occurred.

This first example dates back more than 50 years to when I was working as a construction foreman/engineer on the Navajo and Hopi Indian reservations in Arizona. I had interrupted my university studies in anthropology to earn some money and to gain some practical experience. At that time the conventional view of Indians took two forms: an overly romantic one by those who idolized or romanticized the Indians and another that could only be likened to Social Darwinism (held by the government workers and bureaucrats running the U.S. Indian Service). The problem I faced (which did not seem to bother the other foremen) was that the Navajos simply were not working -- were not

------

structure or grammar. Yet when he would ask, "Can you say this?" the answer would be yes or no to the effect that the form in question was correct or not. I interviewed one of his subjects, a chemist at the American University in Beirut, who expressed his feelings about Ferguson's work as follows: "I could not control the answers or formulate the questions. Ferguson used me like a machine. I had absolutely no control over the data. He got data from me whether I wanted him to or not. I was like a well that must give up its water."

getting their money's worth from the government funds allotted the reservation. I tried the usual persuasive techniques, which relied heavily on the logic of the situation, i.e., there was a fixed amount of money and it was up to the Indians to decide how much they got in water development projects. My words were of no avail.

I didn't know it then but I was about to get my first lesson in how other people's minds work. My persuasive arguments had only made my crews anxious and confused. Talking this over with an older experienced mentor who had grown up on the reservation and who was one of the few people I had ever known who, I felt, really understood Navajo psychology, I discovered that an entirely different approach was called for; Western logic was not only not effective but also counterproductive.12) The key to the Navajo mind lay in spelling out a series of reciprocal tacit arrangements and obligations which the Navajos took for granted and which I fortunately by that time was beginning to understand. The Navajo footdragging was traced to deep concerns as to what the government was going to exact in return for the work it was doing to improve their land. The problem was a matter of comunication, of context, and of meaning. Given these facts it was possible to work out a formula which translated my European logic into a less linear form consistent with Navajo psychology.

From that time on my crews performed well. However, even though I knew and could describe both sides of this cultural equation, no one in the government was interested or could understand that the Navajo mentality was operating from a completely different set of assumptions. Theirs was the logic of interlocking reciprocal obligations. Ours was the linear logic of

12) I later encountered the same pattern in Japan.

abstractions (there is a hidden pool of money there, which you can't see and which you have no control over, but which will provide you with work if you do all the things that the white man says are in your best interests, etc.).

Thirteen years later, following WWII, I found myself on the atoll of Truk in Micronesia in the capacity of adviser to the U.S. Navy in its attempts to govern the Trukese people, who had been "liberated" from the Japanese.13) The cultural distance was vast.

The situation was not unusual. Whenever two cultures try to work things out under such forced circumstances, the entire process can only be described as surreal. Not only were there manifest differences in the tangible realities of the two cultures (work habits, technologies, social and economic relations), but there was the added problem of the "democratic" base that the Navy thought should be imposed without delay on that status-ridden hierarchical society in which chiefs were layered on chiefs.

The U.S. government was anxious to bring democracy to the Trukese. The device chosen was to take advantage of the weekly meetings with chiefs to explain and promulgate the policies of the military governor. The meetings of course were supposed to be "democratic," which meant that everyone was to participate. This did not happen. Information flowed only one way, from the top down, which was consistent with Trukese culture. I was asked to find out what was wrong, as there were beginning to be rumblings of discontent traceable not so much to

---

13) Japanese intervention in the native culture was minimal; as a consequence their administration of the island -- unlike that of the U.S. -- was apparently quite benevolent, with a minimum of disruption of native institutions.

the government's imposed programs as to these newly-instituted and too frequent meetings with the chiefs.14)

Operating on the assumption that no social organism, no matter how authoritarian or autocratic, can function without feedback, I began systematically to trace out (with my counterpart Arty Moses, the atoll chief) the flow of information from the chief down through various channels and then up again to see how individuals who were ostensibly in command were keeping in touch with those whom they were governing. The system revealed was hardly complex, but couldn't have been more different from our own implicit Western system. In the United States, the unstated rule is that only information obtained directly in a face-to-face relationship has legal standing and is binding in court. In the U.S. hearsay is just that and nothing more.15)

The situation on Truk was one in which a high context culture was attempting to cope with a low context one, a distinction to be elaborated below. As it turned out, my particular role was vital to the entire operation, and without my presence (or that of someone like me) in a slot created by the Trukese culture, the link in the feedback chain would have been broken. The hierarchical nature of Trukese society did not allow a subordinate to do anything but agree with a superior when in his presence. Someone on the same level, even though not in the chain of command, was the preferred channel for expression of differences of opinion. I was the channel from Arty Moses to the military governor. When I

---

14) The chiefs were using the opportunity of being together at the meetings to devise methods for exploiting the villagers.
15) Hearsay is permitted in high context cultures such as the French, where the court needs more contexting than the low context American system allows -- a topic worthy of study due to the impact of law on all our lives.

would explain to the governor that there was no way that Arty Moses could possibly disagree with anything he said or wanted (regardless of consequences), the commander, who was normally a mild-mannered man, would explode, pound the desk, and reply, "Goddamn it! Arty Moses <u>told me to my face</u> that everything was ok. And now you are telling me it isn't?"

The protocol of the chief's feasts and other such matters were understood, by the way. Nothing on the covert, informatic level was, however, even when explained with the utmost "logic."

These examples involve behavior and formulations, the rules of which are almost entirely in the unconscious realm. For example, distance-setting within cultures is a powerful communicator of status and mood -- a vital source of feedback as to the state of a given relationship or set of relationships, yet it operates almost totally out-of-awareness.16) Across cultural boundaries, it is also can be, when incorrectly read, a source of miscuing regarding the intentions and rank of the parties concerned.

The two examples above reveal an additional factor that must be explained, namely that ours is a <u>one-way</u> logic.17) We employ it when explaining things to others but not the other way around. Participants who share the same logic set can explain things to each other but our own system of logic breaks down when confronted with another system based on a different set of assumptions (which is simply one more example of the fact that in science one cannot compare two systems based on different models). The primary obstacle, however, is that most if

---

16) Hall 1974, 1968, 1966.
17) Americans are not alone in this. In fact, none of the cultures I have worked with -- literate or preliterate -- has demonstrated any degree of freedom from the chains of its own unique system of logic.

not all informatic systems are apparently blocked on the notion that there really are different models, i.e., the people of the world must contend with the fact that we all inhabit closed systems, the primary and tacit assumptions of which operate out-of-awareness and are not therefore subject to conscious control. Stated differently, cultures and their associated values systems are not just relative but areset up and operate according to assumptions and patterns which are unique to each culture and cannot therefore be compared without translation.

Working with this set of assumptions (a number of which were still implicit -- but not explicit -- in my own thinking) I set about looking for a frame which could be used to analyze more than one culture.18)

It was the result of having to deal with the sheer mass of material, plus an endless variety of contexts, that eventually made it possible for me to tease from the data the outlines of a pattern of the type I was looking for.19) I make it a point when looking for basic patterns to begin by investigating the physiology of human perceptual systems. Significant clues in this case were also available from the studies and insights of the Transactionalists as well as from another group of

---

18) The foundation for my search already existed in <u>The Silent Language</u> in that the primary message systems of culture are universal, as are the formal, informal, and technical modes. Fortunately, there was no lack of cases similar to the two mentioned above. Context as a concern for linguists had proved to be a thorn in the side of many who were involved in research on the intercultural process. Meaning was a chimera. Nothing stayed fixed because of varying contexts.

19) I was aided by chance when a black assistant and I became involved in an intercultural contratemps, the results of which pointed the way to still another category of cultural miscuing. One day in the laboratory, Jessie was working at the drafting table and I was discussing a project I needed his help on. Having

studies (to be described below). The whole basis of perception from the transactional point of view[20] rests on the result of what occurs as the senses process information received from the environment in a transaction with information already stored in the system. At that time there was a wealth of data on context produced by psychologists.[21]

Internal contexting serves many purposes, one of which is to make it possible for human beings to correct automatically for distortions or omissions of information in messages. This mechanism was investigated by Richard and Roslyn Warren,[22] a psychologist-zoologist team. The Warrens excised portions of words on recorded tape and replaced them with background sounds encountered in everyday life, such as a cough. Subjects listening to the tapes the first time were unable to detect the excisions; they still actually "heard" the missing sound and had

------

gotten no conversational feedback from Jessie for several minutes, I asked, "Jessie, are you listening?" He replied, "You listen with your ears." He paused as he thought over what he had just said, and then added, "Man, if you're in the room, I'm listening." Behind this short exchange lay the germ of research on the types of feedback required in order to keep a conversation going ("listening behavior," Hall 1969), requirements that vary from culture to culture. In a flash of insight I understood two things: one source of tension between whites and ethnic blacks is the difference in the expected listening response, and whites need more <u>transmitted</u> information (are lower context) than do ethnic blacks. White transmissions are redundant because blacks are already programmed to the effect that if a significant other is in the room, the interlocutor is automatically tuned in to his or her remarks and therefore does not need to signal, "I hear you." There were dozens of instances of this sort in my research.

20) The moon illusion. Kilpatrick, et al., 1961.
21) The following descriptions of the work of the Warrens, McCulloch, and Gouras and Bishop appeared in slightly different form in Hall 1976.
22) Warren and Warren 1970.

difficulty localizing the cough. At first, the Warrens simply removed one sound -- the central "s" in "legislatures," used in a normal sentence. They were careful to remove enough preceding and following sounds to eliminate traditional cues. Later, the Warrens eliminated an entire syllable -- the "gis" in "legislatures," still used in a sentence -- with the same results. Even when told that a whole syllable was missing from a particular sentence, the subjects were unable to identify even the part of the sentence from which the

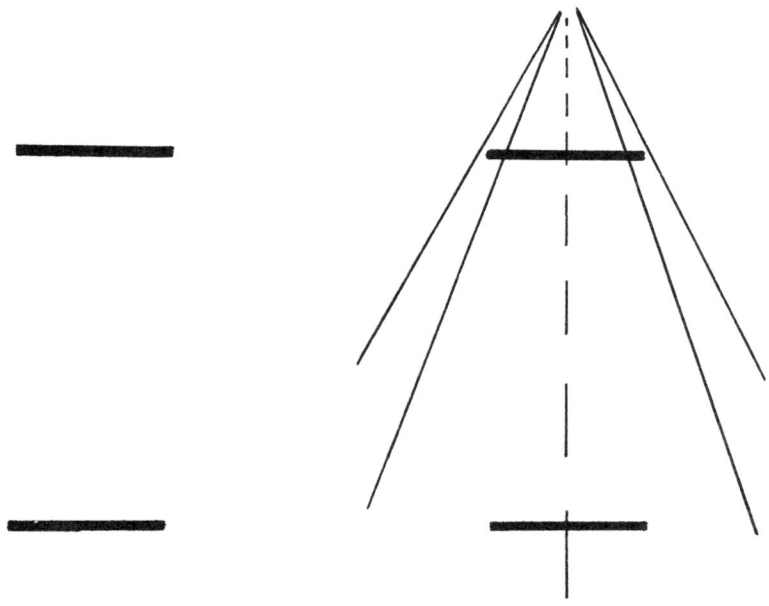

FIG. 2 How Context Affects Perception. All four horizontal bars are the same size.

syllable had been taken.23) The Warrens state:

> Verbal context...<u>can determine completely the synthesis of illusory speech sounds</u>; phonemic restorations are heard <u>when the context is clear</u> but part of the stimulus is absent. Another illusion arises when the stimulus is clear and the context is absent. (italics added)24)

However, when context is absent and a single word is repeated over a three-minute period, nothing remains constant -- the average subject will hear approximately thirty changes involving six different forms. The same word after a time is no longer heard as the same word but as several different words. The word "tress" repeated without pause and heard distinctly is soon transformed into "dress," "stress," "Joyce," "floris," "florist," and "purse."

Contexting is apparently deeply imbedded in processes governing the evolution of both the nervous system and the sensory receptors, particularly the eyes and the ears. An important set of studies, under the direction of Warren S. McCulloch25) of the Massachusetts Institute of Technology, which have been devoted to identifying the information sent to the brain by the eye, reveal that the frog is quite easily fooled because of its <u>low context</u> visual system. Anything that is within a certain size range, is dark, and moves is going to be

---

23) Varying the choice of altered words made no difference.
24) The Warrens' experiments, viewed from the point of view of context theory, provide an alternate explanation for Bateson's hypothesis (1956) that the message itself carries all the necessary information even at lower levels on the phylogenetic scale.
25) Lettvin, Maturana, McCulloch, and Pitts 1959.

treated as a fly by the frog. Live but anesthetized flies that don't move are not eaten, while the burned head of amatch or any small, dark object placed on the edge of a revolving disk will be ingested immediately by any frog within range. As one moves up the phylogenetic scale, more and more contexting (hence greater pattern-recognition capacity) is built into the whole visual apparatus, not only the eye but the brain as well.

Gouras and Bishop reported on the neural circuitry of the retina as it changes in response to increased needs for information as life forms evolve. They state:

> The differences between various vertebrate retinas are not due to new types of synapses but different proportions of the same types, and these differences become most apparent in the output of the retina, the ganglion cells. Ganglion cells of lower order vertebrates are more specific in their stimulus requirements than are those of higher vertebrates. The more specialized a cell becomes, the less potential information it carries, so that by delaying the process of specialization until the more central nervous system -- where more nerve cells are available -- higher vertebrates gain the advantage of extracting more features from the external world. (italics added)[26]

By delaying the specialization of stimulus requirements of ganglion cells as they evolve, animals are also less easily fooled by camouflage and as a consequence have enhanced their survival chances.

26) Gouras and Bishop 1972.

This same observation can be made when comparing high and low context cultures; that is, in many ways the social organism or "culture" can be seen as an extension of the physiology of the perceptual system. Low context cultures tend to be more specialized in the lower (information gathering) echelons, which brings me to the core of this paper, namely observations and results associated with the high-low context model.

Experience with small indigenous communities and cultures reveals that people living in close proximity over long periods of time are thoroughly contexted to each other. In fact, anyone who has moved from a rural setting to an urban cannot fail to notice the difference: the context drops (which is something many country people find attractive about cities -- other people don't know much about them; that is, as context drops, privacy rises). Therefore, it does not seem to be stretching a point to observe that one of the results of what is commonly thought of as "civilization" is a generalized lowering of context. However, some cultures are lower than others (although each culture is unique and may fall on the context continuum from high to middle to low). As recent experiments demonstrate, civilization is just one of a number of factors affecting context. This proved to be the case while I was working with a German and a Japanese publisher over a six-year period. My partner and I interviewed in depth directors of successful business enterprises in France, Germany, the United States, and Japan. Our carefully selected sample included a total of 350 interviews, evenly divided between the four countries. In addition a separate but related informatic analysis was made of each culture to be used as a baseline.

One of our informants, a Frenchman who had briefed himself thoroughly on my thinking and published works, observed that one of his greatest problems in working with the Germans was the matter of context -- they were always "low contexting" him (the meta-language was at work), that is, talking down to him by telling him things he already knew and was deeply familiar with, a hard pill to swallow. Our data corroborated his observations. Not only were the Germans lower on the context scale than the French, which could be observed in virtually everything they did, but this fact infiltrated and influenced almost every aspect of German-French relationships. As the reader may already know, or have gathered from what has been said, the low context Germans quite frequently tell you more than you need to know. One of our favorite examples was generated by our asking directions to a hotel in Hamburg. The man who was directing us not only produced extremely detailed, accurate directions but added that when we crossed the street at the light we should be sure to wait until the red light turned green before crossing!

In the early stages of dealing with a foreign culture, context can be everything you don't know about that system, the patterns and the hidden rules that give the system meaning. In a high context culture practically everything is implicit and taken for granted, whereas in a low context system there will be explicit rules for behavior (e.g., don't flush the toilet after 10:30 PM). The Germans in general have a strong drive to conform to rules, to <u>belong</u> and to be recognized as good citizens. As a consequence they really suffer in France because they will never be French; they will never master the hidden rules of French culture.

Therefore, if the goal is to integrate people into a social system more rapidly, lower the context (which is one reason why armies and bureaucracies are the way they are). Low context cultures will emphasize <u>procedures</u>. In a low context culture, even in a unit such as the family, if something goes wrong there will be an attempt to develop a procedure so the mistake will not reoccur. High context cultures will bypass L/C bureauratic procedures by finding the individual human being who can correct matters. In a high context culture such as Japan and the Pueblo cultures of New Mexico and Arizona, the meaning of life comes from membership in the group. To be expelled or ostracized from the group is social death, which means that laws, when codified, are frequently quite simple. (A codification of Zia Pueblo law, done by the Indians as a guidance instrument for their own people as well as to explain to whites that they did have "laws," was a document of only three single-spaced pages.)

Low context manufacturing involving microchips, the assembly of watches, cameras, and computers, and the production of small machine parts can cross cultural boundaries with fewer complications than can high context activities such as education or the publication of magazines and newspapers. Much less implicit, out-of-awareness information is required in manufacturing. To illustrate: a large German publishing company managed to achieve a thirty to fifty million dollar disaster (estimates vary) when it insisted on German management and control of a magazine venture in the U.S. In contrast, a French magazine that was started by an unusually perceptive German succeeded. His job was to produce something that would appeal to the French. Knowing about the American debacle, he refused to hire German-speaking French staff and, since French was not spoken at company headquarters, management interference was minimal. In the late 19th and early 20th centuries the Chinese decided they should espouse Western education. They sent their young people to European and American universities. However, it wasn't until they established their own universities that they began to produce scientists, physicians, and educators of stature. The context of Western education lacked those reinforcing features necessary for graduates to succeed and fit into the H/C Chinese culture.

Paradoxically, high context systems thrive on detail under certain conditions. It appears that when one is faced with a high context system, as in Japan, new situations can be learned only if they are approached technically and in the greatest detail. On the other hand, those of us in the West who are used to having to struggle with the complexities of L/C systems can, when we are confronted with something new, be quite creative about it and not require an inordinate amount of detailed programming. H/C people can be creative within their own system but have to move to the bottom of the context scale when dealing with anything new, whereas L/C people can be quite creative and innovative when dealing with the new but have trouble being anything but pedestrian when working within the bounds of the old system. To all of this there are limits and exceptions, but it is often necessary in an intercultural situation for the L/C person to have to go into much more detail than he is used to when he is dealing with H/C people. If the L/C person interacting with a high context culture does not really think things through and try to foresee all contingencies, he's headed for trouble.

It is easier to foresee trouble or coming confrontations in L/C cultures than in H/C cultures, because in the L/C culture the bonds that tie people together are somewhat fragile, so that people move away or withdraw if things are not going well. In the H/C culture, according to anthropologist Francis Hsu[27] and others, because the bonds between people are so strong there is a tendency to allow for considerable bending of the system. When the explosion comes, it is likely to come without warning. When the boundaries are overstepped, they must be overstepped so far that there is no turning back. It is sheer folly to get seriously involved with H/C cultures unless one is completely contexted. This is the danger that the West faces in its dealings with the East.

The examples given above are far from comprehensive, and while the principles described here apply to the full range of human transactions, from individuals to business to governments, we can only sketch in the basic patterns. What we know for certain is that there are significant differences in the two ends of the scale. One can only guess at what the total implications will eventually be.

It appears that the information context ratio may be relevant as a guiding principle in the perceptual system at all levels of the phylogenetic scale. One of the consequences of evolution was to increase stored information in the CNS. In humans, this simple and rather obvious relationship seems to have far-reaching consequences depending on where one happens to be in the context scale.

Shifts in context (up or down) represent a meta-language which may be universal. Moving up the scale signifies increasing closeness and moving down the scale indicates the opposite.

In intercultural transactions, it would appear that low context enterprises are easier to transfer from one culture to the next than are high context endeavors. The more difficult high context endeavors would probably

27) Hsu 1970.

benefit from a reduction of interference and management from the headquarters in one country with the new enterprise in another.

It also appears that if people are to be trained for new skills in a minimum of time, lowering the context makes the task much easier. However, there is a price to pay for this type of efficiency: reduced flexibility and reduced response time due to overloading of information channels. Low context systems are much more vulnerable to information overload than are high context systems. And while high context systems tend to be resilient to the pressures of change, either internal or external, change beyond a certain point can be catastrophic. There is no formula for computing the reprogramming of information systems, which does not mean that such an attempt should not be made.

Both high and low context systems are subject to stress from too much or too little information, but they respond to such stresses differently. While there are more generalizations to be made concerning context (such as that monochronic cultures tend to be low context and polychronic cultures tend to be high), the most relevant generalization to a student of information as a cultural process is that "context" and out-of-awareness, tacit culture are synonymous. In general, anthropologists have worked with the transmitted information side of culture whereas informatics is more concerned with the contextual side. It takes both approaches before meaning can be assigned to a series of observations.

Gian-Carlo Rota, who coined the term "informatics," has said, "Scientists always want to show that things that don't look alike are really the same."[28] I want to show that although things may look alike, when context is

---

28) Rota and Sharp 1985.

different they are not the same. When this is understood and the contextual frame is increased as a result, it is possible to envisage a time when they will be the same. And that's the informatic approach to understanding.

BIBLIOGRAPHY AND REFERENCES

Bateson, G., "The Message: This Is Play" in <u>Group Processes: Transactions of the Second Conference</u>, New York: Josiah Macy, Jr., Foundation Publications, 1956.

Caianiello, E.R., "Languages, Hierarchical Structures and Logic," paper presented at Coral Gables Conference, "Orbis Scientiae," January, 1974.

Ferguson, C., personal conversation, early 1950s.

Gouras, P. and Bishop, P.O., "Neural Basis of Vision," Science <u>177</u>, 188-189 (1972).

Hall, E.T. and Trager, G.L., "The Analysis of Culture," Washington, DC: American Council of Learned Societies, 1953.

Hall, E.T., <u>Beyond Culture</u>, Garden City, NY: Doubleday, 1976.

_____, <u>The Dance of Life: The Other Dimension of Time</u>, Garden City, NY: Doubleday, 1983.

_____, <u>Handbook for Proxemic Research</u>, Washington, DC: Society for the Anthropology of Visual Communication, 1974.

_____, <u>The Hidden Dimension</u>, Garden City, NY: Doubleday, 1966.

_____, "Listening Behavior: Some Cultural Differences," Phi Delta Kappan, March 1969.

_____, "Proxemics," Current Anthropology, April-June 1968.

_____, <u>The Silent Language</u>, Garden City, NY: Doubleday, 1959.

Halprin, L., <u>The RSVP Cycles</u>, New York: George Braziller, Inc., 1969.

Hsu, F.L.K., <u>Americans and Chinese</u>, Garden City, NY: Natural History Press, 1970.

Kilpatrick, F.P., ed., <u>Explorations in Transactional Psychology</u>, New York: New York University Press, 1961

Lettvin, J.Y.; Maturana, H.R.; McCulloch, W.S.; and Pitts, W.H., "What the Frog's Eye Tells the Frog's Brain," Proc. Inst. Radio Engrs. <u>47</u>, 1940 (1959).

MacLean, P.D., "An Evolutionary Approach to Brain Research on Prosematic (Nonverbal) Behavior" in <u>Ethology and Nonverbal Communication in Mental Health</u>, ed. S.H. Corson et al., Oxford and New York: Pergamon Press, 1980.

Polanyi, M., <u>The Tacit Dimension</u>, Garden City, NY: Doubleday, 1966.

Rota, G.-C. and Sharp, D., "Mathematics, Philosophy, and Artificial Intelligence," Los Alamos Science <u>12</u> (1985).

Rota, G.-C. and Stein, P.R., "Artificial Intelligence Today," unpublished manuscript, 1985.

Rota, G.-C. and Ulam, S., unpublished recorded conversation, 1982.

Sass, L.A., "Anthropology's Native Problems," Harper's, May 1986.

Shannon, C., <u>A Mathematical Theory of Communication</u>, Urbana: University of Illinois, 1949. The model referred to in the text is known as the Shannon-Weaver model.

Szent-Gyorgyi, A., "Dionysians and Apollonians," Science <u>176</u>, 966 (1972).

Warren, R.M. and Warren, R.P., "Auditory Illusions and Confusions," Scientific American <u>223</u> (1970).

Wiener, N., <u>Cybernetics; or, Control and Communication in the Animal and the Machine</u>, New York: John Wiley & Sons, 1948.

A NON-DETERMINISTIC APPROACH TO ANALOGY, INVOLVING THE ISING MODEL OF FERROMAGNETISM

Douglas R. Hofstadter

Fluid Analogies Research Group, Perry Building
330 Packard Road, Ann Arbor, Michigan 48104

ABSTRACT

A close analysis of several abstract analogies reveals the critical role played by directed links created in the act of perceiving the structures in the analogy. These directed links, having bi-stable orientation properties, are similar to the bi-stable spins of the Ising model of ferromagnetism. The similarity is enhanced by the fact that the links' orientations are not deterministic but stochastic, and the degree of order and disorder in the system is regulated by a formal parameter playing a role analogous to that of temperature in the Ising model. The paper is concerned principally with showing how temperature-controlled flipping of the directed links allows "perceptual Bloch domains" to emerge, thus facilitating discovery of subtle analogies.

In the making of an analogy between two situations, a critical ingredient is how those situations are framed in terms of known concepts. Framing a situation in terms of concepts is much like visual perception of a scene, in which the goal is to attach numerous labels ("chair", "elephant", "Vesuvius") to regions of the visual field. The difference is that a situation is generally abstract rather than visual, and consequently the labels to be attached to parts of it are usually at a higher level of abstraction than those in visual perception.

Our work on analogies[1,2,3] involves highly idealized situations represented by strings of letters of the alphabet. An event in such a situation is a change in the original string. Thus, a typical event would be the changing of the string **abc** into the string **abd**. In fact, we take this event as our prototype event. Our goal is for our computer program "Copycat" to be able to make numerous interesting analogies with that event. For instance, if **abc** -> **abd**, what analogous event should happen in the target situation **ijkl**? There are several conceivable answers, but the most satisfying one for the vast majority of people is: **ijkl** -> **ijkm**. There seems to have been extracted a rule from the prototype event: *"Replace the rightmost letter by its alphabetic successor"*. It seems that this rule is then applied to the target situation, yielding the answer.

On closer analysis, one sees that things are not quite that simple. For instance, consider a different target situation, **iijjkkll**. If the rule cited above were simply applied, as is, to this new target, one would get the following result: **iijjkkll** -> **iijjkkml**, which is contrary to most people's preference on esthetic grounds, which are of major importance in analogy-making. Most people strongly prefer **iijjkkmm** as the outcome. The intuitive explanation for this answer is that generally, the rule should be interpreted a little loosely, and that here in particular, the phrase *"rightmost letter"* should be interpreted according to its new context. In the target situation **iijjkkll**, it seems that doubled letters play the role that single letters played in the prototype, so that the adjusted rule would say something like this: *"Replace the rightmost doubled-letter by its alphabetic successor"*. Of course, that rule yields the appropriate answer.

We call the process whereby a rule is modified according to its new context the translation of the rule. Translation is a key process in the operation of the Copycat program. It is important to see translation occur in a number of different ways, so consider the following new target situation: **srqp**. Application of the "raw" rule (i.e., the untranslated rule) to this target would yield the answer **srqq**, an answer that few people find appealing. Once again, therefore, it seems that translation is called for. However, here we find an interesting split among people. Some prefer the answer **trqp**, while others prefer **srqo**. These answers reveal what translated rules are being created and utilized. In the case of answer **trqp**, it is clear that the operation *replacement by alphabetic successor* was carried out on the *leftmost* letter, rather than the rightmost. On the other hand, in the case of answer **srqo**, the site of the operation remained fixed, but now the operation itself was adjusted into *replacement by alphabet predecessor*.

We have now seen three distinct translations of the raw rule *"Replace the rightmost letter by its alphabetic successor"*. They are summarized below:

Target situation: **iijjkkll**.

Translated rule:
Replace the rightmost doubled-letter by its alphabetic successor.

Resultant answer: **iijjkkmm**.

Target situation: **srqp**.

Translated rule:
Replace the leftmost letter by its alphabetic successor.

Resultant answer: **trqp**.

Target situation: **srqp**.

Translated rule:
Replace the rightmost letter by its alphabetic predecessor.

Resultant answer: **srqo**.

It seems that subtle contextual pressures must be applied to a rule so that it can "flex" or adapt itself to a new situation. In particular, some parts of the rule will stay constant while other parts of it will be modified. The crux of our research project is to determine how such modifications are brought about by the constellation of pressures that arise when the target situation is compared to the prototype situation. But for such a comparison to be made, there must already exist perceptions of the two situations. It is the process of situation perception that we are concerned with in this paper, since that process chronologically precedes and logically underlies the critical process of rule translation.

Because the target situation **srqp** has two distinct answers that appeal to people, it follows that there must be two distinct perceptions of that situation that cause distinct constellations of mental pressures, ultimately leading to the bifurcation in opinions about rule translation. Let us try to characterize the problem. We are comparing prototype situation **abc** with target situation **srqp**. In one view, the letter **s** plays the role of the **c** (i.e., it is the site of change), while in the other view, the letter **p** plays the role of the **c**. Thus, there are really two different mappings going on that give rise to the two different answers:

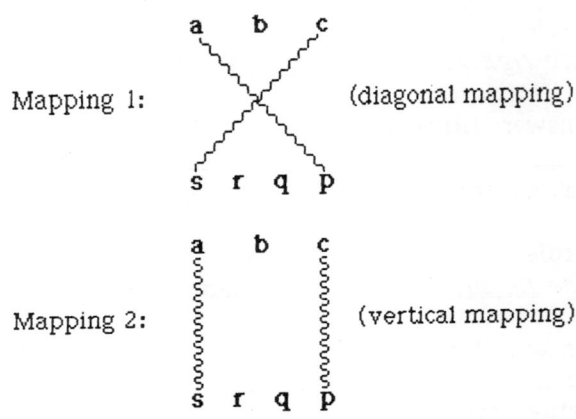

The wiggly lines connecting counterpart entities in a given mapping are called <u>bridges</u>. Bridges play an extremely central role in the Copycat program, for it is they (or more precisely, their <u>pylons</u>) that carry all information about rule translation. However, precisely how bridges and pylons accomplish this need not concern us here. Intuitively, it is clear that the establishment of credible counterparts between situations tells a great deal about how to adapt a statement about the first situation to the second situation.

Clearly the diagonal mapping is the one that gives rise to the answer **trqp**, for in it the **s** is the counterpart of the **c** and the roles of concepts *left* and *right* have been reversed. It is relatively easy to see how these bridges would suggest substituting *"leftmost"* for *"rightmost"* in the rule, thus accounting for the rule translation that leads to the answer **trqp**.

In the vertical mapping, the **p** is the counterpart of the **c** and left and right have not been reversed. Now, these bridges are supposed to give rise to the answer **srqo**. To do so, they would have to suggest substituting *"predecessor"* for *"successor"*, for if that translation of the rule were not carried out, the raw rule would have to be applied, and we would get the distinctly less satisfying answer **srqq**. But nothing in the diagram suggests such a substitution of concepts. There must be more to the vertical mapping than what is shown in the diagram if it is to give us the desired answer. How can we augment our perception of **srqp** to give it the requisite richness?

The answer, curiously enough, is suggested by a more careful examination of the appeal of the diagonal mapping. What is it about **srqp** that suggests a diagonal mapping? After all, it is very unlikely that someone faced with target situation **ijkl** would think of replacing the **i** by another letter. It goes without question that the proper site for change is the **l**. What makes **srqp** different? The answer is that our minds are taking into account the fact that **abc** and **ijkl** are forwards alphabetic sequences, whereas **srqp** is a backwards alphabetic sequence. But our diagrams do not, so far, provide for indications concerning the "internal fabric" of a string of letters. Let us therefore try a simple representation of the internal fabric of strings **abc**, **ijkl**, and **srqp**:

$$a \rightarrow b \rightarrow c$$

$$i \rightarrow j \rightarrow k \rightarrow l$$

$$s \leftarrow r \leftarrow q \leftarrow p$$

Each arrow represents a successorship link between adjacent letters, and we see that in the upper two strings, the arrows all flow to the right, whereas in the lower string, the arrows all flow to the left. If we were to take these arrows as our guidelines for suggesting bridges, they would unequivocally push for the vertical mapping of **abc** onto **ijkl**, and the diagonal mapping of **abc** onto **srqp**. So far, we seem to have only decreased the justification for the vertical mapping of **abc** onto **srqp**! Adding arrows has certainly enriched our representation of what is going on, but hasn't yet solved our puzzle about how to justify the mapping that gives rise to the rule translation that in turn gives rise to the answer **srqo**.

Let us now recall what the desired rule translation was. It was the substitution of *"predecessor"* for *"successor"*. If we wish to get the concept of predecessorship into the picture, it would seem that we would have to have a representation for that concept in our diagrams. And indeed, our diagrams have indubitably manifested a bias towards successorship and against predecessorship. So let us try again with our three strings, now giving equal time to predecessorship.

a <== b <== c

i <== j <== k <== l

s ==> r ==> q ==> p

In these diagrams, an arrow with double thickness represents a <u>predecessorship link</u>. We now realize that each of our three strings has a "bivalence": it can be seen as being composed of either successorship links or predecessorship links. And for each string, switching the type of link defining its internal fabric switches the direction of flow of the arrows.

We now have adequate notation to reexamine the mappings of **abc** onto **srqp**, so let us present enriched diagrams for those situations.

    a --> b --> c   (internal fabric: right successorship)

    s <-- r <-- q <-- p   (internal fabric: left successorship)

---

    a --> b --> c   (internal fabric: right successorship)

    s ==> r ==> q ==> p   (internal fabric: right predecessorship)

Here we have presented **abc** the same way both times, but accompanied by two different visions of **srqp**.

The upper vision frames both **abc** and **srqp** in terms of <u>successorship links</u>, and therefore the flows of arrows in the two strings are opposed: *right* versus *left*. The two starting-points of the flow of arrows (**a** and **p**) are each other's counterparts, as are the two finishing-points (**c** and **s**). This clearly suggests diagonal bridges:

Mapping 1: 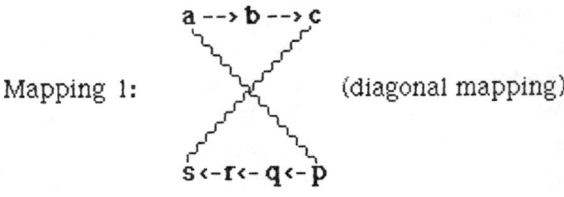 (diagonal mapping)

The pylons of these two bridges, taking into account the <u>identical link types</u> but <u>antiparallel link directions</u>, say that when translation is called for here, the concepts *successor* and *predecessor* should be kept constant, while *left* and *right* should be swapped.

The lower vision, on the other hand, frames both strings in terms of <u>right-moving arrows</u>, and therefore the link types are different: *successorship* versus *predecessorship*. The two starting-points of the flow of arrows (now **a** and **s**) are each other's counterparts, as are the two finishing-points (**c** and **p**). This clearly suggests vertical bridges:

Mapping 2: 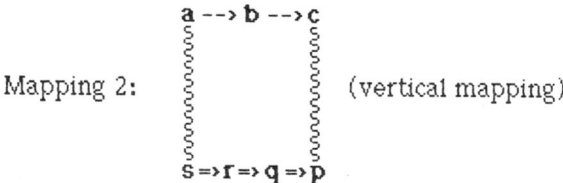 (vertical mapping)

The pylons of these two bridges, taking into account the <u>parallel link directions</u> but <u>opposite link types</u>, say that when translation is called for here, the concepts *left* and *right* should be kept constant, while *predecessor* and *successor* should be swapped. So now we understand not only how the vertical mapping is established, but also how, once it has been set up, the pylons supporting its bridges mediate the proper translation of the rule, so as to produce the answer **srqo**. The insight that allowed us to get to the core of these two answers to the given problem was the notion that arrows, or links, have an intrinsic "bivalence".

This brings us to the central question of this article, namely: "How is it decided, for a given link inside a string of letters, which of its two facets will prevail?" This, after all, is what determines which of the two alternate visions of **srqp** will be chosen, thus determining the mapping, and thereby, the answer to the analogy problem. What makes the **srqp** problem especially interesting is that we wish to have <u>both</u> visions be possible, in principle, on different runs of the program. This would imply that the program need not produce identical answers on distinct runs, and therefore that there is some non-determinism in the program. That is certainly the case, and we now proceed to describe it.

The idea in a nutshell is that each inter-letter link is a bi-stable creature, choosing which way to point according to probabilistic laws rather than deterministically. When any link is first inserted between two letters, it makes such a choice, but that choice is not necessarily final. As the program runs, the link is from time to time given the chance to "change its mind". This makes links sound like capricious creatures, and perhaps gives the impression that utter chaos reigns, with links flipping direction all the time, and no order ever emerging. Such would indeed be the case if all the decisions were unbiased -- in other words, if each link simply flipped a fair coin each time to decide which way to point. But there are time-dependent pressures, both local and global, that bias the coins and tend to ensure that in the long run, order emerges. By local pressure, we mean that each link is somewhat biased to agree with what its closest neighbors are doing. This is roughly analogous to peer pressure. By global pressure, we mean that each link is somewhat biased to agree with the statistically predominant trends in the entire "world". This is roughly analogous to a national mood.

The final factor involved in the emergence of order from this probabilistic chaos is a notion called temperature[4,5,6,7], which serves to regulate the fairness of all the coins at once. A high temperature ("boiling") means that all coins are very nearly unbiased, so that every link can quite capriciously "change its mind", and thus the situation is highly volatile. A low temperature ("freezing") means that every link slavishly follows the latest trend (local or global), and thus the tiniest initial bias is rapidly magnified into an avalanche of conformism from which there is no escape.

The problem with high temperature is easy to see: it is that the system never settles down. The problem with low temperature is subtler; it is that the system is on a hair-trigger and will jump to a conclusion based on only one or two initial coin-flips. An intermediate temperature has some of the good and some of the bad qualities of both these extremes, and is thus not a good compromise. The best solution is to let the system regulate its own temperature, starting it out very high and then lowering it slightly whenever hints of order starts to emerge, and raising it slightly whenever order has not increased recently. The effect of such self-regulation is to "coax" the system gradually into a highly ordered state and to "freeze" it in that state. Of course, to do this, the system has to have a way of measuring its own order. We will describe that shortly.

In order to see how these factors work together to let an orderly vision emerge from an initially inchoate situation, let us follow an example. For the purposes of this discussion, the fact that there are two situations rather than one makes absolutely no difference. In fact, we can combine our two situations **abc** and **srqp** into one longer string, **abcsrqp**. This will serve just as well as the two separate strings to illustrate the ideas of link insertion and link flipping.

Let us look at what happens when the very first link is inserted inside our long string. The choice of site is one of the many non-deterministic aspects of our program; it could take place anywhere inside the string, depending on the output of a random-number generator. Suppose that the locus between the **r** and the **q** is chosen. Then one of the following two diagrams must result:

(1)   a   b   c   s   r <-- q   p

(2)   a   b   c   s   r => q   p

In (1), we have a left-successor link, and in (2), a right-predecessor link. (The scenarios **r** -> **q** and **r** <= **q** represent false statements, and are impossible according to the rules of the game.) Now at the outset, our world is totally unbiased, favoring neither left-pointing nor right-pointing arrows, and favoring neither successorship nor predecessorship. This very first coin flip, therefore, is a 50-50 affair. Let us arbitrarily suppose, then, that fate (in the form of a random-number generator) favors the link in diagram (2). How does that solitary right-predecessor link, once inserted, begin to establish "peer pressure" and a "national mood"?

Locally, this link will tend to bias coin flips affecting its neighbor links (those at the **s-r** and **q-p** loci, although such links don't yet exist). In particular, their coins will be biased to favor rightwards motion and predecessorship. Globally, this link will also tend to bias <u>all</u> coin flips, no matter where they are, towards rightwards motion and predecessorship. Thus this link will tend to say to all links "Be like me!", but it will say it more forcefully to its two neighbors.

Suppose now that the next randomly-selected locus for link insertion is that between **a** and **b**. This link will be under no local pressures (it is too far from the only existent link), but under two distinct global pressures: (1) to be a predecessor-type link, and (2) to point rightwards. The first could be called link-type pressure, and the second could be called directional pressure. Note that in this particular case, these pressures happen to favor opposite outcomes, because the only links consistent with reality are **a** -> **b** (appeasing directional pressure but violating link-type pressure), and **a** <= **b** (appeasing link-type pressure but violating directional pressure). So in this case, the pressures cancel each other out, and consequently we have another 50-50 coin flip. (Strictly speaking, the pressures need not cancel each other precisely, because one of them may be given more weight than the other. But for simplicity's sake, let us right now assume that both pressures are considered equally important.)

Suppose our coin winds up selecting the left-predecessor link (**a <= b**). Now we have two predecessor links, and no successor links. That, in anyone's book, should be read as distinct <u>pressure toward predecessorship.</u> On the other hand, we have one right-moving link and one left-moving link. This is a clear case of cancellation, meaning that there is no pressure towards left or right as of yet. If this early-formed bias favoring predecessorship survives, we will be very likely to settle, as the temperature falls toward freezing, into the following final state:

$$a <= b <= c \quad s => r => q => p$$
$$<====== \quad\quad ==========>$$

This state can be decomposed into two obvious separate parts (**abc** and **srqp**), each with its own uniform internal fabric, shown underneath. Such a region is called a <u>Bloch domain</u>, by analogy with the phenomenon of that name in ferromagnetism, described below.

Obviously, if the second flip had come out the opposite way, namely selecting **a -> b**, there would have been <u>pressure favoring right-moving arrows</u>, while pressures toward successorship and precedecessorship would have canceled each other out, and so the system would have been more likely to settle into the following final state, at low temperature:

$$a \to b \to c \quad s => r => q => p$$
$$-------> \quad\quad ==========>$$

Here again, the same two Bloch domains have emerged, but the one on the left has a different uniform internal fabric.

Just to show that **abc** and **srqp** are not the only possible Bloch domains, here is one other possible final state (although it is certainly less likely to crop up):

$$a <= b \to c \quad s => r => q <- p$$
$$<== \quad --> \quad\quad ======> \quad <--$$

In this state there are three tiny Bloch domains and one medium-sized one.

One might well ask <u>why</u> this particular state would be unlikely as a final state. The answer has to do with how order is measured and how temperature is regulated. We would ideally like final states to have long Bloch domains, because such groupings are similar to what humans tend to perceive. In the above situation, for instance, people would like to see the link between the **q** and the **p** flip, as well as either of the two links inside the **abc** region. Our strategy is thus for the system to <u>lower</u> its temperature slightly whenever a domain having a uniform internal fabric grows <u>longer</u>, and to <u>raise</u> the temperature whenever such a domain becomes <u>shorter</u>. Statistically speaking, the effect of this strategy will be a tendency for long Bloch domains to lock in stably -- but it does not totally prevent short ones from cropping up. And this is important, because there is no guarantee that perceiving the longest possible Bloch domains always provides the best answers to analogy problems.

It is critical that one begin at a high temperature, thus allowing the system to insert its first links relatively unbiasedly and to explore many possible "perceptions" by letting links flip freely, but it is of course important that one finish at a low temperature, thus ensuring that the system has committed itself stably to one unvarying set of links (i.e., to one stable "perception" of the situation).

This is perhaps the appropriate time to bring in the analogy of the Ising model of ferromagnetism[8]. In a ferromagnetic substance, the atoms are arranged in a regular lattice, and each atom is the locus of a spinning electron. For the purposes of this discussion, a spinning electron can be thought of as a tiny magnet capable of pointing in only two directions, usually called <u>up</u> and <u>down</u>. In a real substance, the interactions among spins can be very complex. The Ising model is an idealized model of a ferromagnetic substance, but a very accurate one. Each spin is assumed to be susceptible to "peer pressures" only from its nearest neighbors. (In one dimension, each spin has two nearest neighbors; in two dimensions, four; and in three, eight.) More precisely, this means that each spin "wants" to align itself with the local magnetic field created by its nearest neighbors. Temperature, however, spoils any <u>certainty</u> that it will so conform.

Moreover, if there is a global external magnetic field acting on the substance, every single spin is subject to that "national mood" as well. This external magnetic field H_{global} is added to the sum H_{local} of the magnetic fields of the nearest neighbors, to create a <u>total</u> magnetic field H_{total}.

whose value varies from spot to spot and from moment to moment.

At high temperatures, the Ising model has spins flipping wildly, and forming no global order. At low temperatures, the Ising model settles into disjoint macroscopic regions of atoms whose spins are all aligned. These regions are known (not surprisingly) as <u>Bloch domains</u>, and they are the source of the exceedingly strong intrinsic magnetic fields that substances such as iron characteristically exhibit.

Like our links, Ising-model spins flip probabilistically, biased by the sum of the local magnetic field set up by their neighbors and the global magnetic field (if any exists). When the temperature is high, however, any bias is essentially ignored, so that flips tend to be truly 50-50. As the temperature is lowered, biases receive more and more attention -- and at very low temperatures, a bias becomes effectively an iron-clad rule, so that all randomness is removed.

The process can be expressed quite simply in mathematical terms. To each of a given spin's two possible orientations, there corresponds an energy. Barring coincidences, one of these energies will be lower than the other, and the larger this gap between energies is, the more the spin will want to assume the lower-energy state. The only thing holding it back from doing so is, of course, the temperature. In particular, the probability that a spin will be found in a state of energy E when the temperature is T is proportional to the following expression:

$$e^{-E/kT}$$

where k is the Boltzmann constant. For our purposes, k is irrelevant, so let us simply set it to 1. Let us now suppose that the energies corresponding to the up-state and the down-state are, respectively, E_u and E_d. Then the respective probabilities, p_u and p_d, of the spin being found in those states are (to within a common factor of proportionality):

$$p_u = e^{-E_u/T} \quad \text{and} \quad p_d = e^{-E_d/T}$$

Since the spin must point either up or down, the probability that it will point up is given by the following ratio:

$$\frac{p_u}{p_u + p_d} = \frac{e^{-E_u/T}}{e^{-E_u/T} + e^{-E_d/T}} = \frac{1}{1 + e^{(E_u - E_d)/T}}$$

(In this ratio, the common factor of proportionality cancels out, fortunately.) We can interpret this formula in two ways: if we have a real ferromagnetic substance, then it represents the probability that a given spin will be found pointing up rather than down; if, however, we are computationally simulating such a substance, then it tells us how to bias our coin flip determining the direction of a given spin.

What has not yet been explained is how to calculate the energies attached to spin-up and spin-down states. This is, fortunately, very simple. Suppose a particular spin has value s, and is immersed in a magnetic field of value H. Actually, s can assume only two possible values: +1 (spin up) and -1 (spin down), whereas H can assume any real value, positive or negative. If s and H have the same sign, then s is aligned with H, otherwise s opposes H. Electromagnetic theory tells us that the energy associated with our spin is:

$$E = -sH$$

In particular, for up-spins and down-spins respectively, the two energies are:

$$E_u = -H \quad \text{and} \quad E_d = H$$

What this means in words is that a spin parallel to H has a lower energy than a spin antiparallel to H -- and the larger H is, the bigger that energy discrepancy is. When these two expressions are substituted into the formula for the probability of a coin-flip choosing "up", we get the following expression:

$$\frac{1}{1 + e^{-2H/T}}$$

This has a very simple interpretation. When T is very large, the exponential is close to 1, so that the whole expression is close to 1/2, meaning that the coin flip is essentially unbiased -- just about equally likely to pick "spin down" and "spin up". When H/T is <u>positive</u> and enormous (so that the bias towards "spin-up" should be great), the exponential is very nearly zero, so that the whole expression is very nearly 1, meaning that the coin flip is almost certain to choose "spin up". Finally, when H/T is <u>negative</u> and enormous (so that the bias towards "spin-down" should be great), the exponential is very nearly infinity, so that the whole expression is very nearly zero, meaning that the coin flip will almost never choose "spin up". This is just what we said earlier: high temperature means that coin flips are unbiased so that no global order emerges, whereas low temperature tends to enforce biases very strongly, meaning that spins will line up with their surrounding magnetic field and will form large uniform Bloch domains.

The Ising-model approach lends itself readily to our work, by analogy. The temperature T has already been introduced and, since it is an arbitrary parameter that can be raised or lowered at will, needs no further explanation. Flippable spins are, of course, bivalent links. What, though, corresponds to the magnetic field H? Well, we have seen that H tends to coerce spins to align with it; therefore, H ought to be equated with <u>pressure to conform</u>.

We saw above that in the Copycat world, there are actually two distinct "flavors" of pressure to conform: link-type pressure and directional pressure. To emphasize the analogy to ferromagnetism, let us rename them <u>link-type field</u> H_t and <u>directional field</u> H_d. H_t is a real number telling us how much we should favor successorship links over predecessorship links, and similarly, H_d is a real number telling us how much we should favor right-pointing arrows over left-pointing arrows. (A negative value of H_t means that predecessorship should be favored over successorship, and similarly, a negative value of H_d means we should favor left-pointing arrows.) The following formulas for these quantities suggest themselves:

$$H_t = \#(\text{succ-links}) - \#(\text{pred-links})$$

$$H_d = \#(\text{right-links}) - \#(\text{left-links})$$

Here, a notation such as "$\#(x)-\#(y)$" simply means "Count up the number of current instances of x, and subtract from it the number of current instances of y". These definitions capture the idea that fields are up-to-the-minute fad-measurers or "polls" tracking the popularities of link types and link directions.

Every time a link is about to be inserted or flipped, the bias of the associated coin flip is determined by a calculation using the values of these two fields at that locus. But there is an interesting feedback effect as well: right after each coin flip, the new spin state <u>affects</u> H_t and H_d in turn, because that spin state itself has to be counted if the two polls are to remain up to date.

The fact that there are two types of "magnetic field" reminds us that a link, as well, is actually two spins in one: a "link-type" spin where +1 and -1 mean "successor" and "predecessor" respectively, and a "directional" spin, where +1 and -1 mean "right" and "left" respectively. Thus instead of just one value s associated with a link, we have a pair of such values per link: s_t and s_d. To calculate the energy associated with a given link's state, we multiply each spin by its corresponding field and sum the results:

$$E_t = -s_t H_t \qquad E_d = -s_d H_d$$

$$E_{total} = E_t + E_d$$

We have overlooked one detail -- there are actually two components to each type of pressure to conform: the global component, which doesn't vary from spot to spot, and the local component, which does, since it depends on a locus's immediate neighbors. Therefore we should write:

$$H_{tl} = \#_{local}(\text{succ-links}) - \#_{local}(\text{pred-links})$$

$$H_{tg} = \#_{global}(\text{succ-links}) - \#_{global}(\text{pred-links})$$

$$H_{dl} = \#_{local}(\text{right-links}) - \#_{local}(\text{left-links})$$

$$H_{dg} = \#_{global}(\text{right-links}) - \#_{global}(\text{left-links})$$

The subscript "local" means that one should poll only the immediate neighbors on either side of the given locus, while "global" means that one should poll all links, no matter where they are located.

This means that to make E_{total}, we really should be summing <u>four</u> quantities rather than two:

$$E_{tl} = -s_t H_{tl} \qquad E_{dl} = -s_d H_{dl}$$

$$E_{tg} = -s_t H_{tg} \qquad E_{dg} = -s_d H_{dg}$$

$$E_{total} = c_{tl} E_{tl} + c_{tg} E_{tg} + c_{dl} E_{dl} + c_{dg} E_{dg}$$

The only thing non-obvious thing here is the presence of coefficients, $\{c_{ij}\}$. They are included because it is quite conceivable that one might wish to weight global fields more or less heavily than local fields, and, as was briefly mentioned earlier, link-type fields more or less heavily than directional fields.

For the sake of concreteness, let us take the following situation:

$$a \rightarrow b \rightarrow c \qquad s \Rightarrow r \leftarrow q \Rightarrow p$$

and concentrate on the link in the locus between **r** and **q**. Cursory examination of the situation shows that its two neighbors would very much like it to flip into a right-predecessor link, thus: $r \Rightarrow q$. Globally, the situation is a little more ambiguous, since there is considerable pressure toward right-pointing links, but at the same time, successor links are slightly favored over predecessor links. If all the coefficients $\{c_{ij}\}$ are equal, we would expect that this link would be more likely to flip than to remain as it is.

To check these intuitive conclusions, let us first calculate the values of all the field components at the given locus:

$$H_{tl} = \#_{local}(\text{succ-links}) - \#_{local}(\text{pred-links}) = 0 - 2 = -2$$

$$H_{tg} = \#_{global}(\text{succ-links}) - \#_{global}(\text{pred-links}) = 3 - 2 = +1$$

$$H_{dl} = \#_{local}(\text{right-links}) - \#_{local}(\text{left-links}) = 2 - 0 = +2$$

$$H_{dg} = \#_{global}(\text{right-links}) - \#_{global}(\text{left-links}) = 4 - 1 = +3$$

As things stand right now, $s_t = +1$ (it is a successor link), and $s_d = -1$ (it points to the left). Were it to flip, then we would have $s_t = -1$ and $s_d = +1$. So we need to calculate the energies of these two rival states. We will assume all the $\{c_{ij}\}$ are equal to 1, for simplicity. For the <u>actual</u> state, we have:

$$E_{tl} = -s_t H_{tl} = -(+1)(-2) = 2 \qquad E_{dl} = -s_d H_{dl} = -(-1)(+2) = 2$$

$$E_{tg} = -s_t H_{tg} = -(+1)(+1) = -1 \qquad E_{dg} = -s_d H_{dg} = -(-1)(+3) = 3$$

The sum of these four energies is +6. Now what about the energy if the link were flipped around? Well, this is rather trivial; since both s_t and s_d would change in sign, all four contributions would likewise change in sign. (Note: the four fields stay unchanged until the link actually flips; only at that point are they updated.) Consequently the energy of the hypothetical flipped state would be -6, which is lower by 12 units. Already we can see that our link will be biased towards flipping. The only question is by how much -- and that, of course, depends on the ratio of the energy gap to the temperature. Specifically, the probability of the link's remaining in its current orientation is:

$$\frac{1}{1 + e^{12/T}}$$

If the temperature is, say, 12, then this quantity is $1/(1+e)$, or about 0.27, so that the chance of reversal is almost 75 percent. At double that temperature, this quantity is about 0.38, so the chance of reversal is about 62 percent, which makes sense, since high temperatures tend to reduce biases. At a temperature of 6, this quantity is about 0.12, so that there is almost a 90 percent chance of the link's flipping around! As was mentioned earlier, links tend to "go with the flow" when the temperature is low.

Let us now summarize. Given any situation, a program can easily determine the magnitudes of all four types of pressure ("magnetic fields") at any link-locus. From those values, it calculates the energy gap between the link's two possible orientations, and then, using the current temperature, it determines the bias on the coin it is about to flip. Using a random-number generator, it flips a coin to decide which way to point the current link (and incidentally, whether that link already exists or is about to be created makes no difference to the calculation). Having done this for a given link, it can now go on to another link, and then another, and so on. As it carries out these link-insertion and link-flipping processes, one or more Bloch domains may begin to emerge, and as they grow, the temperature will fall. As the temperature falls, the probability of flipping a link in violation of either the "peer pressure" caused by the Bloch domain it belongs to or the "national mood" caused by the totality of Bloch domains also falls. There is thus a strong tendency toward "locking in" to stable, large Bloch domains. As a result, the Copycat program tends to zero in on highly plausible perceptions of these abstract situations.

It is at this point of Copycat's work that the mapping, or bridge-building, stage takes over from the perceptual, or link-making, stage; once the mapping is done, Copycat goes on to the rule-translation stage, which leads immediately to the construction of its answer to the given analogy problem. And thus ends Copycat's work.

References

Hofstadter, Douglas R. "The Copycat Project: An Experiment in Nondeterminism and Creative Analogies." Cambridge, Massachusetts: M.I.T. Artificial Intelligence Laboratory AI Memo #755, April 1984.

Hofstadter, Douglas R. "Analogies and Roles in Human and Machine Thinking". Chapter 24 of <u>Metamagical Themas</u>. New York, Basic Books (1985): 547-603.

Hofstadter, Douglas R. "Simple and Not-So-Simple Analogies in the Copycat Domain." Unpublished FARG Document (January, 1984), available through Fluid Analogies Research Group, Perry Building, 330 Packard Road, Ann Arbor, Michigan 48104, U.S.A.

Hofstadter, Douglas R. "The Architecture of Jumbo". In <u>Proceedings of the Second Machine Learning Workshop</u>. Monticello, Illinois (1983). Also available through Fluid Analogies Research Group, Perry Building, 330 Packard Road, Ann Arbor, Michigan 48104, U.S.A.

Kirkpatrick, S., C. D. Gelatt, Jr., and M. P. Vecchi. "Optimization by Simulated Annealing". <u>Science</u> **220**, no. 4598 (May 13, 1983): 671-80.

Smolensky, Paul. "Probabilistic Analysis of Inference and Learning in Massively Distributed Parallel Cognitive Systems: The Framework of Harmony Theory". University of California at San Diego, Institute for Cognitive Science Technical Report (1984).

Fahlman, S. E., G. E. Hinton, and T. J. Sejnowski. "Massively Parallel Architectures for Artificial Intelligence: NETL, Thistle, and Boltzmann Machines". In <u>Proceedings of the National Conference on Artificial Intelligence</u> (1983), available through the American Association for Artificial Intelligence, 545 Burgess Drive, Menlo Park, California 94025.

Ziman, J. M. <u>Principles of the Theory of Solids</u> (2nd ed.). Cambridge, U.K., Cambridge University Press (1972): 353-66; 372.

CONSERVATION AND DISSIPATION
IN
NEURODYNAMICS

P.I.M. Johannesma* and A.M.H.J. Aertsen**

* Department of Medical Physics and Biophysics
University of Nijmegen, Nijmegen, the Netherlands
** Max Planck Institut für biologische Kybernetik
Tübingen, B.R.D.

ABSTRACT

Neural interaction is formulated within the frame of population dynamics. Three principles are formulated for information processing in a neural population: local computation, global communication and external stimulation.

The dynamics of a single neuron performing the local computation are based on a simplified form of the Hodgkin-Huxley equations expressed as two coupled first order nonlinear differential equations determined by conservation and dissipation. The global communication is represented by a synaptic connectivity matrix activated by the stochastic action variables of the individual neurons. External stimulation from the sensory environment influences a subset of the population.

A statistical dynamics is derived from the interaction equations. This description is formed by partial differential equations for the development of the distribution of states of the neural population. A stochasticity parameter involved in the generation of the action potentials determines the divergence of trajectories; this parameter may be considered as an inverse temperature.

1. Dynamic Systems

A general form of a dynamic system is sketched in Fig. 1

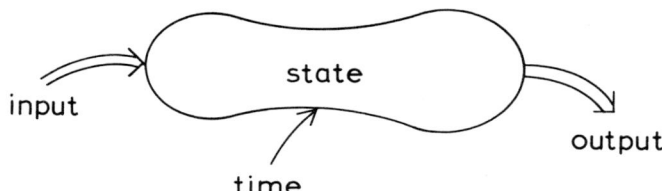

Fig. 1 Dynamic system

The following variables are used:
$x = (x_1, \ldots, x_L)$ represents input,
$u = (u_1, \ldots, u_M)$ represents state,
$y = (y_1, \ldots, y_N)$ represents output.

Equations of motion in state space are
$$\dot{u}_m = A_m(x; u; t) \tag{1.1}$$
$$y_n = B_n(u) \tag{1.2}$$

where A_m and B_n are continuous and differentiable functions of u, x and t
Eq. (1.1) forms a set of coupled nonlinear differential equations describing the change of state as function of state, input and time. Eq. (1.2) are algebraical equations expressing the output as function of state.

Two limitative assumptions are made. Firstly we assume the system to be finite dimensional; this implies that, within the description as a set of coupled differential equations, no time delays are included. Secondly we assume the system to be time-invariant; a diagram is now given in Fig. 2

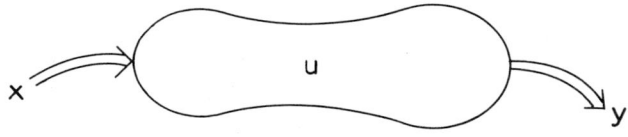

Fig. 2 Diagram of time-invariant dynamic system

The equations of motion are now simplified into

$$\dot{u}_m = A_m(x;u) \qquad (1.3)$$

It should be realised that adaptive, learning and self-organising systems, while changing their parameters in the course of time, do belong to the class of time-invariant systems since time does not appear as a direct or explicit argument. This leads to the conclusion that even if a system is time-invariant and its input is stationary then, under all experimentally realisable tests, both state and output do not necessarily appear as stationary.

2. Reactive and Creative Systems

For nonlinear dynamic systems a distinction can be made between systems which possess and those which do not possess a unidirectional flow of signals inside the system. <u>Reactive</u> systems are characterised by dynamical equations of the form

$$\dot{u}_m = A_m(x;u_1,\ldots,u_m) \qquad m = 1,M \qquad (2.1)$$

Eq. (2.1) states that, by proper choice of the state variable u, its components can be ordered in such a way that the change of component u_m depends only on the preceding components $u_1,\ldots,u_{m-1}$ and on its own value u_m. The influence of u_m on $\dot{u}_m$ should be of a dissipative type.
These type of systems are structurally characterized by convergence and divergence of interaction among elements of the state variable, however, in the network no recurrent connections occurs. Analysis of the processes taking place in reactive systems can be made through the use of correlation functions. If, moreover, the nonlinearities are of a polynomial nature, then measurement of the correlation functions (or Wiener-Kernels) allows computation of the system kernels (or Volterra kernels). For neurophysiological investigations of sensory systems the assumption of reactivity implies that correlation of sensory stimulus and neural events for individual neurons supplies a receptive field which may form a unique and sufficient characteristic of the neuron (Johannesma et al., 1986).

On the other hand <u>creative</u> systems are characterised by the presence of closed nonlinear loops. In contrast with Eq. (2.1) the dynamics are given by

$$\dot{u}_m = A_m(x \; ; \; u_1,\ldots,u_M) \qquad m = 1,M \qquad (2.2)$$

and cannot be reduced to a sequential form. A sketch of reactive and creative system is given in Fig. 3. Circles represent components of the state variable, arrows indicate interaction.

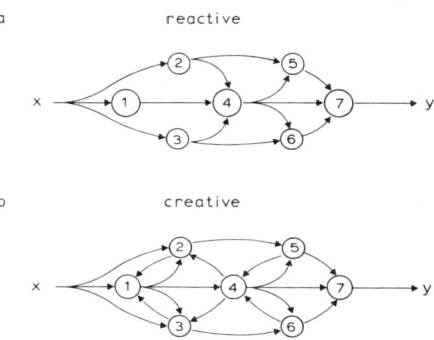

Fig. 3   Reactive and creative system

For linear systems the distinction between reactive and creative is not relevant. Apart from limitations by finite resolution and duration of observation, linear systems can be analysed completely by correlation functions: creative linear systems can be reformulated as reactive linear systems.

Processes occurring in reactive systems compared to those in creative systems differ in a fundamental way. The reactive system is essentially stable, while locally its trajectory may diverge, globally all trajectories converge to a unique stable singular point. This is illustrated in Fig. 4.

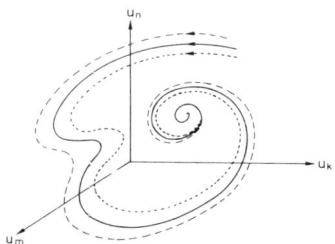

Fig. 4   Trajectories of state variable u in a reactive system

In exceptional cases this point may be located at infinity, however, these systems will usually not be found in nature.

The asymptotic convergence of trajectories in a reactive system implies that solutions will in general not critically depend on initial conditions. As a consequence numerical computations and simulations are not subjected to numerical instabilities.

In creative systems many different type of motion can be present: multiple stable singular points, periodic and quasi-periodic motion as well as chaotic behaviour. Moreover these can occur for the same system depending on differences in initial conditions. An illustration is given in Fig. 5.

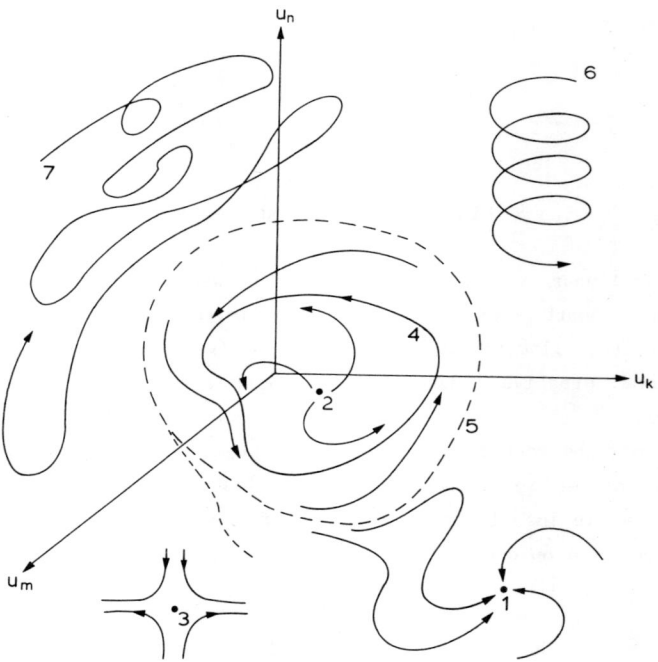

Fig. 5 Illustration of different types of motion in state space of creative system: 1. stable singular point, 2. unstable singular point, 3. saddle point, 4. limit cycle, 5. separatrix, 6. quasi-periodic behaviour, 7. chaotic motion

3. Statistical Dynamics

In creative systems trajectories may converge in some dimensions and diverge along other dimensions, this behaviour being again dependent on the position in state space. This phenomenon of "small causes, large effects" impedes a general solution for the dynamics of motion, the underlying cause being the lack of the superposition principle which holds for linear systems as well as the absence of global stability which applies to reactive systems. These difficulties are not limited to the mathematical aspects, also computationally there are consequences. Simulations of creative systems may be numerically unstable; moreover a single result is normally not representative for comparable situations with a small shift in initial conditions or parameters.

In order to generalise our point of view we look not toward a single system with precise initial conditions, but to an ensemble of systems. In any natural situation this will apply because the state of a system can only be measured with finite resolution. We define $\rho(u,t)du$ as the probability that the system is in the volume $(u, u+du)$ of state space at time t. The rate of change of the probability density ρ is now focus of interest.

The change of ρ as function of time is related to its local value and to its spatial gradient by the equation

$$\dot{\rho} = - \dot{u} \cdot \nabla \rho - \rho \nabla \cdot \dot{u} \tag{3.1}$$

where

$\rho = \rho(u,t)$ is density of state in u at time t,

$u = (u_1, \ldots, u_M)$ is position in state space,

$\dot{u} = (\dot{u}_1, \ldots, \dot{u}_M)$ is motion in state space

$\nabla = (\frac{\partial}{\partial u_1}, \ldots \frac{\partial}{\partial u_M})$ is spatial derivative in state space,

$\nabla \rho = (\frac{\partial \rho}{\partial u_1}, \ldots \frac{\partial \rho}{\partial u_M})$ is gradient of density,

$\nabla \cdot \dot{u} = (\frac{\partial \dot{u}_1}{\partial u_1}, \ldots \frac{\partial \dot{u}_M}{\partial u_M})$ is divergence of motion.

Insertion of the equations of motions for u as given in Eq. (1.3) into Eq. (3.1) leads to

$$\dot{\rho} = - A \cdot \nabla \rho - \rho \nabla \cdot A \tag{3.2}$$

forming a partial differential equation describing the flow in state space.

If a stationary asymptotic distribution exists, it has to obey the equation

$$A \cdot \nabla \rho + \rho \nabla \cdot A = 0 \qquad (3.3)$$

Eq. (3.3) implies that the uniform distribution, where all states are equally probable, will only be an asymptotic solution if

$$\nabla \cdot A = 0 \qquad (3.4)$$

Study of Eq. (3.2) is possible by computational methods; on the other hand we may try more formal approaches. A possible method may be given by the assumption that ρ belongs to a certain class of distributions characterized by a set of parameters (e.g. moments or cumulants). Substitution of this hypothetical distribution into Eq. (3.2) may then lead to an equation for the parameters. The substitution into Eq. (3.2) of

$$\psi = -\ln \rho \qquad (3.5)$$

leads to

$$\dot{\psi} + A \cdot \nabla \psi - \nabla \cdot A = 0 \qquad (3.6)$$

Since ρ may, in a number of situations, be approximately exponential Eq. (3.6) can be a suitable starting point for an approximative procedure (Victor and Johannesma, 1986).

4. Population Dynamics

In the preceding parts a general sketch was presented for description and properties of nonlinear systems. This approach, however, is so general that no specific results can be expected. Therefore we focus now on a more specific description of population dynamics. For this goal three principles are assumed: local computation, global communication and external stimulation.

We assume that the dynamic system consists of a number (K) of elements, the elements are essentially identical but may differ in parametric values. Between the elements there is an interaction or communication based on the principle "to whom it may concern": each element emits messages identified by the source, each element is able to receive all messages but accepts these on the base of a specific sensitivity determined by emitter and receiver. Influence from the environment is present in the form of stimulation. Computation, communication and stimulation are nonlinear but their effects add in a linear way.

I. Local computation

The transformation of signals within a given element (k) is based on the dynamical interaction of conjugated state variables: u_k and v_k. A minimal model is based on two state variables; this leads to

$$\dot{u}_k = f_k(u_k, v_k) + \ldots$$

$$\dot{v}_k = g_k(u_k, v_k) + \ldots \quad (4.1a)$$

II. Global communication

The communication between the elements consists of a sum of dyadic interactions

$$\dot{u}_k = \ldots + \sum_l p_{kl}(u_k, v_l) + \ldots$$

$$\dot{v}_k = \ldots + \sum_l q_{kl}(v_k, u_l) + \ldots \quad (4.1b)$$

III. External stimulation

The input or stimulus has an additive effective on the change of state variables, however, this effect may be state-dependent

$$\dot{u}_k = \ldots + r_k(u_k, x_k)$$

$$\dot{v}_k = \ldots + s_k(v_k, x_k) \quad (4.1c)$$

Combination of these three partial equations gives

$$\dot{u}_k = f_k(u_k,v_k) + \sum_l p_{kl}(u_k,v_l) + r_k(u_k,x_k)$$

$$\dot{v}_k = g_k(u_k,v_k) + \sum_l q_{kl}(v_k,u_l) + s_k(u_k,x_k) \qquad (4.1)$$

as our fundamental equations for population dynamics. A diagram is sketched in Fig. 6.

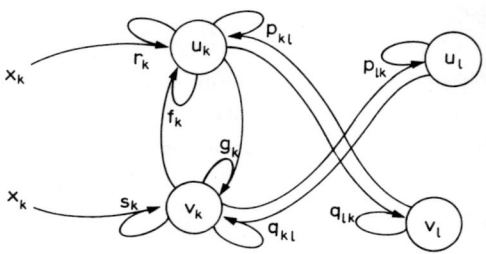

Fig. 6 Signal flow diagram of population dynamics

In the frame of population dynamics different interpretations can be given for the preceding set of equations. Examples are found in epidemiology (spread of infection), communication between natural or artificial entitities (animal vocalisations) and interaction between neurons. In several of these situations it makes sense to discriminate u_k and v_k and consider u_k as a receptive variable while v_k forms the effective variable. As a consequence we assume q_{kl} and s_k to be zero. This allows a reduction of Eq. (4.1) leading to

$$\dot{u}_k = f_k(u_k,v_k) + \sum_l p_{kl}(u_k,v_l) + r_k(u_k,x_k)$$

$$\dot{v}_k = g_k(u_k,v_k) \qquad (4.2)$$

A simplified diagram representing Eq. (4.2) is given in Fig. 7.

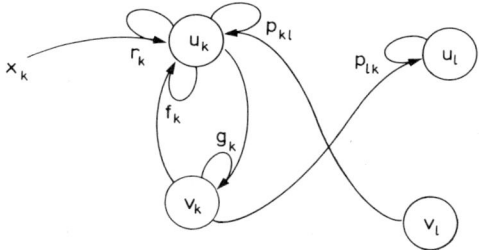

Fig. 7 Reduced signal flow diagram of population dynamics

The equations for the dynamical behaviour of a population as given in Eq. (4.2) form a basic type of description. Their character is that of coupled relaxation oscillators. The single oscillator as given in Eq. (4.1a) may be linear or nonlinear, including the Rayleigh-van der Pol equations.

5. Neurodynamics

For an application to the dynamics of interacting neural populations Eq. (4.1a) may be considered as a simplification of the 4th-order Hodgkin-Huxley equations into the form of 2nd-order Bonhoeffer-van der Pol equations as presented by Fitz-Hugh (1969).

The interaction $p_{kl}(u_k, v_l)$ introduced in Eq. (4.1b) can be made more specific

$$p_{kl}(u_k, v_l) = W_{kl}(u_k) \cdot z_l(v_l) \tag{5.1}$$

where
   v_l = effective variable of emitter l,
   u_k = receptive variable of receiver k,
   z_l = activity generated by emitter l,
   W_{kl} = sensitivity of receiver k for activity generated by emitter l.

The activity z_l may be considered as the action potential generated by neuron l. For simplicity we assume it to be dependent only on v_l and not on u_l. Two types of dependency of z_l on v_l may exist: deterministic or stochastic. In the deterministic case we may simply write

$$z_1 = z_1(v_1) \qquad (5.2)$$

The form of the relation may be semi-linear, sigmoidal, stepwise or exponential. The choice of the form of Eq. (5.2) has far-reaching consequences for the general dynamic properties of the population. In the stochastic case we assume that $\underline{z_1}$ is a stochastic variable asuming the values 0 and 1 with a probability distribution determined by v_1. Examples are sigmoidal form

$$P(z_1|v_1) = \frac{\exp z_1 v_1}{1+\exp z_1 v_1} \qquad z_1 = 0, 1 \qquad (5.3a)$$

or exponential

$$P(z_1|v_1) = \exp z_1 v_1 \qquad z_1 = 0, 1 \qquad (5.3b)$$

For a more extensive and intensive discussion see Johannesma and van den Boogaard (1985).

For neural interaction Eq. (4.2) become

$$\dot{u}_k = f_k(u_k,v_k) + \sum_l W_{kl}(u_k) z_l(v_l) + r_k(u_k,x_k)$$
$$\dot{v}_k = g_k(u_k,v_k) \qquad (5.4)$$

A sketch is given in Fig. 8, where the circles represent the neurons, while the arrows indicate the action potentials travelling between the neurons.

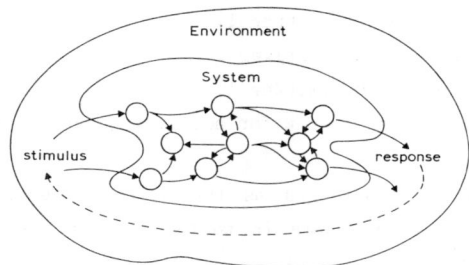

Fig. 8 Sketch of a system as a set of elements and relations in interaction with its environment.

6. Conservation and Dissipation

In assumption I preceding Eq. (4.1a) we introduced conjugated variables u_k and v_k e.g. position and momentum, or current and potential. Here we introduce conjugated functions determining the dynamics. For a single element, neglecting interaction and input and omitting the index k, Eq. (4.1a) becomes

$$\dot{u} = f(u,v)$$
$$\dot{v} = g(u,v) \qquad (6.1)$$

We now define $C = C(u,v)$ and $D = D(u,v)$ such that

$$f = +\frac{\partial C}{\partial v} + \frac{\partial D}{\partial u}$$
$$g = -\frac{\partial C}{\partial u} + \frac{\partial D}{\partial v} \qquad (6.2)$$

From Eq. (6.2) follows

$$\Delta C = \frac{\partial f}{\partial v} - \frac{\partial g}{\partial u}$$
$$\Delta D = \frac{\partial f}{\partial u} + \frac{\partial g}{\partial v} \qquad (6.3)$$

where

$$\Delta = \frac{\partial^2}{\partial u^2} + \frac{\partial^2}{\partial v^2}$$

If C and D are prescribed on a finite or infinite closed boundary in the (u,v)-plane, then C and D are completely determined by f and g.

As a consequence Eq. (6.1) becomes

$$\dot{u} = +\frac{\partial C}{\partial v} + \frac{\partial D}{\partial u}$$
$$\dot{v} = -\frac{\partial C}{\partial u} + \frac{\partial D}{\partial v} \qquad (6.4)$$

The description in Eq. (6.4) allows the derivation of some properties of this computational element. If an arbitrary function K of the state of the element is given by K(u,v) then its change in the course of time induced by the dynamics of the element is

$$\dot{K} = \dot{u}\frac{\partial K}{\partial u} + \dot{v}\frac{\partial K}{\partial v} \tag{6.5}$$

Substitution of Eq. (6.4) given

$$\dot{K} = \{K,C\} + \nabla D \cdot \nabla K \tag{6.6}$$

where $\{K,C\} = \frac{\partial K}{\partial u}\frac{\partial C}{\partial v} - \frac{\partial K}{\partial v}\frac{\partial C}{\partial u}$ and $\nabla = (\frac{\partial}{\partial u}, \frac{\partial}{\partial v})$

As a consequence

$$\dot{C} = \nabla C \cdot \nabla D \tag{6.7}$$

$$\dot{D} = \{D \cdot C\} + (\nabla D)^2 \tag{6.8}$$

This leads to the conclusion
- if dissipation D = 0,
 then C is a dynamical invariant (constant of motion)
- if conservation C = 0,
 then D increases toward a (local) maximum value.
or: D = 0 ↔ system purely consersative,
 C = 0 ↔ system purely dissipative.
Therefore we may give somewhat more loosely the interpretation
 C represents that what tends to be preserved,
 D represents that what tends to be maximised.
We should realise however that while C and D are additive in Eq. (6.4) they have mixed influences in the dynamics of motion.

For the distribution in state space we find analogously to Eq. (3.2)

$$-\dot{\rho} = \dot{u}\frac{\partial \rho}{\partial u} + \dot{v}\frac{\partial \rho}{\partial v} + \rho\frac{\partial \dot{u}}{\partial u} + \rho\frac{\partial \dot{v}}{\partial v} \tag{6.9}$$

Substitution of the dynamics of motion as expressed in Eq. (6.4) into Eq. (6.9) gives

$$-\dot{\rho} = \{\rho,C\} + \nabla\rho\cdot\nabla D + \rho\Delta D \tag{6.10}$$

Substitution of $\psi = -\ln\rho$ into Eq. (6.10) gives

$$\dot{\psi} = -\{\psi,C\} - \nabla\psi\cdot\nabla D + \Delta D \tag{6.11}$$

From Eq. (6.10) follows directly that for a conservative system for which $D = 0$, we have

$$\dot{\rho} = -\{\rho,C\} \tag{6.12}$$

Eq. (6.11) leads to the conclusion that for a conservative system a stationary distribution exists and is given by

$$\rho_s(u,v) = \rho_s(C(u,v)) \tag{6.13}$$

All points in state space compatible with the given value of C have equal probability density.

For a dissipative system, $C = 0$, we have

$$-\dot{\rho} = \nabla\rho\cdot\nabla D + \rho\Delta D \tag{6.14}$$

which gives the conclusion that in this case no finite stationary distribution will exist. This conforms our expectations since the dissipative systems will evolve toward singular forms located in the maxima of D.

Asymptotic stability of the dynamics is related to the properties of C and D for large values of $|u|$ and $|v|$.

7. Simple C-D systems

The general equation for the two-dimensional C-D system given by Eq. (6.3) is repeated here

$$\dot{u} = +\frac{\partial C}{\partial v} + \frac{\partial D}{\partial u} \quad , \quad \dot{v} = -\frac{\partial C}{\partial u} + \frac{\partial D}{\partial v} \tag{7.1}$$

A diagram of the general form of this relaxation oscillator is given in Fig. 9A. It is rather complicated because conservation C and dissipation D depend on both u and v. If this dependency is additive

$$C(u,v) = C_1(u) + C_2(v)$$

$$D(u,v) = D_1(u) + D_2(v) \tag{7.2}$$

then Eq. (7.1) becomes

$$\dot{u} = c_2(v) + d_1(u) \quad , \quad \dot{v} = -c_1(u) + d_2(v) \tag{7.3}$$

and the diagram simplifies into that of Fig. 9B.

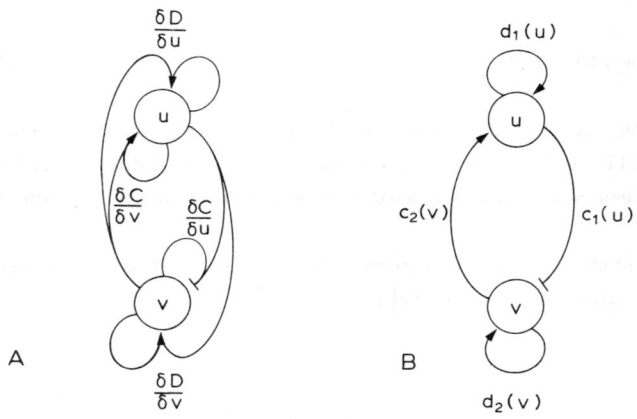

Fig. 9 Complete (A) and simplified (B) diagram of relaxation oscillator with two degrees of freedom.

If $D = 0$, then the system has an invariant of motion. If its state space is two-dimensional and bounded then the motion is periodic. For the general conservative oscillator the equation reads

$$\dot{u} = +\frac{\partial C}{\partial v} \quad , \quad \dot{v} = -\frac{\partial C}{\partial u} \tag{7.4}$$

for arbitrary $C \in C^\infty$. Its diagram is given in Fig. 10A.

If $C = 0$, then the system is described by

$$\dot{u} = +\frac{\partial D}{\partial u} \quad , \quad \dot{v} = +\frac{\partial D}{\partial v} \tag{7.5}$$

The state variable moves toward its local maximum (hill-climbing) and comes there at rest. The diagram is shown in Fig. 10B.

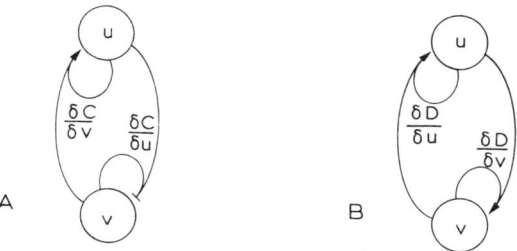

Fig. 10 Diagram of two-dimensional purely conservative (A) and purely dissipative (B) system.

Specific examples

1. Conservative harmonic oscillator

$$C(u,v) = \frac{1}{2} cu^2 + \frac{1}{2} v^2 \quad , \quad D(u,v) = 0$$

with equations of motion

$$\dot{u} = v \quad , \quad \dot{v} = -cu \tag{7.6}$$

or $\quad \ddot{u} + cu = 0$

2. Damped harmonic oscillator

$$C(u,v) = \frac{1}{2} cu^2 + \frac{1}{2} v^2 \quad , \quad D(u,v) = -\frac{1}{2} dv^2$$

gives equations of motion

$$\dot{u} = v \quad , \quad \dot{v} = -cu - dv \tag{7.7}$$

or

$$\ddot{u} + d\dot{u} + cu = 0$$

3. Conservative relaxation oscillator

$$C(u,v) = C(u) + \frac{1}{2} v^2 \quad , \quad D(u,v) = 0$$

with equations of motion

$$\dot{u} = v \quad , \quad \dot{v} = -c(u) \tag{7.8}$$

or $\quad \ddot{u} + c(u) = 0$

$C(u)$ is the potential function. The Duffing equation belongs to this class.

4. Relaxation oscillator

$$C(u,v) = C(u) + \frac{1}{2} v^2 \quad , \quad D(u,v) = -D(V)$$

leads to equations of motion

$$\dot{u} = v \quad , \quad \dot{v} = -c(u) - d(v) \tag{7.9}$$

or $\quad \ddot{u} + d(\dot{u}) + c(u) = 0$

where $d(u)$ is the nonlinear friction and $c(u)$ the nonlinear elastic force. This equation forms for $c(u) = c \cdot u$ a generalisation of the Rayleigh equation. In a comparable way the van der Pol equation can be formulated, as well as a generalised Duffing-van der Pol equation of the form

$$\ddot{u} + b(u)\dot{u} + c(u) = 0 \tag{7.10}$$

Also the Bonhoeffer-von der Pol equation is of this type. This equation forms a quite realistic representation of the information processing inside a neuron including the generation of the action potential. A large class of equations, including many well known, representing different types of harmonic and relaxation oscillators is given by Eq. (7.1) where C and D are polynomials in u and v

$$C(u,v) = \sum_{m,n} C_{mn} u^m v^n$$
$$D(u,v) = \sum_{m,n} D_{mn} u^m v^n \tag{7.11}$$

For many models of elements of a biological population Eq. (7.11) will give an acceptable description for C and D polynomials of less than or equal to 4th degree. As a basic reference for nonlinear dynamical systems see Guckenheimer and Holmes (1983).

8. Three examples

In this section three examples will be shown of two-dimensional C-D systems. The general form of the dynamical equation as given in the previous section is repeated here

$$\dot{u} = +\frac{\partial C}{\partial v} + \frac{\partial D}{\partial u} \quad , \quad \dot{v} = -\frac{\partial C}{\partial u} + \frac{\partial D}{\partial v} \tag{8.1}$$

For the conservation function C a quadratic function of u and v is taken in all three cases

$$C(u,v) = a u^2 + b v^2 \tag{8.2}$$

The dissipation D is taken as a fourth order degree algebraic function of u and v:

$$D(u,v) = \prod_{m=1}^{4} (u-u_m) + \prod_{n=1}^{4} (v-v_n) + c\, uv(1 - \frac{1}{2} uv) \tag{8.3}$$

where c = 0 in the first example,

   c = 1 in the second and third example.

In all three examples the range for u and v is (-2, +2).

Since we want to display both the change $\dot{u}$ of u and $\dot{v}$ of v in one picture we have used a chromatic coding for $(\dot{u},\dot{v})$. The colour-key for this code is shown in the upper right hand corner of each figure. White corresponds with $(\dot{u},\dot{v})$ = (0,0), dark blue with $(\dot{u},\dot{v})$ = (1,0), purple red with $(\dot{u},\dot{v})$ = (0,1), yellow with $(\dot{u},\dot{v})$ = (-1,0) and light blue with $(\dot{u},\dot{v})$ = (0,-1). The more saturated the colour, the larger the components of $\dot{u}$ and/or $\dot{v}$.

Since C and D are polynomials in u and v it follows from Eq. (8.1) that the dynamic equations can also be written in teh general form

$$\dot{u} = \sum_{m,n} A_{mn} u^m v^n \qquad (8.4)$$

The coefficients A_{mn} and B_{mn} are indicated at the right hand side of each figure under the colour-key.

In the first example the choice for C and D is

$$C = \frac{1}{2}(u^2 + v^2)$$
$$D = 0.05\, u^2(1-3/2\, u^2) + 0.1\, v^2(1-v^2) \qquad (8.5)$$

which gives the explicit form

$$\dot{u} = v + 0.1\, u - 0.3\, u^3$$
$$\dot{v} = -u + 0.2\, v - 0.4\, v^3 \qquad (8.6)$$

The picture is shown as the upper one in Fig. 11. A singular point $(\dot{u}=\dot{v}=0)$ is clearly seen in the origin, study of the colour distribution indicate that this singular point is stable: trajectories converge toward the origin. Also at the boundaries we see that the flow of trajectories is converging. Returning to Eq. (8.5) this can be understood since D becomes monotonically decreasing for large $|u|$ and $|v|$ indicating a negative feedback of the state-variable (u,v) onto itself resulting in a convergence of trajectories. While the two first order equations given in Eq. (8.6) are relatively simple, a combination of these two into a single second order equation leads to an unwieldy expression.

In the second example C and D are taken as

$$C = \frac{1}{2} u^2 + v^2$$
$$D = -u(u-1/2)(u-1)(u-3/2)-v(v-1/2)(v-1)(v-3/2) + uv - \frac{1}{2} u^2 v^2$$
(8.7)

The polynomial form is

$$\dot{u} = 0.75 - 5.5\ u + 3\ v + 9\ u^2 - uv^2 - 4\ u^3$$
$$\dot{v} = 0.75 - 5.5\ v + 9\ v^2 - u^2 v - 4\ v^3$$
(8.8)

Both this form and the form of D show that for $|u|$ and $|v|$ large trajectories will converge; as a consequence global stability is assumed. The dynachroom for this system is given in the middle part of Fig. 11. Two singular points can be clearly seen on the horizontal axis for positive u; both appear to be stable.

In the third example

$$C = \frac{1}{2} u^2 + v^2$$
$$D = -u(u-1/2)(u+1)(u-3/2)-v(v+1/2)(v-1)(v+3/2) + uv - 1/2\ u^2 v^2$$
(8.9)

or alternatively

$$\dot{u} = -0.75 + 2.5\ u + 3\ v + 3\ u^2 - uv^2 - 4\ u^3$$
$$\dot{v} = + 0.75 + 2.5\ v - 3\ v^2 + u^2 v - 4\ v^3$$
(8.10)

Again global stability is assured through the choice of the highest powers of u and v in D. The resulting dynachroom is shown in the lowest part of Fig. 11. While the chromatic resolution is not sufficient to observe clearly all singular points there appear at least five to be present. For investigation of local stability different resolutions for $(\dot{u},\dot{v})$ have to be choosen.

Fig. 11(a) Dynachromes of three C-D systems.

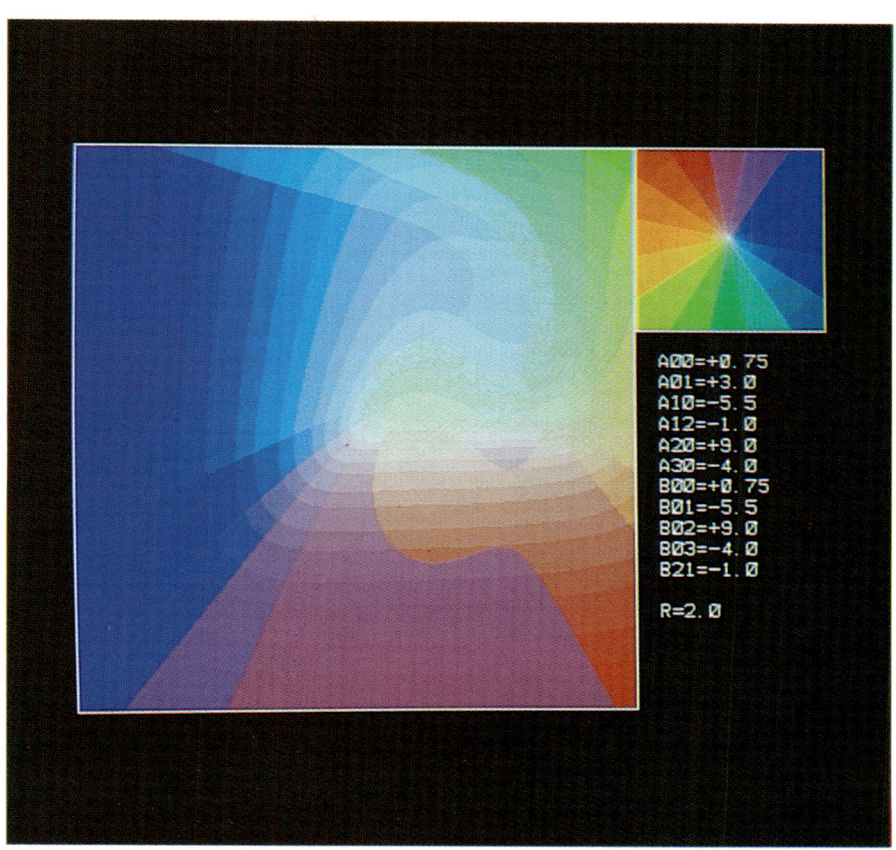

Fig. 11(b)

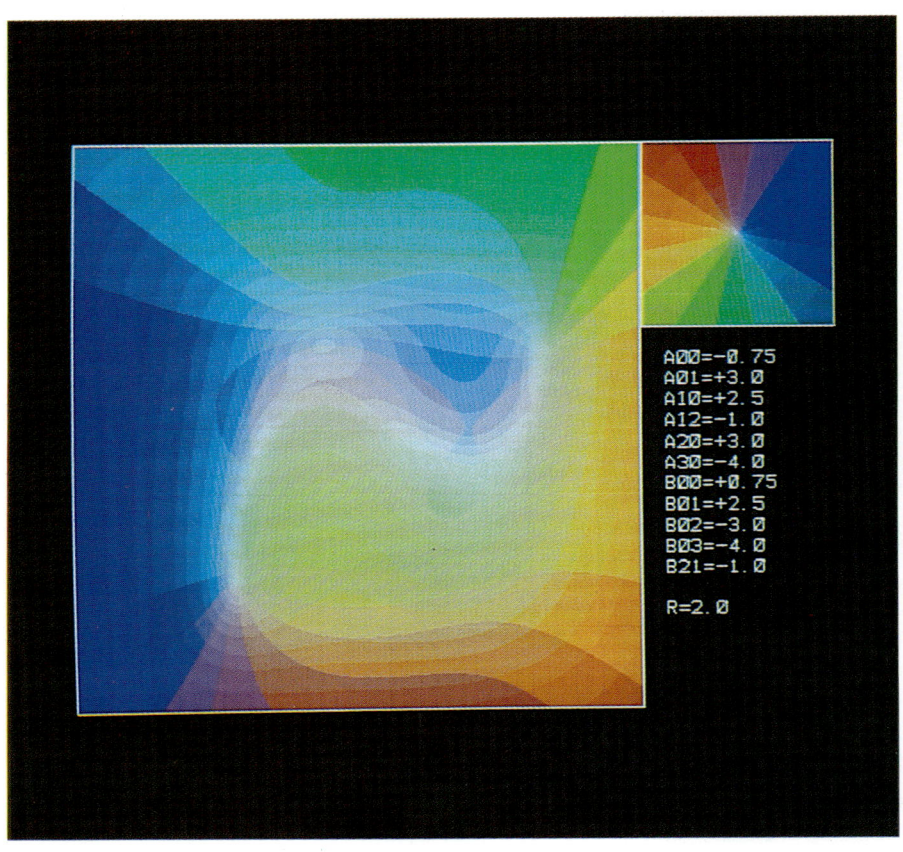

Fig. 11(c)

9. A special class of C-D systems

For a specia class of nonlinear oscillators we start again from Eq. (7.1). Now however we assume a relation between dissipation D and conservation C. The dissipation does not depend directly on the state variable (u,v) but instead is a function of C

$$D = -G(C) \tag{9.1}$$

where G is a continuous and differentiable function of C which may have a single or more zero-points. Examples are a polynomial expression or a sinuoidal one.

The dynamical equations can now be written as

$$\dot{u} = +\frac{\partial C}{\partial v} - g(C)\frac{\partial C}{\partial u}$$
$$\dot{v} = -\frac{\partial C}{\partial u} - g(C)\frac{\partial C}{\partial v} \tag{9.2}$$

where $g(C)$ is the derivative of $G(C)$ with respect to C. The change in the conservational function C now becomes

$$\dot{C} = -g(C)\cdot(\nabla C)^2 \tag{9.3}$$

showing that the "energy" exchange between system and environment depend on $g(C)$

$$\begin{aligned} g(C) &> 0 \rightarrow \text{emission of energy by the system} \\ &= 0 \rightarrow \text{conservation of energy} \\ &< 0 \rightarrow \text{absorption of energy by the system} \end{aligned} \tag{9.4}$$

For $g(C) = 0$ Eq. (9.2) reduces to the Hamiltonian form

$$\dot{u} = \frac{\partial C}{\partial v} , \quad \dot{v} = -\frac{\partial C}{\partial u} \tag{9.5}$$

indicating that (u,v) follows a periodic trajectory determined by $C(u,v)$ invariant.

This trajectory with an invariant of motion will be stable if $G(C)$ is a minimum and unstable if $G(C)$ is a maximum. So we come to the conclusion that for $g(C) = G'(C) = 0$ the system has a periodic solution which for $g'(C) = G''(C) > 0$ is unstable (separatrix) and for $g'(C) = G''(C) < 0$ is stable (limit cycle).

As an example we may take

$$C(u,v) = \frac{1}{2}(u^2 + v^2)$$

$$G(C) = \sin C$$

(9.6)

then we find an equidistant concentric set of circular limit cycles interspersed by concentric separatrices.

For a set of oscillators again under the assumption of D depending on C two quite different situations may arise. We assume in both cases

$$C = \sum_k C_k(u_k, v_k) \tag{9.7}$$

In the first case according to the assumption of local computation we take

$$D = \sum_k D_k(u_k, v_k) \quad \text{and} \quad D_k = -G_k(C_k) \tag{9.8}$$

The dynamical equations become now

$$\dot{u}_k = +\frac{\partial C_k}{\partial v_k} - g_k(c_k)\frac{\partial C_k}{\partial u_k}$$

$$\dot{v}_k = -\frac{\partial C}{\partial u_k} - g_k(c_k)\frac{\partial C_k}{\partial v_k}$$

(9.9)

and the elements are uncoupled. Each element has its own dynamic behaviour characterised by stable or unstable singular points, limit cycles and separatrices.

In the second case however we loosen the assumption of local computation and take

$$D = -G(C) \tag{9.10}$$

leading to the dynamic equations

$$\dot{u}_k = + \frac{\partial C_k}{\partial v_k} - g(C) \frac{\partial C_k}{\partial u_k}$$
$$\dot{v}_k = - \frac{\partial C_k}{\partial u_k} - g(C) \frac{\partial C_k}{\partial v_k}$$
(9.11)

For $g(C) = 0$ each element again has a periodic solution. However, a disturbance of the state of a single element leads to $G(C) \neq 0$ which influences the behaviour of all elements including the position of their limit cycles. The elements will "search" for a cooperative mode of behaviour such that again $g(C) = 0$ with $g'(C) < 0$. The limit cycle determines the individual behaviour, but the population determines the limit cycle. In population dynamics this situation may well arise if there exist some general non localised form of interaction: thermal or chemical, availability of light or food.

10. Population dynamics and neurodynamics in C-D form

The formulation of the equations for population dynamics results from the combination of Eq. (4.2), Eq. (5.1) and Eq. (6.2)

$$\dot{u}_k = + \frac{\partial C}{\partial v_k} + \frac{\partial D}{\partial u_k} + \sum_l W_{kl}(u_k) z_l(v_l) + r_k(u_k, x_k)$$
$$\dot{v}_k = - \frac{\partial C}{\partial u_k} + \frac{\partial D}{\partial v_k}$$
(10.1)

Because of the principle of local computation

$$C(u,v) = \sum_k C_k(u_k, v_k)$$
$$D(u,v) = \sum_k D_k(u_k, v_k)$$
(10.2)

From here onwards two ways are open. The first is toward a slightly more abstract version. We define a functional for the communication inside the population by

$$E(u,v) = \sum_{k,l} W_{kl}(u_k) z_l(v_l)$$
(10.3a)

with

$$w_{kl}(u_k) = \frac{\partial}{\partial u_k} W_{kl}(u_k)$$

and a functional for the external stimulation by

$$R(u,x) = \sum_k R_k(u_k, x_k) \qquad (10.3b)$$

with

$$r_k(u_k, x_k) = \frac{\partial}{\partial u_k} R_k(u_k, x_k)$$

Combination of Eqs. (10.1)-(10.3) gives

$$\begin{aligned} \dot{u}_k &= +\frac{\partial C}{\partial v_k} + \frac{\partial D}{\partial u_k} + \frac{\partial E}{\partial u_k} + \frac{\partial R}{\partial u_k} \\ \dot{v}_k &= -\frac{\partial C}{\partial u_k} + \frac{\partial D}{\partial v_k} \end{aligned} \qquad (10.4)$$

as an abstract formulation of population dynamics.

We observe that we need four functionals: C and D comprising 2 K arguments, E with K^2 arguments and R with K x L arguments. The system dynamics are determined by the local derivatives of these functionals. The important aspect is the additivitiy of the contributions of the elements to the total functional; for C, D and F the sum is over the K elements, for the interaction E the sum is over all pairs of elements.

A second approach is directed toward a more concrete form of the equations, especially for information processing in a neural population. We start again from Eq. (10.1). Now the simplification is made that the effect of interaction w_{kl} is independent of the emitter l. This implies that a neuron has identical dynamic sensitivity for all its incoming synaptic connections. As a consequence

$$W_{kl}(u_k) = E_k(u_k) \cdot w_{kl}$$

and

$$E(u,v) = \sum_{k,l} E_k(u_k) w_{kl} z_l(v_l) \qquad (10.5)$$

Since our interest is directed upon the information processing in a population of neurons we neglect the state dependent sensitivity of a neuron for the incoming signals from other neurons, this implies the approximation

$$e_k(u_k) = \frac{dE_k}{du_k} = 1 \quad \text{or} \quad E_k(u_k) = u_k \tag{10.6}$$

In neurons the action potential z_k is an all-or-none phenomenon which depends in a complex and partly stochastic way on the current and potential in a neuron. We approximate the description of the action potential by a binary stochastic variable z_k assuming the values 0 and 1 with probability

$$g(z_k|v_k) = \frac{\exp b\, z_k(v_k - a_k)}{1 + \exp b(v_k - a_k)} \qquad z_k = 0,1 \tag{10.7}$$

Combining the previous assumptions and approximations the equations for dynamical interaction in a neural population are written in vector notation

$$\begin{aligned}
\dot{\vec{u}} &= \frac{\partial C}{\partial \vec{v}} + \frac{\partial D}{\partial \vec{u}} + W \cdot \vec{z} + \vec{r} \\
\dot{\vec{v}} &= -\frac{\partial C}{\partial \vec{u}} + \frac{\partial D}{\partial \vec{v}} \\
g(\vec{z}|\vec{v}) &= \Lambda^{-1}(\vec{v}) \exp b\, \vec{z} \cdot (\vec{v} - \vec{a}) \\
\Lambda(\vec{v}) &= \sum_{\vec{z}} \exp b\, \vec{z} \cdot (\vec{v} - \vec{a})
\end{aligned} \tag{10.8}$$

The parameter b in the probability of generation of the action potential determines the amount of stochasticity; for small b the probability is a smooth function of v, for large b a steep function approaching a deterministic threshold. The parameter b may be interpreted as the inverse of a 'temperature' of the population. The parameter $\vec{a}$ represents the transition level or threshold for firing.

The neurodynamics as represeted by Eq. (10.8) forms a generalisation of the McCulloch-Pitts description (se Caianiello, 1986) with respect to both the inclusion of internal neural dynamics as well as stochastic aspects. Moreover it contains as special cases the purely conservative statistical neurodynamics introduced by Cowan (1967) as well as the purely dissipative description given by Hopfield (1982).

11. Conclusion and speculation

In this paper we have formulated neural interaction in the framework of population dynamics at the same time using a description inspired by and related to statistical physics. The definition of an element based on an interaction of two conjugated variables supplies the minimal unit with individual dynamics. The introduction of the conservation and dissipation function is intended as a relation with biological concepts as well as a discrimination of different types of dynamic behaviour. Conservation is related to periodic or ergodic type of behaviour with equations which are time reversible, dissipation leads to irreversible developments. We are aware of the fact that the presence of both conservation and dissipation combined with the nonlinear form of the equations does lead to a mixture of these two influences in the actual behaviour. However, by singling out special cases (as in section 9) or by perturbation methods, it may be possible to acquire a better insight. Moreover the concepts of conservation and dissipation can be of use in the formulation of the dynamics of biological processes.

The approach to neurodynamics as presented is intended as an intermediate level of precision between the complexity of detailed processes inside a single neuron and the global level of stochastic automata. This level of description is expected to be useful for analysis and interpretation of multi-unit recording of stimulus-evoked activity of neural populations. It does suggest forms of representation of the experimental data. Because of the contribution of the autonomous component to the total dynamics it gives room for a creative description of the neural base of perception.

Acknowledgement

The authors express their appreciation for constructive comment by Dr. H.J.F. Knops (dept. of physics) and Prof.Dr. A.H.M. Levelt (dept. of mathematics) from the university of Nijmegen and Dr. G. Palm from the Max Planck Institut für biologische Kybernetik in Tübingen.

References

Caianiello E.R.: Neuronic equations revisited and completely solved. In: Brain Theory (Eds. G. Palm and A. Aertsen) p. 147-160; Springer, 1986

Cowan J.D.: Mathematical theory of central nervous activity. Doct. Disseration, London Univ. 1967

FitzHugh R.: Mathematical models of excitation and propagation in nerve. In: Biological Engineering (Ed. H.P. Schwann) p. 1-85; McGraw-Hill, 1969

Guckenheimer J., Holmes P.: Nonlinear oscillations, dynamical system, and bifurcations of vector fields. Appl. Math. Sc. 42; Springer, 1983

Hopfield J.J.: Neural networks and physical systems with emergent collective computational abilities. Proc. Nat. Acad. Sci. (USA) 79; 2554, 1982

Johannesma P.I.M., Boogaard H.F.P. van den: Stochastic formulation of neural interaction. Acta Math. Appl. 4; 201-224, 1985

Johannesma P., Aertsen A., Boogaard H. van den, Eggermont J., Epping W.: From synchrony to harmony. Ideas on the function of neural assemblies and on the interpretation of neural synchrony. In: Brain Theory (eds. G. Palm and A. Aertsen) p. 25-47. Springer, 1986

Nadal J.P., Toulouse G., Changeux J.P., Dehaene S.: Networks of formal neurons and memory palimpsests. Europhysics Letters 1(10); 535-542, 1986

Victor J.D., Johannesma P.I.M.: Maximum-entropy approximations of stochastic nonlinear transductions: an extension of the Wiener theory. Biol. Cybernetics 54; 1-12, 1986

SELF-ORGANIZED SENSORY MAPS AND ASSOCIATIVE MEMORY

Teuvo Kohonen

Helsinki University of Technology
Department of Technical Physics
Rakentajanaukio 2 C, SF-02150 Espoo
Finland

ABSTRACT

In the control of behavior, the following two functions of the brain can be distinguished: 1. Internal representation of sensory information, especially of its abstractions. 2. Storage and recall of the above representations. - This paper shows that a simple physical self-organizing process is able to compress sensory information and to represent its main features in low-dimensional (say 2-D) maps that approximately preserve the metric and topological relations between the observations. In this view the generalized associative memory is one which is able to automatically form such topological, abstract maps and to store these representations in distributed networks.

1. SENSORY MAPS IN THE BRAIN

There are many enormous obstacles to the explanation of information processing in the brain. One of them is lack of clear physiological correlates of long-term memory traces, and another the collective and global nature of those neural phenomena which are associated with high-level cognitive processes.

It is indeed not possible to understand the structures of the brain without knowing for what purposes they are meant. If we now set aside the neural functions which are necessary for the centralized control of circulation, respiration, etc., there is little doubt about that the most important task of the brain is the control of behavior on the basis of sensory information. To this end the brain has to forecast the behavior of the environment. In order to make this possible, the brain must be able to form simplified internal models or representations of the exterior world and its history. The multitude of these models, and the mechanism which implements them, may then be called memory in the most general sense.

From above it may be obvious that the two main functions of memory are:

1. A mechanism which collects sensory information and transforms it into various internal models or representations.

2. A mechanism which "blindly" interrelates the signal processes in these representations and stores them as collective state changes of the neural network. The same mechanism is then applied to the recall of information from memory according to the association laws.

Most of the earlier and contemporary neural modelling work has concentrated on the second aspect. It is self-evident, however, that the first one, the "front-end" problem of memory must be solved first.

A few attempts (cf., e.g.[1]) have been made to transform feature values of sensory signals into spatial patterns on the cortex. In 1981, this author[2] succeeded in "isolating" those functional laws which seem to be essential and sufficient for the formation of such internal models. It turned out that the structure of an internal model can be a two-dimensional "map" or coordinate system which represents topological relationships between various sensory occurrences. In effect, such a map "flattens" a high-dimensional signal space in a nonlinear way into two dimensions, keeping only the most important feature dimensions in the representation.

The cortex of the brain is essentially a two-dimensional sheet, too, and many areas of it are specialized to different sensory modalities. In many areas, first of all in the primary sensory areas, it is possible to find subareas in which the various cells respond to distinct abstract qualities of the sensory stimuli in an orderly fashion. For instance, in the auditory areas there is a "scale" for different acoustic frequencies, and in the visual areas one can find a "color map", maps for orientation of line segments, etc. In some animals for which hearing is an important means of orientation (e.g. owls) there have been found maps of the acoustic space. Bats have coordinate systems relating to echo delays on their cortex, etc. It seems as if two-dimensional geometric mapping of sensory feature dimensions were one of the most central principles for the organization of the "internal representations".

A computational model which has an ability to form geometric "maps" for various abstract feature dimensions will be derived below.

2. THE BASIC NEURAL NETWORK MODEL

The most important types of neural signal are the trains of action pulses which are propagated down the axons, the output branches of the neural cells (neurons). We hold the view that a single pulse has a negligible information value, whereas it is the intensity of the pulse train, or the average frequency of neural pulses which defines the signal. This frequency is a positive real number denoted by ξ_{ij}

or η_i in the following. An approximation of the "transfer function" of a single cell may then be expressed, e.g., by the functional

$$\eta_i = \sigma(\sum_{j=1}^{n} \mu_{ij} \xi_{ij} - \Theta_i) \tag{1}$$

where $\sigma(.)$ is a "sigmoid" function, having saturation limits at zero and some positive value $\eta_{i\,max}$; Θ_i is a threshold or bias, and the μ_{ij} can be positive or negative, corresponding to excitatory and inhibitory synapses (synaptic connection strengths), respectively.

It is nowadays generally agreed that the elementary state variables of memory must be identified with the synaptic connection strengths between the cells. Whatever the underlying biochemical or anatomical changes in them are, the strength values are assumed to result from one or several parallel synapses. For the presynaptic (sending) cell i and postsynaptic (receiving) cell j, let this strength be denoted by μ_{ij}. There is a huge number of such state variables in the human brain; if the number of neural cells in the whole brain were of the order of 10^{11}, the number of synaptic connections would be of the order of 10^{15}. It is neither clear which types of synapse are the most modifiable ones. The fringe area of the cerebral cortex, the hippocampus has synapses which exhibit strong but relatively short-lasting changes, whereas the more permanent changes may be found in other brain areas where the magnitude of these changes is smaller, almost unnoticeable. We shall propound the view that even though individual synaptic changes were extremely small, they may collectively manifest themselves by summation of a large number of mutually correlating changes.

A neural cell, as implicitly stated above, never operates alone. It is usually embedded in a network with abundant feedback connections. The operation of feedback systems is in general very complex, even when one would have only few feedbacks and although the transfer functions were linear. The number of feedback loops in the human brain, as mentioned, may be of the order of 10^{15}, and the transfer functions of the neural cells are nonlinear. Moreover, the system parameters μ_{ij} are variable, each one being described by several widely different time constants. Is there any hope for analytical treatment of such systems? At least, depending on the approximations made, there is plenty of leeway for different views, theories, schools, and fashions.

For a theoretical discussion, the neural network must be simplified structurally as well as functionally. A hypothetical structure which may be found anywhere in the brain and which is believed to contain the most essential types of connection is shown in Fig. 1.

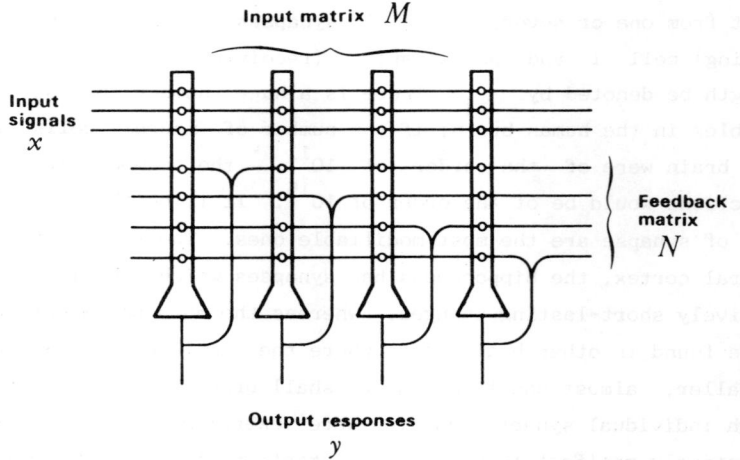

Fig.1. The basic structure of neural circuits used in brain models.

The larger symbols stand for neural cells, and the small circles for synapses, respectively. Each cell receives two types of connection, either from external sources, or by feedback from the other cells of the same system. The external signals may also come from other similar systems; if the interaction between systems were taken into account, we would have to discuss the global behavior on a higher hierachical level, too. Although such hierarchical phenomena are typical for the living brain, their discussion must be postponed until the one-level systems are understood well enough.

The present author, during a long course of research, has acquired a view that the above two systems of synaptic connections within the same cells must serve different purposes. The weighted summation of the external input signals at every cell must perform some kind of feature detection, or decoding of the relative signal values converging upon this cell, with the result that each cell produces a selective response to a particular signal combination. The cell acts like a "lock", and the set of input signals corresponds to a "key". The input synapses may adaptively change their strengths to produce a wanted "code" for the "lock", as seen below. The second synaptic system, associated with the feedback connections, seems to implement the basic associative memory function. If the connection strength between two cells adaptively changes into a value which is somehow proportional to the correlation of activity values relating to cells at both sides of the connection, then a "holographic" memory can be rendered possible. We do not mean holography literally, implemented by phased wavefronts, but only a collective transformation of signals, such that the memory traces are disseminated over the network. This network can then be shown to possess a similar reproducing ability as a hologram when an input "key" pattern excites the network. Its activity state will develop towards a recollection of a "memorized" pattern, and the recall selectivity obeys the classical laws of association, as seen below.

The system model of Fig. 1 can be described by the following system of differential equations. Many different functional forms can be chosen for $f(.)$, $g(.)$, and $h(.)$. Most works on "massively parallel computer network" models only concentrate on Eqs. (2) and (4).

$$dy/dt = f(x,y,M,N) \qquad (2)$$
$$dM/dt = g(x,y,M) \qquad (3)$$
$$dN/dt = h(y,N) \qquad (4)$$

Here the inputs to the cells are denoted by the vector x and the outputs of the cells by the vector y. M and N are connection matrices relating to the different synaptic connections between the cells.

The first differential equation, Eq. (2) relates the incoming signals, the feedback signals, and the connection matrices. This equation mainly describes relaxation of electrical activities in the network within a time of the order of 100 ms. The parametric equation for the incoming signals, Eq. (3) describes a much slower phenomenon (changes occurring within weeks) based on changes in proteins and macroscopic structures. This equation is responsible for the formation of the feature maps.

The third equation, Eq. (4) relates to the feedback connections which are responsible for the associative memory. Here we have the difficulty that the memory traces should be formed rapidly whereas they should be permanent. The accurate treatment of this equation is not possible here and we should also include the control circuits for attention which modulate the plasticity of the neural connections.

3. RELAXATION OF THE ACTIVITY

We will first tentatively consider the relaxation effect, Eq. (2) in a memoryless case assuming that the interaction function between the cells is like the "Mexican hat" function seen in Fig. 2 a). In higher mammalia at least, at short distances up to about half a millimeter the feedback is positive, and from there maybe up to one millimeter negative. The role of this feedback is to enhance the signal activity in the net. At still longer distance there is a weak positive feedback which is presumed to relate to associative memory. The function shown here is first assumed constant in time.

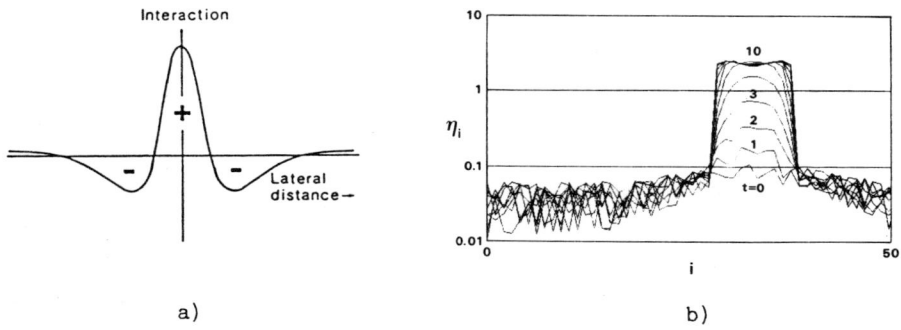

Fig. 2. a) The "Mexican hat" function.
b) Formation of "bubbles" of activity over a one-dimensional network

In Fig. 2 b) there is a simple example of the effect caused by this function. Input to the network has a random spatial distribution with a slight maximum in its average (see curve t=0). Curve t=10 gives the activity after relaxation; it has concentrated around the initial maximum. We call this area of activity the "bubble".

Now it is necessary to notice that the real effective input to the network is mainly determined, according to Eq. (1), by the inner products of the input vector x with the weight vectors m_i which are the columns of matrix M (cf.Eq. (2) and Fig. 1.)

It may also be necessary to note that the bubbles can be annihilated if the inhibitory input to the network is momentaneously increased, which is known to be the case with fluctuating attention. So the "bubbles" are only temporary phenomena.

The bubbles can be formed in the same way in two- or

multidimensional nets. They have also been measured, e.g., on the cortices of several animals[3].

4. SELF-ORGANIZATION OF THE INPUT CONNECTIONS

Let us next concentrate on the adaptive changes in the input connectivity matrix M. We are restricting to rather small pieces of network, with dimensions of the same order of magnitude as the outer diameter of the "Mexican hat" function; in other words, on the average only one bubble is formed. The self-organizing process to be described below can there be shown to create an ordered feature extraction mapping in the incoming synaptic connections over such an area. Assume first, for simplicity, that the input vector x is connected in parallel to all the cells of a two-dimensional net. In the following discussion, the input weights are assumed modifiable. Assume that bubble formation for slowly varying inputs has happened first. Now we can write the adaptation equations for the weights μ_{ij}, or the elements of M in many ways, of which one is

$$d\mu_{ij}/dt = \alpha \eta_i \xi_j - \beta(\eta_i) \mu_{ij}. \tag{5}$$

The first term corresponds to the classical assumption about the synaptic plasiticity, the so-called Hebbian law[4] which states that the changes of the connection are proportional to the product of the incoming signal and the activity of the cell. The second term is an additional nonlinear, activity-dependent forgetting term which we have found useful to stabilize the memory traces[5]. The parameter α is a positive constant.

The process can be simplified considerably by making use of bubble formation and assuming that the constant term in the Taylor expansion of $\beta(\eta_i)$ is zero, which is equivalent with assuming that the forgetting is active:

$$d\mu_{ij}/dt = \text{const.} \ (\xi_j - \text{const.} \ \mu_{ij})$$
$$\text{inside the bubble,}$$

$$d\mu_{ij}/dt = 0 \qquad \text{outside the "bubble"}. \qquad (6)$$

Since the scaling of signals in relation to the connection strengths is arbitrary, we may put the second constant (in the parentheses) equal to unity.

It can be shown[5] that for a rather general choice of the form $\beta(.)$ the norms of the vectors tend to constant, identical values. Because of this, matching on the basis of maximum inner products yields the same location for the "bubble" as if the matching were based on the minimum of the vectorial differences between x and the m_i. Combining this result and Eq. (6) leads to the simplest self-organizing algorithm: Starting with random initial connections, the following two steps for each learning pattern presented to the network are computed:

Bubble formation:

If $||x - m_c|| = \min_i \{||x - m_i||\}$, then N_c is defined as the set of cells corresponding to the bubble with fixed radius, centered at cell c.

Adaptation:

$$dm_i/dt = \alpha (x - m_i) \quad \text{for} \ i \in N_c$$
$$dm_i/dt = 0 \qquad \qquad \text{for} \ i \notin N_c, \qquad (7)$$

where α is the "adaptation" gain ($0 < \alpha < 1$).

As shown by numerous experiments, for best organizing results the radius of the "bubble" and the parameter α should be made time-variable. Experimentally established values for $\alpha = \alpha(t)$ and $N_c = N_c(t)$ can be found from Ref. 5. There may exist biological counterparts of these effects, associated with time-variable neural plasticity and control of lateral inhibition, which we do not discuss here, however.

We have demonstrated by a number of experiments[5,6,7] that the above process is able to form two-dimensional "maps" of input signals of very different sensory modalities such that the most important feature dimensions that are present in the statistical distribution of x will be displayed as coordinates in the map. Since this would perhaps remain obscure otherwise, the following result is used for illustration.

To find out the usefulness of a neural model, it must be tested with natural signals which have a certain stationary density function of the input denoted p(x). The "maps" have been tested by using natural speech as input[7]. The input signals for the map were obtained by simulating the effect of the inner ear by computing short-time spectra of the acoustic waveform every 10 milliseconds from samples over 25 milliseconds. The responses from the "map" to different acoustic spectra were labeled according to the location of the centroid of the "bubble" and the corresponding phoneme present in speech. The map was shown to learn the responses to the different phonemes in an orderly fashion (Fig. 3). The cells were organized in a hexagonal lattice corresponding to the letters.

Fig. 3. Two-dimensional map of Finnish phonemes. The double labels mean mapping of different phonemes onto the same location.

We have used such phoneme maps for the recognition of continuous speech, and a practical microprocessor equipment has already been constructed[8].

One would, of course, like to have a clear analytical explanation and mathematical proofs for this self-ordering process. In principle, this is a Markov process, although a rather complicated one. It is not possible to show here the proof even in the simplest case[5]. Intuitively one may become convinced that Eqs. (5) represent a smoothing process and the m_i tend towards a monotonic two-dimensional sequence. The distribution of the asymptotic m_i values, however, can be shown to be a delicate function of the x values; in fact, the set of the m_i can be shown to approximate the density function $p(x)$, and the map represents a hierarchical clustering graph of the x values. In other words, the various cells of the network correspond to various feature detectors which will be distributed into the input space optimally with respect to the statistics of the input signals.

5. RELATIONSHIP OF SENSORY MAPS TO THE DISTRIBUTED MODELS OF ASSOCIATIVE MEMORY

It is now interesting to see how the three systems described by the differential equations, Eqs. (2)-(4) may cooperate in the network. The input synaptic system determines a set of features from the input signals. These features are assumed constant in the following discussion and this results in a set of fixed input weights for each cell. In the feedback there exist, however, two different factors: the time-invariant "Mexican-hat" type of feedback enhancing the activity in the net, and the long-range time-variable feedback responsible for the associative memory. The latter can be modeled e.g. by the cross-correlation matrix of the input-output pairs stored in the memory[9,10,11,12]

$$N = \sum_{k=T_0}^{T} y_k y_k^T, \qquad (8)$$

where the y_k are the output patterns which we associate with themselves. This matrix can be used in the differential equation for the outputs, Eq. (2) in many ways but it can be simply added to the activity-enhancing feedback to show the effect of the memory. This can be shown with the following simple example. We have again the linear array of 50 cells but now we have added to the "Mexican-hat" feedback function a feedback function which corresponds to memorization of three patterns (y_1, y_2, y_3). Each pattern consists of two "bubbles" (at locations 7 and 28, 15 and 35, and 22 and 42, respectively; see Fig. 4 a) created with the same relaxation as before. When a key pattern containing only one slight maximum (at cell 15 in Fig. 4 b) is presented to the network, first a "bubble" is formed at that location. A little later (after t=5) another "bubble" rises and this corresponds to the "bubble" (at cell 35) associated with the key in the memory.

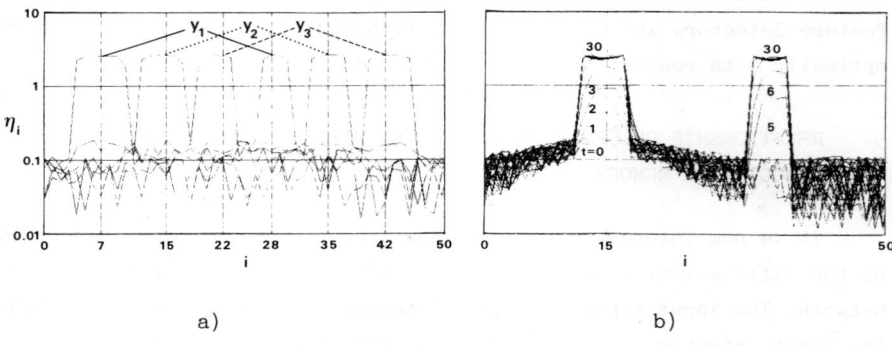

a)   b)

Fig. 4. a) The three "bubble" pairs stored in the memory.
b) Recollection of a two peak pattern using a one peak key.

The state of the network, i.e., the set of interconnections represented by the matrix N is thus changed, and these changes represent the memory traces stored in the network. Reading of this memory can only be made associatively, for instance, one may start with some partial activation of the network (key) which then spreads into other parts (recollection). The recall process is a relaxation process itself which is described by the "relaxation equation", Eq. (2). This possibility was already pointed out long ago[11,12]. This "relaxation" can be shown to develop into a direction which initially starts by multiplying the key vector y by matrix N (the "perturbation approximation"). The exact treatment is cumbersome (cf., e.g., the matrix Bernoulli equations discussed in [5]. Although the process represented by the "relaxation equation" amplifies the recollection and strongly enhances the contrasts, finally leading to binary patterns of activity, its "perturbation approximation" already contains the essential information that is recalled from memory. A demonstration of the selectivity of recollections with respect to different key patterns was also demonstrated earlier[12]. The illustration on the left in Fig. 5 represents one sample of 500 different photographic images used as the patterns y. Here each dot corresponds to one cell in the network, and the size of the dot is proportional to its activity value. The middle figure is the key activation, and the rightmost one the perturbation approximation of recollection which was obtained after spreading of activation into the inactive cells. Notice that the memory traces left by all the 500 images were really superimposed on the same network!

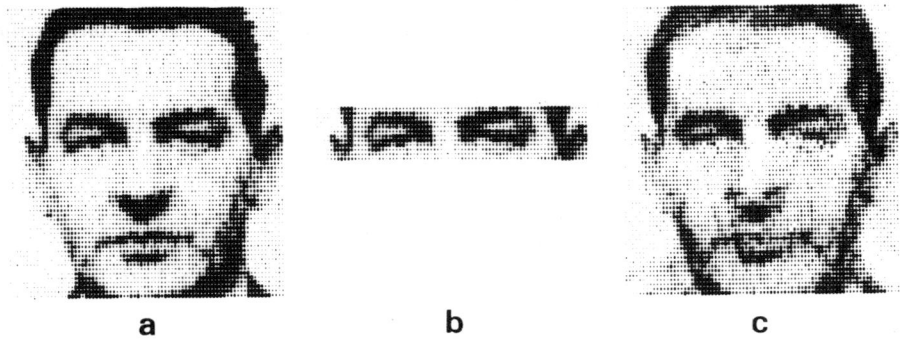

      a   **b**   **c**

Fig. 5. Demonstration of associative recall in a distributed network.
    a) One of the 500 images stored
    b) Key pattern used for excitation
    c) Recollection

6. CONCLUSIONS

The various phenomena associated with biological memory are obviously organized hierarchically, and there are no chances of describing the complete brain network analytically. The present approaches, at their best, must only be considered as descriptions of the lowest-level functions.

For even the network models to be valid, it is necessary that the structural features, the transfer functions, and the adaptation equations all be qualitatively right. There exists nowadays some experimental anatomical and electrophysiological evidence for each partial assumption. For instance, the feedback connections at short distance, corresponding to the "Mexican-hat function", have amply been documented in anatomical literature; for their electrophysiological measurement, cf.[13]. On the other hand, the spreading of axons at a longer distance which has also been known for a long time has recently been illustrated by convincing staining studies by Gilbert[14]. A couple of works (e.g.[15,16]) have accumulated experimental evidence to the Hebbian law of synaptic plasticity. Nonetheless, in relation to the importance of this problem, one would need much more convincing experimental evidence especially for the transfer functions and modifications of the Hebbian law.

REFERENCES

1. Malsburg, Ch. v.d., Kybernetik 14, 85-100 (1973)
2. Kohonen, T., in Proc. 2nd Scand. Conf. on Image Analysis (Oja, E. and Simula, O., eds) Suomen Hahmontunnistustutkimuksen Seura, Helsinki, pp. 214-220 (1981)
3. De Mott, D.W., Medical Research Engineering, 23-29 (1966)
4. Hebb, D., Organization of Behavior, Wiley, New York (1949)
5. Kohonen, T., Self-Organization and Associative Memory, Springer-Verlag, Berlin-Heidelberg-New York-Tokyo (1984)

6. Kohonen, T., in Proc. 6th Int. Conf. on Pattern Recognition, Munich, FRG, pp. 114-128 (1982)
7. Kohonen, T., Mäkisara, K., and Saramäki, T., in Proc. 7th Int. Conf. on Pattern Recognition, IEEE Computer Society Press, Silver Spring, MD, pp. 182-185 (1984)
8. Torkkola, K. and Riittinen, H., Helsinki University of Technology Report TKK-F-A591 (1986)
9. Kohonen, T., Helsinki University of Technology Report TKK-F-A130 (1970)
10. Kohonen, T., IEEE Trans. Comp. $\underline{C-21}$, 353-359 (1972)
11. Kohonen, T., Lehtiö, P., and Rovamo, J., Ann. Acad. Sci. Fenn. A.V. Med., 164 (1974)
12. Kohonen, T., Lehtiö, P., Rovamo, J., Hyvärinen, J., Bry, K., and Vainio, L., Neuroscience (IBRO) $\underline{2}$, 1065-1076 (1977)
13. Creutzfeld, O.D., Kuhnt, U., and Benevento, L.A., Exp. Brain. Res. $\underline{21}$, 251-274 (1974)
14. Gilbert, C.D., TINS-April 1985, 160-165 (1985)
15. Rauschecker, J.P. and Singer, W., J. Physiol. (Lond.) $\underline{310}$, 215-239 (1981)
16. Levy, W., in Synaptic Modification, Neuron Selectivity, and Nervous System Organization (Ed. Levy, W.B., Anderson, J.A., and Lehmkuhle, S.), pp. 5-33 (1985)

A NEW, ALGEBRAIC, CALCULUS OF THE IDEAS IMMANENT IN NERVOUS ACTIVITY.

by Francesco E. Lauria

Universita di Napoli

Dipartimento di Informatica e Sistemistica

Via Claudio 21, 80125 NAPOLI

ABSTRACT.

A concurrent architecture based on a McCulloch and Pitts network (MPN) is presented toghether with a kind of functional language which admits, at the syntactical level, a free group structure. The simulation of such an architecture on a INMOS's Trasputers network has been possible thanks to the algorithm, here presented, to automatically allocate resources in (MPN). As application some programs are discussed to illustrate the way our (MPN) version is supposed to work and the role played by that meta-linguistic device usually called memory.

0. INTRODUCTION.

We present here the peculiar algebraic properties of a functional language associated to a kind of a data flow network whose nodes compute substantially the same threshold functions introduced about forty years ago by McCulloch and Pitts (1) in an epoch marking paper. Toghether with the language we give an universal algorithm to automatically allocate the physical resources, that is, the nodes' coupling values needed to compute each elementary instructions and thus any arbitrarily given program. Moreover we propose the network as a trainable control system uninterruptedly interacting with the environment. And whose outputs can be purposefully modified exploiting either the algebraic properties of the associated language or its capability to learn by experience.

The primary context into which this work can be seen is constituted by those AI researches as represented by Weyhrauch's FOL (2), by Stefik (3), by McCarthy and Hayes (4) and by Moore (5). I. e., by works in which reasoning about reasoning to attain purposeful goals is at lenght discussed and, in some cases, implemented in expressly built high-level languages. Here the equations we can write taking advantage of the free group structure, which at the syntactical level characterize algebraically our language, introduce in a quite natural way both the rules and the procedures needed to reason about programs. Moreover to store in the high speed memory we need only to alter the nodes' coupling values as a function of their past workload, so we have another handy way to modify, automatically or at user's will, the functions computed by the machine.

Because of the way the network is supposed to be

operated we refer also to the Wilkins's problem-independent planning proposal (6).

A particular mention is due to the very interesting Levesque's paper (7). Its functional approach to auto-referentiality is based on the auto-epistemic operator K belonging to the language. Thus to alter the meta-rules he needs to change at least the said operator. In our case instead it is sufficent to choose another set of equations, within the same language, or to store the desired alterations, into the structure of the machine. To this end an high-level language was purposefully avoided with the aim to have a simpler presentation and a better understanding of the said procedures' interplay. In any case, should the need arise, it is not an impossible task to identify a suitable subgroup, or subset, of our group as the needed formal structure for the required higher-level language.

The version of the machine here presented could be defined as data-flow oriented and it is so engineered that singling out parallel executable instructions becomes quite a natural process. Some of the notions here developed have been suggested in the discussions found in the literature (8,9,10,11,12,13,14,15) about the languages associated with distributed architectures.

Now we shall discuss the identification of the network's nodes with some of the neurophysiological structures found in a mammalian central nervous system. In the McCulloch and Pitts's paper the nodes were identified as nerve cells or neurons, i. e., as the elementary components of any ganglion belonging to a nervous system. The elementary instructions defining our language are each computed by a set of nodes. Thus if we continue to identify a node with a neuron then the execution of any one instruction could be thought of as accomplished by some nervous structure more complex than a single neuron. Perhaps by a kind of columnar structure as the ones discovered by Mountcastle in the acoustic cortex, extensively studied in the visual cortex by Hubel and Wiesel (e. g., see 16 and the literature there quoted) and which seem to be present in some basal ganglion as well (17). In this respect it is worth to note how much the parallelism is developped in a nervous system but how this is in contrast with purposeful behaviour which seems to exhibit some kind of parallelism only in extreme circumstances. E. g., the algorithms so far developped by mankind are essentially sequential in character notwithstanding that man has a brain split into two hemispheres and a pair each of: eyes, arms and hands. The sequential execution of operations, in any event a non intrinsic feature of them, contrasts with the streams of data which if independent are processable in parallel, almost by definition. Perhaps this can help explain some of the difficulties encountered in developping parallel languages when the emphasis is primarily on their algorithmic nature and not, as in our case, on the flows of data to be processed.

To conclude let us remind how the present work can be seen as a small step in the direction shown some century ago by Raimondo Lullo with its revolving wheels.

1. THE MACHINE AND THE ASSOCIATED LANGUAGE.

a) Our Version Of A McCulloch And Pitts (MP) Architecture.

In accordance with a suggestion found in (8, 13) and to simplify the nodes we opted for the threshold function given in (18) as the standard computation carried out by the processing element (PE) associated with the generic node of the synchronous net. Given $D(x)=0$ if $x<=0$ and $D(x)=1$ if $x>0$ then the time discrete threshold function $f(h,t)$, the one computed by the h-th node at time t, is given by:

$f(h,t)=D(ADD(k,l)\{C(h,k,t-ltau)f(k,t-ltau)-T(h)\})$,

where t is the present time, tau is the smallest interval, i. e., a quantum of time. The quantity $C(h,k,t-ltau)$, the coupling coefficient or C, stand for the coupling between node k and node h at ltau before present time and $T(h)$ is the threshold value of node h; it is assumed $h>0$. Finally $ADD(k,l)\{X(h,k,l)\}$ it means that we must take the sum of X for all the k and l values.

From now on we shall use the nodes' identifiers h and k

both as physical devices' names and as, respectively, input and output, either internal or external, data flows' names. This is a time saving choice because almost by definition to each MP's node it corresponds an input and an output data flow and viceversa. And the same is true of an MP network: the context shall always help in deciding whathever of the two cases we are considering.

b) The Functional Language.

As elsewhere discussed [19,20] we consider the alphabeth or set of elementary instructions :

i(an,am) or identity

s(an,am) or successor

p(an,am) or predecessor

j(K(a1),L) or jump

e(L,K(a1)) or end of jump,

where an and am are independent variables on the set {a0, a1, ...} of data flows' names, K(x) is a true or false propositional calculus's expression (only "and", "vel" and "not" are allowed) and L is a variable defined on {N1, N2, M1, M2, M3}, the set of jump's parameters.

A program is a finite word in the given alphabeth if whenever it contains j(K(an),L) then it contains also e(L,K(an)) and viceversa. The two are said to be coupled or to be a coupled pair.

The interpretation of a program runs as follows:

s(p){i}: with an unitary delay the standard signals'

number belonging to the output data flow an is an increasing (decreasing) (identical) function of the standard signals' number belonging to the input data flow am. Because in any given interval of time there is a maximum number of standard signals then we shall adopt the convention that the predecessor of an empty sequence is equal to the successor of one containing the maximum signals number.

j(e): with an unitary delay the control's flow jumps to the coupled e (next) instruction whenever K(a1) is true, to the next (next) instruction otherwise. In other words to the input data flow of a j instruction it corresponds, with an unitary delay, an identical input data flow of either the coupled e or the next instruction accordingly to the K(a1) truth or falsity.

The elementary instructions' sequence comprised in between a pair of coupled instructions is said to be their "scope". A jump is said to be forward if the j precedes its coupled e, to be a loop otherwise. Two pairs of coupled jump instructions are said to be nested if one pair belongs to the other's scope, chained if the j of one pair, and thus the other's e also, belongs to the other's scope. Unconditional jumps are those with an always true condition as argument, the others are said to be conditional.

Because we need to know how they are coupled the jump and the end of jump instructions contained in a program, where they can be both intermingled in any conceivable way and coupled both forward and bakward,then we shall adopt a convention on the use of the names of the variables as their arguments. It is based on the parenthesis algorithm (19, pg. 143, Th. I) and it sounds like:

CONVENTION ON THE VARIBLES OF THE JUMP INSTRUCTIONS.

The second, first, variable of a backward coupled jump, end of jump, can be either N1 or N2. The second, first, variable of a forward coupled jump, end of jump, can be either M1 or M2 or M3.

The above quoted theorem prove the adequacy of the choice to write strings containig any number of jump instructions however intermingled the forward and backward coupled pairs are.

From now on, whenever possible, the argument L of the jump instructions' pairs shall be substituted by the same data flow's name as the one which appears as the first (second) argument of the instruction to be sequentially executed before (after) j (e). Or of an input (output) data flow, if no such instruction exists. Accordingly to the above mentioned results (19) such a substitution can take place unless there are nested, or chained, jump pairs with identical K(x) and L. However in such a case we can always rewrite the K(x) so that each jump pair has a different one and thus to solve the problem.

c) The Resources Allocating Algorithm.

Let us now turn to one of our goals, namely to find the coupling coefficients' values which implement the said elementary instructions. First we must decide how much redundancy has to be built in the system, that is, how many nodes have to compute any given elementary instruction: let us suppose R be such a number. Then a two entries table has to be built: we shall write in the left hand column the names of both the internal and external output data flows, that is, of all the elementary instructions first arguments which are found to be different. Then for each an found in the left hand column we enter, in the right hand one, first the input data flows, that is, all those second arguments am which correspond to a first argument an, with an different from am. Followed by as many pairs of (elementary instruction's name, R-tuple of nodes' identifiers) as elementary instructions there are with an as first argument. Let us suppose R=1, then for i, s and p we have respectively:

$$\text{output flows! R-tuple}$$

$$\text{an} \quad ! \text{ am, (i, 1)}$$
$$\text{an} \quad ! \text{ am, (s, 1)}$$
$$\text{an} \quad ! \text{ am, (p, 1)}.$$

If the threshold is T we have:

$C(1, am) > T$ for the i instruction;
$C(1, am, (t)) = C(1, am, (t-tau)) > T$ for the s instruction;

$$C(1, am, (t)) = C(1, am, (t-tau)) < T$$

and

$C(1,am,(t)) + C(1,am,(t-tau)) > T$ for the p instruction.

To simplify our presentation we shall give separately the cases for j and e: the reader should keep in mind that the two must be either forward coupled or they have to form a loop. We have to calculate the Cs for the two constructs:

loop	forward case
.	.
.	.
.	.
x1(f(a2,a3),a1)	x3(L,a5)
e(L,K(a2))	j(K(a2),L)
x2(a4,L)	x4(a6,f(a2,a7))
.	.
.	.
.	.
x3(L,a5)	x1(g(a2,a3),a1)
j(K(a2),L)	e(L,K(a2))
x4(a7,g(a2,a6))	x2(a4,L)
.	.
.	.
.	.

where x1, x2, x3 and x4 are variables on the set of elementary instructions, and f(x) and g(y) are logical combinations of their actual arguments. Whenever any of the formers is lacking then it is intended to be substituted by an output, input, data flow with the same name as its first, second, argument.

We can now complete the table, where to the e's output data flow it corresponds two inputs: the normal one and the other arising whenever the jump has been successful executed.

output flows	R-tuples
f(a2,a3)	a1,(x1,1)
L	1,(e,2);K(a2) & 5,(j,5),(e,2)
a4	2,(x2,3)
.	
.	
.	
K(a2)	a5,(x3,4),(j,5)
a7	g(a2,a6),(x4,6)
.	
.	
..	

The Cs are respectively:

   C(1,a1)>T if x1=i,

   C(1,a1,(t)) = C(1,a1,(t-tau))>T if x1=s, etc. etc.;

   C(2,1)>T

and

   C(2,K(a2))<T

and

$$C(2,5) < T$$

and

$$C(2,5) + C(2,K(a2)) > T$$

for the e instruction;

$$C(3,2) > T \text{ if } x2=i,$$

$$C(3,2,(t)) = C(3,2(t-tau)) > T \text{ if } x2=s, \text{ etc. etc.};$$

$$C(4,a5) > T \text{ if } x3=i, \text{ etc. etc.};$$

$$C(5,4) > T \text{ for the j instruction.}$$

For the instruction following the jump one we have, where abs(x) is the x's absolute value:

$$C(6,5) > T$$

and

$$C(6,K(a2)) < 0$$

and

$$abs(C(6,K(a2))) > T.$$

In the Appendix an introductory evaluation on the available storage in a MP architecture is discussed.

2. THE LANGUAGE'S META-RULES.

a) The Algebraic Structure At The Syntactical Level.

In (19,20) it was discussed the free group structure which, at syntactical level, it is possible to recognize in our language. There the associated machine was of a sequential kind so we shall outline here how to adapt the said discussion to the previously given MP architecture.

Let us now consider the set of unbounded strings, or sequences, of symbols belonging to the union of:

and
$$F = \{ s, j \}$$
and
$$F^{-1} = \{ p, e \}$$
$$\{ i \}.$$

Let L be the subset of the said set containing all those strings in which either do not appear j or e or whenever they contain a j, e, then they contain also its inverse, namely an e, j, and thus in each of them there are as many j as e. In other words for each j, e, belonging to a squence of L there is one and only one e, j, which is its inverse and belongs to the same sequence. The elements of each pair of inverse jump operating codes belonging to a same string are said to be coupled: thus in this paper all the jump instructions are coupled with an end of jump, and vice versa. We shall suppose the sequences written in accordance with the Convention given in (19, pg. 153). Here as there chained, nested, pairs it means that the j or, and, the e of a coupled pair falls in between those of another coupled pair. Here the second, first, variable f a jump, end of jump, stands for the ADD and LOC parts referred to in (19). Eventually if we follow verbatim the chain of definitions and of congruences' relations given in (19, pg. 156-159) we arrive again at the L-group whose elements are the programs of our network, considered as syntactical objects, i. e., as sequences of symbols and not as sequences of transformations to be carried out by the machine. We note that F, and thus also F^{-1}, is an alphabeth with a finite number of letters. The finiteness of the network implies a finite number of distinct values for h and k. The finiteness of its life span put an upper bound on the set of values assumed by l. Eventually the Convention for the variables of the jump functions proves the finiteness of F whenever the associated network is supposed to be finite.

One possibility to obtain a trainable machine depends on our ability to make it able to react to its environment. I. e., the laws governing the environment and thus also the environment-machine interactions cannot be ignored any longer. If something like a "WRITE" command must be executed using an hand then the correct sequence of accelerations must be generated, by no means an easy task. Then if the output programs drive the machine as to do not violate the laws characterizing the environment, this must be taken into account when causally linking them to the input. Here the algebraic structure associated with our language shall come into play. Actually a group can be defined as the smallest algebraic structure in which the equation

$$ax=b$$

admits always one and only one solution whenever a and b are elements of the group.

It is possible to write other equations in a group but in general they do not admit a solution. E. g., axaxa=i is an homogeneous equation which does not admit a solution whenever a is one of the generators, or one of their inverses, of a free group. In the case of the L-group: if a belongs either to F or to F^{-1}. Instead axax=1, still in a free group, admits as solution $x=a^{-1}$ for any a belonging to the group. From now on we shall consider only homogeneous equations: the ones whose right hand member is the identity. In so doing no generality is lost: we can always multiply both members of a non homogeneous equation for a suitably chosen factor to obtain, as required, an homogeneous one.

In our machine an otput program can be found as the value a variable has to assume to satisfy a given condition. E. g., that the output be such to preserve the steady state in the environment-machine interaction. This condition is obviously formalized by an homogenous equation when the unknown represents the output program, the coefficents represent the inputs. And the costants represent preassigned macroinstructions: either built-in by the manufacturer or stored in the memory as result of past experience. In both cases actually recordered in the wiring linking the nodes.

Because the output programs must take into account the environment's constraints then the equations' coefficents,

e. g., the input searching strategies, cannot be arbitrary. This in turn requires input devices so devised as, once the input data are into the machine, to trigger the execution of the input programs and built-in macroinstructions which, feeded as coefficents into the relevant equation, identify as the solution, and elicit the execution of, the desired output program. Or, in another jargon, the execution of the input programs and built-in macroinstructions must result in flows of data such that they elicit the execution of output programs, which are the correct ones with respect to the chosen equation. I. e., if the output program x is the response the machine should give to the data stream d then both the input devices must compute the program a and the built-in macroinstructions m must be such that, e. g., the condition:

$$amx=i$$

is satisfied whenever a starts on d.

To conclude this chapter we observe that, once found, the solution of an equation is an easily mechanizable formula.

b) The Memory As A Linguistic Meta-rule.

To store information in our network we choose to alter, in accordance with a given law, the coupling coefficients' values. This can be done because the way we intend to use the machine requires more storage facilities to record programs than data. I. e., we need a rule to alter the stored programs so that they can be adapted to the unpredictability of the laws governing the environment. Because the control depends upon the data flows it means the data must be seen just as signals coming in and eliciting the elaboration: the data records must be kept in external devices like the graffiti on a wall or a file in a text editor (the choice depending on the technological maturity of the environment). At most the internal data recording capability should consist of temporary buffer stores, needed to complete some computations.

It should be clear that the choice of the law governing the alteration of the coupling coefficients' values it is the crucial one for the evolution of the machine's behaviour. We plan to try some of them in a simulation to be reported in a next paper. Here we shall suppose the existence of a function of time $M(t)$, or memory function, which changes the coupling coefficients on a time scale far greater than tau. But so chosen as to do not decrease the number of input patterns which are known, i.e., recognized to correspond to the same object in accordance with the following definitions:

DEFINITION I.

The programs p', p'', belonging to the L-group, are M-equivalent, $p'Mp''$, if whenever the machine executes one of the two with the input data d and stops on the result r, then the same is the case for the other.

DEFINITION II.

The network MP "knows" or "has knowledge" or "recognize" that the input (p,d) is the same as the input (p',d') with respect to the equivalence M iff (r,b) and (r',b') are the corresponding outputs of MP and rMr' is the case, i. e., iff the respective output programs are M-equivalent.

A Poisson like coupling coefficients' multiplicative factor, with values comprised in between 1 and a maximum to be chosen taking into account the threshold' value, it could be a choice for M(t). The effect of such a factor, on C(h,k,t-1tau), have to be triggered iff the nodes h and k, in the order, assume value one with a delay 1.

The given rule seems to be of an automatic kind, i. e., it acts whenever the said conditions are meet. Actually the matter does not stop here. As it seems to be the case for a vertebrate brain (22), we can always have learning programs. I. e., programs whose effct is to repeat again and again the same task so to reinforce, with respect to the others, the relevant coupling coefficients. In other words, the old dictum "repetita iuvant" suggests a linguistic way to exploit the meta-rule to alter the previously vired in programs and therefore to record new ones.

3. THE PRODUCT'S IMPLEMENTATION AS AN EXAMPLE.

a) The Prerequisites.

We shall discuss what role they must play both the interactions with the environment and the previously learned skills in a machine executing a simple computation like a product.

Our network shall be equipped with an unbounded bidimensional sheet, probabily of paper, divided in squares each of which is either empty or marked with a sign chosen amid the followings: a standard one or an addition, a +, or a multiplication, a x, or an equal, a =, sign. No square can be in any other state. The numbers are coded in the unary alphabeth: the number n is represented as n+1 adjacent squares marked with standard signs in an horizontal row.

The machine looks through a window 2px1 squares wide: the execution of the output program L, R, U, D, causes the window to be shifted of 2p or respectively of one square in a left, right, or respectively upper, down direction.

To look through the window there is an eye which can scan one square at the time and is able to tell its state. The output programs l and r move the eye one square at the time in a left, right, direction within the boundaries set by the window: otherwise the eye and the window are jointly moved. To erase or mark the observed square there is an arm: the programs to execute the desired action are called e, m, add, mul, eq.

We shall suppose the machine already trained in the execution of the "cpy (X)" program, i.e., a program to copy the content of the observed square into the square (X), with X a word in the alphabeth {L, R, U, D}. It is not too difficult to see how it could be obtained the program cpy(X). To this end we shall suppose the eye to be a mutipixel gadget, i. e., a parallel processing one with many output channels. Then it is possible to suppose, without any loss of generality, the existence of so many sets of nodes as many states can have a square. In this respect it should be noted how the said sets do not need to be all distincts: their intersection is not required to be empty. The desired program is the one which executes the sequence of the window's movements as indicated by X and then elicits the arm to compute the appropriate action chosen amid the previously given five. Because under certain circumstances the signs +, x and = are to be interpreted as names of programs to be executed, this possibility has to be excluded whenever cpy(X) is running: a comand to this end shall be inserted into the respective programs.

If the machine has to compute as output some preassigned programs then the input ones can be determined by taking advantage of the possibility to write equations in our language. E. g.,

$$x \; b \; eq = i,$$

where the programs must be executed in a left to right direction and b, possibly the identity, is a costant. it gives us:

$$x = eq^{-1} \; b^{-1},$$

with x the desired input program

For the moment being let us stress how each, out of the five possible outputs, can be obtained by means of a "convolution array" I. e. a net whose nodes have positive coupling coefficents with the relevant inputs and negative with all the others This correspond to a program with five jump instructions each of which is not executed only if the relevant input is present and all the others are absent

Two more programs are supposed to be known to the machine: ADD1 and ADD2. The difference between the two is simply the input device used: the former is elicited when the eye is scanning a square with a + and the cpy is not running The latter when somebody whisper the word ADD into the machine's ear, and the cpy is not running. ADD2 start the eye scanning in a leftward direction until a square either empty or marked with a + is reached. At this stage both ADD1 and ADD2 write on the observed square a standard sign, after this the eye start scanning in a rightward direction until it finds an empty square Then it erase two squares toward the left and the process terminate. Moreover the cpy and both ADD1 and ADD2 must contain conditionals to take care of the case when the eye reach the window's boundary. We note that for these programs the actions' sequences depend on the detection of boundaries between sets of squares marked with standard signs and those not so marked. In other words conditions easily implemented in convolution arrays and easily formalized with conditional jump instructions

To complete the chapter we give the program ADD2: to this end let k be the data stream associated with the detecton of an unmarked square and h', h", the one associated wth the detection of the left, right, window boundary.

Let it be:
$$c = j(h",M1)Rr^p j(h',M2)e(M1,h")Ll^p e(M2,h')$$

Then the program ADD2 is:
$$e(N',k)lcj(k,N')me(N',k)rcj(k,N')(1ce)^2$$

In both the programs the jumps are, not executed whenever the condition stated by the jump's first subscript is, not satisfied: that is, whenever the standard signals are, not, present in the appropriate data flow

b) The Algorithm.

As in the addition case two are the algorithms and the programs to compute a product namely MUL1 and MUL2. The first is the one activated when the eye scans a square marked with a x. The second when somebody whispers MULTIPLY in the machine's ear

Both are executed whenever cpy is not running. For both the initial conditions must be: two integers written on one row and the eye scanning the square at the left of the rightmost marked one. For MUL1 the two integers must be separated by a x sign. For MUL2 either by a x or by an empty square. Let us start with the MUL2 algorithm bearing in mind that the arm writes or erases only onto the square currently observed by the eye.

M1. The eye starts scanning in a leftward direction until a square marked with a x or an empty one is found. The arm writes an x sign. (From now on MUL1 and MUL2 are identical.)

M2. The eye scans in a leftward direction until a square marked with an = or an empty square has been reached. It goes one square back toward the right. The window is shifted down a square and the arm writes an = on the observed square.

M3. The window is backshifted one square up. The eye goes one square right and cpy(D) is executed. M3 is repeated until an empty square has been copied. The arm writes a +.

M4. The eye start scanning in a leftward direction until a square marked with a + or a x sign is detected.

M5 The eye goes back one square toward the right then the arm erases a square if it is a standard sign, otherwise the eye goes back toward the right until an empty square is scanned, it goes one square to the left and then step M6 is executed. Otherwise the eye goes in a rightward direction until an empty square is detected and then the arm writes a standard sign. The eye starts anew in a leftward direction until an empty square is detected, then the arm writes the standard sign previously erased. Eventually the step M5 is repeatedly executed.

M6. The eye scans in a leftward direction until it finds a square marked with an x or an empty one. In the first case MUL1 is automatically activated. In the latter one the algorithm is terminated and the eye goes rightward until it finds either an empty square, in which case it stops or a square marked with a + sign, in which case ADD1 computes the sum until the task is completed.

Now some comments on the algorithm just given. To simplify the exposition we did not explicitly mention the inhibitions which shall be included within the program. Then we shall observe how little parallelism have been introduced. But here we must distinguish because of the ambiguity inherent in the notion of parallelism. As observed in the introduction the transformations, i. e., the sequences of operations to be carried on are by theyrself neither sequential nor parrallel. Only the streams of data to be processed can be seen as one or a multiplicity of entities. If the data constitute a multiplicity then they can be either independently or concurrently processable. E. g., the cpy progam can be easily transformed in such a way so to copy at once, onto as many squares, the content of all those falling within the window's boundary. This obviously requires the existence of many eyes, arms and so on, with not so many insurmountable problems. But once we have such a wealth of peripheric devices then it is all but an easy task to have them operated indipendently of each others. It cannot be said that the uninterrupted interactions with an unpredictable enwironment renders unworkable such a system. But the easiness of syntactical and above all semantical errors, with the ensuing unsolvable deadlocks, deprives it of any interest under the said conditions.

In conclusion to suppose to have just one or a fixed number of input and output gadgets it is not so fundamentally different so we opted for the simpler expository solution, leaving for the interested reader an easy exercise to solve.

c) The Program.

We recall that the first, second, subscript of the jump, end of jump, instruction shall represent the condition upon which the jump has to take place. Thus j(add,M1) it means that the forward jump has to be executed if there it is a + on the square scanned by the eye. Or e(N1,=ve) it is the end of a backward jump which has to be executed until the eye does not scan a square either empty or marked with a =.

I. e., for forward jumps the given condition must be satisfied for the jump to take place. For loops they continue to be executed until the stated condition, usually an action to be computed by the peripheric devices implicitly referred to in the scope, has been satisfied. The conditions shall be either programs' names, known to the machine, or input states connected by the logical connectives. For greater clarity the program shall be broken in the six steps of the previously given algorithm: there are two main differences.

The first is represented by a set of jump instructions which inhibit the executions of the MUL and ADD programs whenever a + or a x sign is seen from the eye, e. g., during the execution of a cpy program. To help readability it has been indicated, in square brackets, the algorithm's step which contains the corresponding end of jump. The second difference, the one occurring in the step M6, shall be discussed later on. And now the program:

[M3]	j(cpy(D),M1)
[M3]	j(add,M1)
[M4]	j(H,M1)
[M5]	j(H',M1)
[M5]	j(H",M1)
[M5]	j(H"',M1)
[M5]	j(K,M1)
M1.	e(N1,xve)lcj(xve,N1)mul

M2.   e(N1,=ve)lcj(=ve,N1)rDeq

M3.   e(N1,e)Urccpy(D)e(M1,cpy(D))j(e,N1)adde(M1,add)

M4.   e(N1,+vx)s(H,N1)lcp(N1,H)j(+vx,N1)e(M1,H)

M5.   e(N1,e)s(H',N1)rcp(N1,H')e(M1,H')j(m,M1)e(N1,e)

    s(H",N1)rcp(N1,H")e(M1,H")j(e,N1)lcj(m,M2)e(M1,m)e

    e(N1,e)s(H"',N1)rcp(N1,H"')e(M1,H"')j(e,N1)me(N1,e)

         s(K,N1)lcp(N1,K)e(M1,K)j(e,N1)j(e,N1)

M6.   e(M2,m)e(N1,xve)lcj(xve,N1)e(N1,ev+)rcj(ev+,N1).

Now some observation on the program. First let us examine M6; to conform to the given algorithm it should have read:

     e(M2,m)e(N1,xve)lcj(xve,N1)j(e,M1)MUL1e(M1,e)

           e(N1,ev+)rcj(ev+,N1)j(e,M1)ADD1e(M1,e).

The difference is simply accounted for by the fact that we are not dealing with a stored program machine. The activation of MUL1, or ADD1, in our case it depends on the eye reading a square marked with x, or with +. In M4 to avoid starting anew MUL1, or ADD1, we had to inhibit them by means of the j(H,M1) ... e(M1,H) comand: in M6 the avoidance of such a comand automatically starts the relevant program once the input is present. This drive us to point out the existence of an interesting mechanism, i. e., how the machine can learn universal skills from a few particular cases. It is sufficent to train it taking into account the intrinsic property of a data flow machine: namely the automatic activation of procedures.

Now let us consider the couples s ... p which are present in M4 as in M5: they just give a name to a flow of data. Name which we found in the subscripts of the aforementioned jump instructions: the choice is due to its brevity and readability. When the program is running on an actual machine its own activity can be, if we so choose, the source of the various H, H'... data flows.

4. CONCLUSIONS.

Let us examine the given programs and algorithms. As previously noted they are primarily sequences of conditional jumps which can be implemented in the net by means of convolution arrays. Yet this conclusion is little more than to say we can have positive and negative coupling coefficients. Then let us look a little more carefully to the said conditionals. Trivially they are all conditional upon the happening of something like: a boundary between two different kinds of signs read from the eye, with the difference being specific or generic, or the execution of some program, i. e., the existence of activity somewhere in the net. All this can plainly be catalogued under a few headings, i. e., the conditions to be satisfied are chosen amid few selected ones. And this is not the particular case for the aforegiven programs and algorithms. If we consider the case for the selection function, the substitution and minimalisation schemata, which the reader should admit are only boring exercises, then we have more of the same diet:

i. e., jumps depending on specific or generic input conditions or on activity going on somewhere in the net.

Thus if we start with a machine which has a base of coupling coefficients such that it can be trained to execute recursive functions it shall be a quite universal one. Otherwise it shall have some intrinsic limitations on the skills it can learn.

To conclude we should give some suggestions on how to train the machine. This is an empirical ground which is especially explored at antropomorphic level: there are many available techniques, e. g., the punishement and reward one. In our case this can become an automatic, or trainer driven, diminishement and reinforcement of coupling coefficents, locally or at large. In any case it seems always useful to bear in mind that old dictum: repetita iuvant.

And now a concluding remark. If we put ourself from a mechanistic point of wiev it means, almost by definition, to algebraize reasoning: to break down reasoning in a discrete stream of, often unrelated, steps. Thus in a system to which we must told "what", as opposed to "how", to do it is essential the role played by the memory. From Turing machine onward every time we ask a machine to execute an instruction we are just telling "what" to do. The "how" to do bothered the designer, perhaps the manufacturer: the user does not need to know such minutiae.

The natural automata, usually called vertebrate nervous systems, apparently store macroinstructions chosen accordingly to the past workload to learn new tasks: this is the trik we plan to simulate.

Acknowledgements: the work here presented was supported in part through both the Progetto ASSI of the MPI's 40% fund and a research contract of the CNR's Mathematical Committee.

THE END

APPENDIX.

The amount of storage space provided by our network can be estimated once some details have been fixed. Following the text we say $C(i,j)$ is the value of the generic coupling coefficient which connects the node j with the node i. In other words $C(i,j)$ is a variable on the set $\{C(i,j,t-l\tau)\}$, for $0<=l<=L$.

Let be i' and i" two distint nodes. We shall call "common" the set of all those nodes such that if j belongs to the common set then from

$$C(i',j)<>0$$

it follows:

$$C(i',j)=C(i",j).$$

We hall call "different" the set of all those nodes such

that if j belongs to the different set then from

$$C(i',j,t-l'tau) <> 0$$
$$C(i'',j,t-l''tau) <> 0,$$

for at least an l', l", it follows either:

$$C(i'',j,t-l'tau) = C(i',j,t-l'tau),$$
$$C(i',j,t-l''tau) = C(i'',j,t-l''tau),$$

or:

$$C(i'',j,t-l'tau) = 0,$$
$$C(i',j,t-l''tau) = 0.$$

Obviously the common set is a subset of the different set.

Let us consider the effect of a learning process, governed by the memory function M(t), upon the coefficient C(i,j). Let it be k', k", the smallest number of addends for which:

$$ADD(j)\{C(i,j,t-ltau)-T(i,t-ltau)\} > 0$$

after it has taken place the learning process of the computation whose name is k', k". Moreover we shall suppose k', k", to be the computation which involves the smallest, greatest, number of coupling coefficients, that is, we suppose k'<k". For simplicity's sake we shall consider their median value k. In other words k is the number of coupling coefficients nedeed to store a generic computation. Let n be the total number of C(i,j,t-ltau) for a given i, that is, n should be the memory capacity of the ram associated to each node in a simulation: in general

$n > j$.

Thus $\binom{n}{k}$ is the number of computations which can be stored exploiting the coupling coefficients impinging on i. It is worth recalling here that $\binom{2n}{n}$ is a quite fast increasing function of n: of the order of $(n3.14)^{-.5} 4^n$ [21, pg.71 exer. 46]. It seems safe to conclude that the total storage capacity of an N nodes MPN is given by:

$$N\binom{n}{k}.$$

Such a conclusion however does not take into account redundancy. In other words if the node i fails then the information stored into the coefficients C(i,j) is gone.

Thus let us consider the different and the common sets of a set of m nodes, the generic of which we shall continue to call i. By definition all the C(i,j,t-1tau), for j belonging to the common set and l fixed, have the same value. Let it be n' the number of distinct coefficients which couple the common set's nodes with i. Thus n' is also the number of the C(i,j,t-1tau) which are independent of each other. Next let it be m(h), h=>2, the nodes belonging to the different set such that each of them originate n(h) independent C(i,j,t-1tau). Then

$$n" = ADD(h)(m(h)n(h)) < n,$$

with m(1)=1, it is the total number of independent coupling coefficients impinging upon node i. Thus to the redundancy built in the proposed kind of diffuse memory it corresponds the reduced capacity of $\binom{n}{k}''$ for any set of m nodes, against the original value:

$$m\binom{n}{k}.$$

BIBLIOGRAPHY.

(1) McCulloch, W.S. and Pitts,W.-"A logical calculus of the ideas immanent in nervous activity." Bull. Math. Bio. 5 (1943) 115-133.

(2) Weyhrauch, R.W.-"Prolegomena to a theory of mechanized formal reasoning." AI 1 (1980) 133-170.

(3) Stefik,M.-"Planning and Meta-planning (MOLGEN: part 2). "Reading in AI. Webber and Nilsson (eds.). Tioga, Palo Alto (Ca). (1981) 272-286.

(4) McCarthy,J. and Hayes,P.J.-"Some philosophical problems from the standpoint of artificial intelligence." Reading in AI. Webber and Nilsson (eds.). Tioga, Palo Alto (Ca). (1981) 431-450.

(5) Moore, R.C.-"Reasoning about knowledge and action." Reading in AI. Webber and Nilsson (eds.). Tioga, Palo Alto (Ca). (1981) 473-477.

(6) Wilkins,D.-"Domain independent planning: representation and plan generation." AI 22 (1984) 269-301.

(7) Levesque, H.J.-"Foundations of a functional approach to knowledge representation." AI 23 (1984) 155-212.

(8) Dennis, J.B.-"Data-flow supercomputers."Computer (1980)

48-56.

(9) Gurd, J. - "Developments in data-flow architecture." Preprint Un. of Manchester U.K. (1982).

(10) Gottlieb,A. and al.-"The NYU supercomputer. Designing an MIMD shared memory parallel computer." IEEE Trans. on computers. Vol. C-32, No. 2 (1983).

(11) Treleaven, P. C. and Lima, I. G. -"Fifth generation computing: logic or control flow?" Atti del Cong. AICA Vol. II (1983) XXXIII-XL.

(12) Hudak, P. and Goldberg, P.- " Experiments in diffused combinator reduction. " To appear in "Proceedings of 1984 Symp. on LISP and functional programming."

(13) Hudak, P. - " Distributed applicative processing systems: project goals, motivation and status report." Technical report YALEU/DCS/TR-317. May (1984).

(14) Hillis, D. W. - "The connection machine." MIT (1985).

(15) Hoare, C. R. A.- "Communicating sequential processes." Com. of the ACM. 21 (1978) 666-677.

(16) Hubel, D. H. and Wiesel, T. N.- "Brain mechanisms of vision." Sci. Am. 241, 3, (1979), 130-144.

(17) Jones, E. G. and al. -"Thalamic basis of place and modality specific columns in monkey somatosensory cortex: a correlative anatomical and physiological study." J. Neurophysiol. 48, 2, Aug. (1982) 545-568.

(18) Caianiello, E. R. -"Outline of a theory of thought-processes and thinking machines." J. Theoret. Biol. 2

(1961) 204-235.

(19) Lauria, F. E. - "An addressable machine as the interpretation of a free group." Information sciences. 28 (1982) 131-160.

(20) Lauria, F. E. - "Toward an algebraic gnoseology." Systems and Cybernetics. 16 (1985).

(21) Knuth, D.E.-"The art of computer programming." Addison Wesley. II ed. Vol. I (1969).

(22) Nottebohm,F. "Learning, Forgetting, and Brain Repair." in "Cerebral Dominance" Geschwind and Galaburda eds. Harvard Un. Press (1984) Chp. 7, 93-113.

(23) INMOS-"OCCAM programming manual." Prentice Hall (1984)

To my unforgettable teachers:

E. R. CAIANIELLO AND V. BRAITENBERG.

Amalfi, June 17-18, 1986.

WHAT KIND OF UNCERTAINTY ATTACHES TO HUMAN ACTIONS?

Donald M. MacKay

Department of Communication and Neuroscience, University of Keele
Keele, Staffs. ST5 5BG
England

ABSTRACT

Two quite different kinds of 'uncertainty' may attach to human action. One arises from the stochastic aspect of the physical processes in the brain. The other, which is inescapable for embodied agents and those in dialogue with them, arises from the non-existence of a complete future specification of a cognitive mechanism, unconditionally binding on the agent whose mechanism it is.

1. STOCHASTIC INFORMATION-PROCESSING

In the late '40s, when trying to sort out different quantitative aspects of the concept of information[1,2], it occurred to me that in order to handle and generate information in the way that human beings do, it would be advantageous to have an essentially stochastic information-processing system, in which the probabilities attached to representations could be represented by physical variables that determined their statistical influence on the stochastic process. In such a mechanism, propositions or other representations that were poorly supported by evidence would be embodied in structures that were (physically) weakly coupled to the process to which they were relevant, whereas those with stronger support (higher probability) would be embodied in structures more strongly coupled. In 1949 I aired some of these ideas in a paper "On the combination of digital and analogical techniques in the design of analytical engines", which was first tried out on a few friends, then circulated as an appendix to a paper[3] at our second London Information Theory conference in 1952, and eventually published as an appendix to a later paper[4]. The reception of this notion among my fellow information theorists and computer scientists at the time was hardly encouraging, though a few physiologists, like J. Z. Young and Ralph Gerard, and especially the ever-generous Warren McCulloch, gave it a more friendly hearing. At the Macy meeting on Cybernetics in March 1951[5] the idea aroused no enthusiasm, the suggestion that stochastic exploratory strategies might sometimes have advantages over deterministic ones, especially in concept-formation, being authoritatively dismissed by Savage (see pp. 204-207 of ref. 5).

None of the objections I met, however, seemed to rule out the possibility that a system on these stochastic lines might not only generate 'mind-like behaviour', but also show in a natural way (rather than by special contrivance) many of the characteristics that we regard as specifically human. In particular, such systems seemed capable in principle of transcending some of the limitations of digital computers which were then being pointed out by philosophers. A meeting of the recently formed British Society for Philosophy of Science (BSPS) in 1950 gave me a chance to explore this aspect[6], leading later to a full-scale philosophical debate[7]. At this stage we were talking mainly about the powers of artefacts rather than about human brains: how genuinely <u>creative</u> could an artefact be? Could it <u>generate</u> information, as well as merely processing information fed to it? Must all its behaviour be in principle predictable? It will probably be obvious nowadays, when associative stochastic automata, "Boltzmann machinery" and the like are commonplace, how readily such questions could be met[3,4,7,8,9].

2. SYSTEMATIC UNPREDICTABILITY

All this, however, led to a second question. If a stochastic mechanism could so naturally meet many tests for human-like behaviour, might it not be that the human brain itself was just such a mechanism[10])? If so, what would be the implications for our ideas of human creativity and human freedom?

In this connection Popper's classic paper[11]) entitled "Indeterminism in quantum physics and in classical physics" had already demonstrated that no discrete-state computer could predict completely the future of a universe that included itself. (The argument was, briefly, that any such computer would need information about the results of its calculations before it could complete those calculations.) Popper did not himself consider the human brain to be a computing machine; but he suggested that "In so far as man is a predictor, my modest results are ... applicable to man and the human society" (loc. cit. p. 193). I do not wish here to defend (or attack) Popper's conclusion: my concern is with what does, and does not, follow logically if we accept it.

3. 'DETERMINATE' VS. 'PREDICTABLE' VS. 'INEVITABLE'

In the first place, what the argument demonstrates ought strictly to be called unpredictability, rather than indeterminism. If granted, it shows that even if our brains were computers governed by classical physical laws, we could not compute precisely our own future, nor that of any system such as our own human society to which we are sufficiently tightly coupled. It does nothing, however, to disprove the existence of a determinate specification of our future with an unconditional claim to our assent - i.e. such that whether we know it or not, it states definitively what we would be correct to believe and mistaken to disbelieve. To see the importance of this point, consider what we mean by calling the times of future events, such as solar eclipses, 'determinate' in the system of classical astronomy. We do not necessarily mean by this that we (or anyone else) can predict them. We mean that whether or not it can be calculated by us, there exists (now) a detailed solution of the deterministic equations of state that govern the solar system, for the future time in question, such that (whether we know it or not, or like it or not) that specification has an unconditional claim to our assent. In this sense, the occurrence of the eclipse at that future time is <u>inevitable for us</u>. Whether we know it or not, it has already a specification 'written in heaven', which lays down what we <u>would</u> be correct to believe and mistaken to disbelieve about it. Thus, the incalculability-by-us of a future event does nothing to disprove its inevitability-for-us.

Secondly, Popper's argument (even if accepted) did nothing to rule out the possibility that a sufficiently informed 'super-observer' of our brain-machinery, carefully isolated from us so that his activities could not significantly influence us, might accurately predict our immediate future. No doubt this possibility is as remote in practice as the exploits of Laplace's famous demon; but merely practical considerations can hardly allow us to ignore it for philosophical purposes. In short, Popper's conclusion, even if granted, left open the question whether my future, despite its unpredictability-by-me, might not still be both predictable-for-others and inevitable-for-me in the foregoing sense.

4. RANDOMNESS

My attachment to stochastic models of brain activity had led me from the outset to emphasize the possibility that our brain activity might be not just unpredictable-by-us but unpredictable-by-anyone. Indeed, if the 'randomness' of the stochastic process in question originated at the level of Heisenberg uncertainty, it could be argued (on one interpretation of that uncertainty) that its future was strongly indeterminate, in the sense that no uniquely determinate solution of its state equation would exist now, specifying its states uniquely at a later time[6]. On the other hand, although such indeterminacy might well underlie the 'freedom' we recognize in human creativity, it seemed less than satisfactory as the basis of human responsibility[12]. If a brain event were caused by some Heisenberg-indeterminate transition, it should surely be regarded as something that happens to me rather than something I achieve: something I suffer rather than something for which I am responsible? A random transition might make some act of mine unpredictable-for-anyone; but its effect would still be inevitable-for-me, and so not rationally creditable-to-me in the way that a sober choice could be.

What then of normal rational non-capricious decision-making? Popper had shown that this could be <u>unpredictable</u> for the agent himself; but was there any rational basis for our common intuition that until we make up our mind on an issue of choice the outcome is in some sense <u>undetermined</u>? To clarify this question I considered the extreme theoretical case of a human brain in which Heisenberg-indeterminate events were as negligible as they are in a well-functioning digital computer, so that every state-transition was a determinate function of prior states and inputs. It turned out that even in this extreme case (and <u>a fortiori</u> if less physically-deterministic assumptions are made) there is a clear and non-trivial sense in which our future brain activity is not merely unpredictable-by-us but indeterminate-for-us[17].

5. I-STORY AND BRAIN-STORY

This superficially implausible conclusion is in fact a direct consequence of the working assumption of mechanistic brain science, that every fact of our conscious experience has its correlate in the physical state or activity of our brain, in the sense that no change could take place in the data of conscious experience without a corresponding physical change taking place in the brain. This working hypothesis is schematized in Fig. 1, where the lefthand column represents the kinds of statement we could make in bearing truthful witness to our prime data about ourselves - the facts of our conscious experience: I see-X; I hear-Y; I believe-Z etc. Collectively, let us call these the 'I-story'. It is not suggested that all the data of our conscious experience can be specified completely in words. (Think for example of the tumultuous experience of feeling seasick!). The purpose of Fig. 1 is merely to illustrate the working assumption that for

Figure 1

I-STORY	BRAIN-STORY
I see-X	Brain state x
I hear-Y	Brain state y
I believe-Z	Brain state z
etc.	etc.

every fact of our conscious experience there is a corresponding entry in the 'brain-story' that an ideally informed super-observer could tell about our brain. (Note that we have no need at the moment to accept or reject this assumption. Our purpose is to see what would (and would not) follow if it were true).

6. THE PHYSICAL COSTS OF KNOWING

Among the facts to which our I-story bears witness are facts about what we know or believe to be the case. Mechanistic brain science presupposes that all our knowledge, whether more or less certain, whether articulate or tacit, is represented somehow in the physical structure of our brains - perhaps in terms of interneuronal couplings of variable strength forming an associative network that determines our total state of conditional readiness to reckon with the nature and contents of our world in all relevant circumstances[13,14]). Let us call this physical structure the 'cognitive mechanism', or CM for short, remembering that it is defined operationally as that (postulated) subsystem or aspect of our brain that must change significantly if a significant change takes place in what we know or believe. Consider now the complete physical specification of your cognitive mechanism at a given time. Assuming the extreme case in which the physical laws governing its structure and behaviour are deterministic, the specification of its state now will imply a specification of its immediate future as a function of the inputs it will receive. Assuming for the sake of argument that those inputs are also completely determinate, there must exist now a unique solution to the state equation of CM as a physical system. For any time t in the immediate future this will imply a complete specification CM(t) of the state of CM.

7. LOGICAL RELATIVITY

Let us now ask what may seem a strange question: <u>who would be correct to believe CM(t)</u>? It might seem obvious that if CM(t) is implied by the relevant physical state equation, then it is something that anyone and everyone would be correct to believe and mistaken to disbelieve, just as in our earlier case of a solar eclipse. Certainly it is what a sufficiently remote super-observer (i.e. one sufficiently uncoupled from the situation) would be correct to believe: indeed whether such a super-observer knows it or not, CM(t) has an unconditional claim to his assent. But nevertheless - is it what <u>you</u> would be correct to believe? Certainly not; for (you remember) CM is precisely that little part of the physical universe that <u>represents</u> what you believe, and so must <u>change</u> if any change takes place in what you believe. The calculated CM(t) implies a complete description of CM now. It follows directly that if CM(t) is a correct inference in the case where you do not hear of it or believe it, it cannot (logically cannot) have a well-founded claim to <u>your</u> assent. Indeed it offers a description of your future that you would certainly be <u>in error</u> to believe. For CM(t), if embodied in your cognitive mechanism (CM), would necessarily be falsified in at least one particular - namely, in respect of the changes that its embodiment in CM must have brought about. In other words, no complete specification of the immediate future of your CM could be equally accurate whether or not you believed it.

Notice that we are not here considering the case where somebody actually tries to offer you CM(t), and so interferes with your cognitive process. (We shall glance at that special case later). The question we are asking is a logical one, namely: Assuming that nobody ever tries to tell you about CM(t), so that our imaginary super-observer is correct to believe it, does CM(t) lay down what you would be correct to believe (and mistaken to disbelieve), although you happen not to know it? To this logical question we have seen that the answer is 'no'. With respect to descriptions of the present or immediate future of anybody's CM the situation is inescapably relativistic: The description the super-observer is correct to believe is something you would be mistaken to believe[17]. What you would be correct to believe is therefore necessarily something different. No complete present and future specification of your CM exists with an unconditional logical claim to your assent as well as that of everyone else. In this respect, even on the most deterministic of classical physical assumptions, future states of cognitive mechanisms are startlingly different from future states of the solar system, or indeed of any other part of the physical universe.

8. SELF-FULFILLING PREDICTIONS?

By now (if not before) you may be impatient to suggest an obvious move. Why should the super-observer not adjust the calculation of CM(t) to allow for the effects of your believing it? It is not in fact obvious that this can be done, especially if the relevant changes in CM occasioned by changes in belief are discrete. For the sake of argument, however, let us suppose that some relaxation method could enable our non-participant super-observer O to come up with a revised specification of the immediate future of your CM - call it CM'(t) - which he can now say has a well-founded logical claim to your assent. O has carefully adjusted it so that the changes (in your CM) occasioned by your believing CM'(t) would be just what are required to make it correct. But what follows from this achievement? Admittedly (if it can be done) it enables O to assert that there exists a specification of the future of your CM that you would be correct to believe if only you knew it. CM'(t) can claim to be beliefworthy-by-you. On the other hand, since CM'(t) was calculated so as to be correct if (and only if) you believed it, it follows that unless you believe it, CM'(t) will be false, and <u>nobody</u> would then be correct to believe it! In other words, CM'(t) is a description of the future of your cognitive mechanism which you would be <u>equally correct to believe or disbelieve</u>. Its claim to your assent, so far from being unconditional, is entirely <u>open</u>, in that if you were to believe it, it would be true; if you were to disbelieve it, it would be false. It does not at all have a claim to assent that is inevitable for you. On the contrary, in this strict and explicit sense, the correctness of CM'(t), and so the correct specification of your future, is <u>up to you</u>.

So this clever move on the part of the super-observer (which has been repeatedly proposed since the middle '50s in would-be refutation of the present argument) turns out only to reinforce the main point. The best that deterministic brain science could show to exist by way of a specification of the immediate future of your CM is a description that you would be <u>equally correct to accept or reject</u>. No specification of CM's immediate future (nor of any future brain events sufficiently dependent on it) exists, unknown to you, with an unconditional claim to your assent. In this sense, and to this extent, you are correct to regard your immediate future as <u>indeterminate</u>; even if predictable-by-others, it is still <u>open</u> for you to determine by your cognitive agency.

9. THE EMBODIMENT OF MIND

Talk of 'determination' here may seem to smack of interactionism - the idea that the brain is subject to quasi-physical 'influences' from the 'world of the mind' on which it exerts reciprocal 'influences'[18]. In a (philosophically desperate) move to escape this conclusion, some 'physicalists' would argue that 'brain' and 'mind' are merely two ways of talking about the same thing, and thinking is merely a name for something the brain does. As I have argued elsewhere[19,20,21,22], this last move is not so much false as categorically inept. It is not brains, but people, that think. Thinking is a datum of the 'I-story'. There is however a realistic alternative to both interactionism and

physicalist reductionism, in the view that our conscious agency is
<u>embodied in</u> (rather than interactive with, or identical with) the
activity of our brain-and-body as an information-system. On this view,
our thinking can determine the form of our action because it is
embodied in our brain activity, even though every physical event in our
brain were determined by physical laws. Much the same, after all, can
be said in the case of a computer solving an equation. The equation
determines the form of the computer's action because it is embodied in
it, even though every significant physical event in the computer is
determined as such by the laws of physics.

10. OPENNESS VS. UNCERTAINTY.

Notice now the contrast between the 'openness' thus established
and the 'uncertainty' that stems from inability to predict. Whereas
the latter rationally entails a 'wait and see' attitude of suspended
judgment, and requires our minds to be made up by additional <u>data</u>, the
former rationally entails <u>action</u>: the action (which we all know how to
attempt) of 'making up our own mind' by the process we call thinking,
valuing and choosing. We have our being as cognitive agents in this
reflexive process of internal self-supervisory agency, whereby we each
act as our own goal-selector[6,15,16].

It is equally important to realize that the 'logical
indeterminacy' so established in no way excludes the predictability of
our choices by others causally uncoupled from us. The appearance of
conflict or contradiction here arises only if we forget the
relativistic point made earlier. The specification $CM(t)$ that the
superobserver O would be correct to believe is precisely what the agent
would be mistaken to believe. It is at best only locally-valid-for O
(and others like him, causally uncoupled from A). It is systematically
invalid-for-A (and others in reciprocal dialogue with him). As in
physical relativity theory, what we must ask is how views from the two
standpoints (of super-observer and agent) are related. There is no
suggestion that the agent A is free to believe anything he likes,

regardless of what O may know and believe. What A rationally ought to believe about his own future must differ from what O rationally ought to believe, but it is not <u>arbitrarily</u> different. Even in the extreme case we have considered, what A is correct to believe is that his future is open for him to determine, and has no uniquely definitive specification (unknown to him) that he ought rationally to accept as inevitable if only he knew it. This in no way conflicts with the existence of a definitive prediction that O (causally uncoupled from the events in question) would be correct to believe. Indeed A could perfectly well accept without inconsistency that such a prediction exists, as long as he recognizes that it is not a prediction he would be correct to believe if he knew it.

11. THE RETROSPECTIVE VIEW

Perhaps the easiest way to see this is to imagine that O (having successfully predicted the future state of A's CM) waits until after the event and then visits A with a complete record of his predictive performance. Obviously, A can now be rationally convinced that CM(t) was what O was correct to believe; but from O's own records, it will be equally evident that A would have been mistaken to believe the same. Thus in retrospect, so far from O's evidence proving that A was under an illusion due to his ignorance of CM(t), it would actually show that A was <u>correct</u> in believing that there existed no specification of CM definitive-for-him until he (A) determined the state of his CM by his cognitive activity.

12. MANIPULATIVE PREDICTION

This is perhaps the point at which to consider the special case mentioned earlier, where O interferes with A by actually offering him a prediction of CM. If A is sufficiently counter-suggestible, he can doubtless give O a hard time; but assuming (for the sake of argument) that O is able to allow for the effects of his interference so that his prediction is self-fulfilling, we can readily imagine cases in which what A is induced to believe turns out to be correct. But what follows from that? Does it follow, for example, that if A _had_ believed something else, he would have been mistaken? Obviously not, since in that case his CM would not have been in the state which the prediction (by definition) was designed to bring about! To object that in this case it was predictably impossible for A to believe anything else would be to miss the point. The question is not whether A was psychologically obliged to accept CM(t) (this is granted, and there are of course lots of ways of psychologically obliging people to accept predictions), but whether he was (or could be) rationally obliged to do so. Since (ex hypothesi) O's data showed that CM(t) would have been false unless A accepted it, they clearly demonstrate no rational obligation on A's part to do so.

13. CONCLUSION

Having (I hope) cleared the air by discussing an extreme case, it is time to return to reality and sum up. We have seen that in real brains, subject to all the physical fluctuations recognized in thermodynamics and quantum mechanics, there is room for uncertainty of the sort we might call 'conjectural'. Not all brain activity, even when macroscopically specified, need be determined uniquely by current state variables and input since, even without invoking Heisenberg's principle, some neural state-transitions can be expected to show dynamic instability leading to 'deterministic chaos'. I have argued, however, that this kind of uncertainty is more defensibly associated with such human features as caprice and creativity than with the 'freedom' of responsible choice.

Of greater interest is an 'uncertainty', or rather 'openness', of a quite different order, which attaches to our future brain states regardless of the degree of physical indeterminacy they exhibit. This, which we have distinguished as 'logical indeterminacy', arises from the fact that no complete (present or future) specification of our cognitive mechanism exists, even unknown to us, with an unconditional claim to our assent. This indeterminacy, which is not incompatible with predictability-by-detached-observers, attaches to any human actions sufficiently causally dependent on the cognitive mechanism in question. Unlike thermodynamic or quantum uncertainty, it is not to be seen as limiting our knowledge of what is the case, but rather as offering an opportunity for us to decide what is the case. Because the 'openness' thus established is inseparable in principle from embodied cognitive agency, it affords, I believe, an inescapable measure of responsibility for our actions as cognitive agents.

REFERENCES

1. MacKay, D.M. "Quantal Aspects of Scientific Information", Phil. Mag., 41, 289-311 (1950).

2. MacKay, D.M. "Complementary Measures of Scientific Information-Content", Methodos, 7, 63-90 (1955).

3. MacKay, D.M. "Generators of Information", Communication Theory, (W. Jackson ed.), Butterworths, 475-485 (1953).

4. MacKay, D.M. "Operational Aspects of Intellect", Mechanization of Thought Processes, (N.P.L. Symp. No. 10, 1958, London, H.M.S.O.) 37-52 (1959).

5. MacKay, D.M. "In Search of Basic Symbols", Proc. 8th Conf. on Cybernetics (H. von Foerster, ed.), Josiah Macy Jr. Found., New York, 181-221 (1951).

6. MacKay, D.M. "Mindlike Behaviour in Artefacts", Brit. J. Phil. of Sci., 2, 105-121 (1951).

7. Wisdom, J.O., Spilsbury, R.J. and MacKay, D.M. "Mentality in Machines", Men and Machines, (Aristot. Soc. Suppt., 26), London, Harrison, 1-86 (1952).

8. MacKay, D.M. "The Epistemological Problem for Automata", Automata Studies, (C.E. Shannon and J. McCarthy, eds.), Princeton University Press, 235-251 (1956).

9. MacKay, D.M. "From Mechanism to Mind", Brain and Mind, (J.R. Smythies, ed.), 163-200, and Comments on pages 129-131, Routledge and Kegan Paul (1965).

10. MacKay, D.M. "Towards an Information-Flow Model of Human Behaviour, Brit. J. Psychol., $\underline{47}$, 30-43, 1956.

11. Popper, K.R. "Indeterminism in Quantum Physics and in Classical Physics", Brit. J. Phil. of Sci., $\underline{1}$, 117-133 and 173-195 (1950).

12. MacKay, D.M. (correspondence) Brit. J. Phil. of Sci., $\underline{3}$, 352-353 (1953).

13. MacKay, D.M. "Operational Aspects of Some Fundamental Concepts of Human Communication", Synthese, $\underline{9}$, 182-198 (1954).

14. MacKay, D.M. "The Place of 'Meaning' in the Theory of Information", Information Theory, (E.C. Cherry, ed.), Butterworths, London, 215-225 (1956). Reprinted in MacKay, D.M., Information, Mechanism and Meaning, M.I.T. Press, Cambridge, Mass. (1969).

15. MacKay, D.M. "Cerebral Organization and the Conscious Control of Action", Brain and Conscious Experience, (John C. Eccles, ed.), 422-445, Springer-Verlag, New York (1966).

16. MacKay, D.M. "Conscious Agency with Unsplit and Split Brains", Consciousness and the Physical World, (B.D. Josephson and V.S. Ramachandran, eds.), Pergamon, Oxford, 95-113 (1980).

17. MacKay, D.M. "On the Logical Indeterminacy of a Free Choice", Mind, $\underline{69}$, 31-40 (1960).

18. Popper, K.R. and Eccles, J.C. "The Self and Its Brain", Springer-Verlag, Berlin, Heidelberg, London, New York, (1977).

19. MacKay, D.M. "The Use of Behavioural Language to Refer to Mechanical Processes", Brit. J. Phil. of Sci., 13, 89-103 (1962).

20. MacKay, D.M. "A Mind's Eye View of the Brain", Cybernetics of the Nervous System, (Norbert Wiener and J.P. Schade, eds.), Progress in Brain Research, 17, 321-332, Elsevier (1965).

21. MacKay, D.M. "Ourselves and our Brains: Duality Without Dualism", Psychoneuroendocrinology, 7, 285-294 (1982).

22. MacKay, D.M. "Mind Talk and Brain Talk", Handbook of Cognitive Neuroscience, (M.S. Gazzaniga, ed.), Plenum, New York, 293-317 (1983).

THE BIOLOGICAL FOUNDATIONS OF SELF-CONSCIOUSNESS AND THE PHYSICAL DOMAIN OF EXISTENCE

Humberto R. Maturana

1. PURPOSE

My purpose in this essay is to explain cognition as a biological phenomenon, and to show, in the process, how language arises and gives origin to self-consciousness, revealing the ontological foundations of the physical domain of existence as a limiting cognitive domain. In order to do this I shall start from two unavoidable experiential conditions that are at the same time my problem and my explanatory instruments, namely: a), that cognition, as is apparent in the fact that any alteration of the biology of our nervous system alters our cognitive capacities, is a biological phenomenon that must be explained as such; and b), that we, as is apparent in this very same essay, exist as human beings in language using language for our explanations. These two experiential conditions are my starting point because I must be in them in any explanatory attempt, they are my problem because I chose to explain them, and they are my unavoidable instruments because I must use cognition and language in order to explain cognition and language.

In other words, I propose **not to take** cognition and language as given unexplainable properties, but to take them as phenomena of our human domain of experiences that arise in the praxis of our living, and that as such deserve explanation as biological phenomena. At the same time, it is my purpose to use our condition of existing in language to show how the physical domain of existence arises in language as a cognitive domain. That is, I intend to show that the observer and observing, as biological phenomena, are ontologically primary with respect to the object and the physical domain of existence.

2. THE PROBLEM

I shall take cognition as the fundamental problem, and I shall explain language in the process of explaining cognition.

We human beings assess cognition in any domain by specifying the domain with a question and demanding adequate behaviour or adequate action in that domain. If what we observe as an answer satisfies us as adequate behaviour or as adequate action in the domain specified by the question, we accept it as expression of cognition in that domain, and claim that he or she who answers our query knows. Thus, if someone claims to know algebra, that is, to be an algebrist, we demand of him or her to perform in the domain of what we consider algebra to be, and if according to us he or she performs adequately in that domain, we accept the claim. If the question asked is not answered with what we consider to be adequate behaviour or adequate action in the domain that it specifies, the being asked disintergrates or disappears, it loses its class identity as an entity existing in the operational domain specified by the question, and the questioner proceeds henceforth according to its nonexistence. In these circumstances, since adequate behaviour (or adequate action) is the only criterion that we use and have to assess cognition, I shall take adequate behaviour or adequate action in any domain as this is specified by a question, as the phenomenon to be explained when explaining cognition.

3. NATURE OF THE ANSWER

I am a biologist, and it is from my experience as a biologist that in this essay I am treating the phenomenon of cognition as a biological phenomenon. Further more, since as a biologist I am a scientist, it is as a scientist that I shall provide a biological explanation of the phenomenon of cognition. In order to do this: a), I shall make explicit what I shall consider as adequate in the context of what I consider is a scientific explanation (section 4), so that all the implications of my explanation may be apparent to the reader and he or she may know when it is attained; b), I shall make explicit my epistemological standing with respect to the notion of objectivity (section 5), so that the ontological status of my explanation may be apparent; c), I shall make explicit the notions that I shall use in my explanation by showing how they belong to our daily life (section 6), so that it may be apparent how we are involved as human beings in the explanation that I shall provide; and d), I shall make explicit the nature of the biological phenomena involved in my explanation (section 7) so that our involvement as living systems in the explanation as well as in the phenomenon of cognition itself may be apparent. Finally, in the process of explaining the phenomenon of cognition as a biological phenomenon, I shall show how it is that scientific theories arise as free creations of the human mind, how it is that they explain human experience and not an independent objective world, and how the physical domain of existence arises in the explanation of the praxis of the observer as a feature of the ontology of observing (sections 8-11).

4. THE SCIENTIFIC DOMAIN

We find ourselves as human beings here and now in the praxis of living, in the happening of being human, in language languaging, in an a priori experiental situation in which everything, that is, everything that happens, is and happens in us as part of our praxis of living. In these circumstances, whatever we say about how anything happens takes place in the praxis of our living as a comment, as a reflexion, as a reformulation, in short, as as explanation of the praxis of our living, and as such it does not replace or constitute the praxis of living that it purports to explain. Thus, to say that we are made of matter, or to say that we are ideas in the mind of god, are both explanations of that which we live as our experience of being, yet neither matter nor ideas in the mind of god constitute the experience of being that they are supposed to explain. Explanatons take place operationally in a metadomain with respect to that which they explain. Furthermore, in daily life, in the actual dynamics of human interactions, an explanation is always an answer to a question about the origin of a given phenomenon, and is accepted or rejected by a listener who accepts or rejects it according to whether it satisfies or not a particular implicit or explicit criterion of acceptability that he or she specifies. Due to this, there are as many different kinds of explanations as there are criteria of acceptability of reformulations of the happening of living of the observers that the observers specify. Accordingly, every domain of explanations as it is defined by a particular criterion of acceptability, constitutes a closed cognitive domain as a domain of acceptable statements or actions for the observers that accept that criterion of acceptability. Science, modern science, as a cognitive domain is not an exception to this. Indeed, modern science is that particular cognitive domain which takes what is called the scientific explantion as the criterion of validation(acceptability) of the statements that pertain to it. Let me make this explicit.

4.1 Scientific Explanations

Scientists usually do not reflect upon the constitutive conditions of science. Yet, it is possible to extract, from what modern scientists do, and operational (and hence, experential) specification of what constitutes a scientific explanation as the criterion of validation of what they claim are their scientific statements. Furthermore it is possible to describe this criterion of validation of scientific statements as a reformulation of what is usually called the scientific method.

A. Different domains of human activities entail different intentions. Thus, as the intention of doing art is to generate an aesthetic exp-

erience, and the intention of doing technology is to produce, the intention of doing science is to explain. It is, therefore, in the context of explaining that the criterion of validation of a scientific explanation is the conjoined satisfaction in the praxis of living of an observer of four operational conditions, one of which, the proposition of an ad hoc mechanism that generates the phenomenon explained as a phenomenon to be witnessed by the observer in his or her praxis of living, is the scientific explanation. And, it is in the context of explaining that it must be understood that the scientific explanation is the criterion of validation of scienific statements. Finally, it is also in the context of explaining that it must be recognized that a modern scientific community is a community of observers (henceforth called standard observers) who use the scientific explanation as the criterion of validation of scientific explanations, which are:
a) The specification of the phenomenon to be explained through the stipulation of the operations that a standard observer must perform in his or her praxis of living in order to be a witness of it also in his or her praxis of living.
b) The proposition, in the domain of operational coherences of the praxis of living of a standard observer, of a mechanism, a generative mechanism, which when allowed to operate gives rise, as a consequence of its operation, to the phenomenon explained to be witnessed by the observer also in his or her praxis of living.This generative mechanism, which is usually called the explanatory hypothesis, takes place in the praxis of living of the observer in a different phenomenal domain than the phenomenal domain in which the phenomenon explained is witnessed, and the latter as a consequence of the former stands in an operational metadomain with respect to this. Indeed, the phenomenon explained and its generative mechanism take place in different nonintersecting phenomenal domains in the praxis of living of the observer.
c) The deduction, that is, the computation, in the domain of operational coherences of the praxis of living of the standard observer entailed by the generative mechanism proposed in (b), of other phenomena that the standard observer should be able to witness in his or her domain of experiences as a result of the operation of such operational coherences, and the stipulation of the operations that he or she should perform in order to do so.
d) The actual witnessing, in his or her domain of experiences, of the phenomena deduced in (c) by the standard observer who actually performs in his or her praxis of living the operations stipulated also in (c).

If these four operational conditions are conjointly satisfied in the praxis of living of the standard observer, the generative mechanism proposed in (b) becomes a scientific explanation of the phenomenon brought forth in (a). These four operational conditions in the praxis of living of the observer constitute the criterion of validation of scientific explanation, and science (modern science) is a domain of statements directly or indirectly validated by scientific explanations. Accordingly, it follows from what I say that there are no such things as scientific observations, scientific hypotheses or scientific predictions, and that there are only scientific explanations and scientific statements. It also follows that the standard observer can make scientific statements in any domain of his or her praxis of living in which he or she can make scientific explanations.

B. According to A a scientific statement is valid as a scientific statement only within the community of standard observers that is defined as such because they can realize and accept the scientific explanation as the criterion of validation of their statements. This makes scientific statements consensual statements, and the community of standard observers a scientific community. That in principle any human being can belong to the scientific community is due to two facts of experience: one is that it is as a living human being that an observer can realize and accept the scientific explanation as the criterion of validation of his or her statements and become a standard observer, the other is that the criterion of validation of scientific statements is the operational criterion of validation of actions and statements in daily life, even if it is not used with the same care in order to avoid confusion of phenomenal domains. Indeed, these two experiential facts constitute the fundament for the claim of universality made by scientists for their statements, but what is peculiar in scientists is that they are careful to avoid confusion of phenomenal domains when applying the criterion of validation of scientific statements in the praxis of living.

C. Scientists and philosophers of science usually believe that the operational effectiveness of science and technology reveal and objective independent reality, and that scientific statements reveal the features of an independent universe, of an objective world. Or, in other words, many scientists and philosophers of science believe that without the independent existence of an objective reality, science could not take place. Yet, if one makes, as I have done above, a constitutive, and ontological analysis of the criterion of validation of scientific statements, one can see that scientific explanations do not require the assumption of objectivity because scientific explantions do not explain an independent objective reality. Scientific explanations explain the praxis of living of the observer, and they do so with the operational coherences brought forth by the observer in his or her praxis of living. It is this fact which gives science its biological foundations and which makes science a cognitive domain bound to the biology of the observer with characteristics that are determined by the ontology of observing.

4.2 Science

In conclusion, the operational description of what constitutes a scientific explanation as the criterion of validation of scientific statements, reveals the following characteristics of scientific statements in general, and of scientific statements in particular.

A. Scientific statements are consensual statements valid only within the community of standard observers that generate them, and science as the domain of scienific statements does not need an objective independent reality nor does it reveal one. Therefore, the operational effectiveness of science as a cognitive domain rests only on the operational coherence that takes place in the praxis of living of the standard observers who generate it as a particular domain of consensual coordinations of actions in the praxis of their living together as a scientific community. Science is not a manner of revealing an independent reality, it is a manner of bringing forth a particular one bound to the conditions that constitute the observer as a human being.

B. Since the members of a community of standard observers can generate scientific statements in any phenomenal domain of the praxis of living in which they can apply the criterion of validation of scientific statements, the universality of a particular body of scientific statements within the human domain, will depend on the universality in the human domain of the standard observers that can generate such a body of scientific statements. Finally, scientific statements are valid only as long as the scientific explanations that support them are valid, and these are only valid as long as the four operational conditions that must be conjointly satisfied in their constitution, are satisfied for all the phenomena that are deduced in the praxis of living of the standard observers in the domain of operational coherences specified by the proposed generative mechanism.

C. It is frequently said that scientific explanations are reductionist propositions, aducing that they consist of expressing the phenomena to be explained in more basic terms. This view is inadequate. Scientific explanations are constitutively non-reductionist explanations because they consist of generative propositions and not in expressing the phenomena of one domain in phenomena of another. This is so because in a scientific explanation the phenomenon explained must arise as as result of the operation of the generative mechanism, and cannot be part of it. In fact, if the latter were the case, the explanatory proposition would be constitutively inadequate and would have to be rejected. The phenomenon explained and the phenomena proper to the generative mechanism constitutively pertain to non-intersecting phenomenal domains.

D. The generative mechanism in a scientific explanation is brought forth by a standard observer from his or her domain of experiences in his or her praxis of living as an *ad* *hoc* proposition that in principle requires no justification. Therefore, the components of the generative mechanism, as well as the phenomena proper to their operation, have a foundational character with respect to the phenomenon to be explained and as such their validity is in principle accepted *a priori*. Accordingly, every scientific domain as a domain of scientific statements is founded on basic experiential premises not justified in it, and constitutes in the praxis of living of the standard observer a domain of operational coherences brought forth in the operational coherences entailed in the generative mechanisms of the scientific explanations that validate it.

5. OBJECTIVITY IN PARENTHESIS

If one looks at the two shadows of an object that simultaneously partially intercept the path of two different lights, one white and one red, and if one has trichromatic vision, then one sees that the area of the shadow to the white light that receives red light looks red and that the area of the shadow to the red light that receives white light looks blue-green. This experience is compelling and unavoidable, even if one knows that the area of the shadow to the red light should look white or gray because it receives only white light. If one asks how it is that one sees blue-green where there is white light only, one is told by a reliable authority that the experience of the blue-green shadow is a chromatic illusion because there is no blue-green shadow to justify it as a perception. We live numerous experiences in our daily life that we class like this as illusions or hallucinations and not as perceptions, claiming that they do not constitute the capture of an independent reality because we can disqualify them by resorting to the opinion of a friend whose authority we accept, or by relying on a different sensory experience that we consider as a more acceptable criterion. In the experience itself, however, we cannot distinguish between what we call an illusion, a hallucination or a perception: illusion, hallucination and perception are experientially indistinguishable. It is only through the use of a different experience as a metaexperiential authoritative criterion of distinction, either of the same observer or of somebody else subject to similar restrictions, that such a distinction is socially made. Our incapacity to experientially distinguish what we socially call illusion, hallucination or perception, is constitutive in us as living systems, and not a limitation of our present state of knowledge. The recognition of this should lead us to put a question mark on any perceptual certainty.

5.1 An Invitation

The word perception comes from the latin expression <u>per capire</u> that means through capture, and carries with it the implicit understanding that to perceive is to capture the features of a world independent of the observer. This view assumes objectivity, and, hence, the possibility of knowing a world independent of the observer, as the ontological condition on which the distinction between illusion, hallucination and perception that it entails is based. Therefore, to question the operational validity in the biological domain of the distinction between illusion, hallucination and perception, is to question the ontological validity of the notion of objectivity in the explanation of the phenomenon of cognition. But, how to proceed? Any reflection or comment about how the praxis of living comes about is an explanation, a reformulation of what takes place. If this reformulation does not question the properties of the observer, if it takes for granted cognition and language, then it must assume the independent existence of what is known. If this reformulation does question the properties of the observer, if it asks about how cognition and language arise, then it must accept the experiential indistinguishability between illusion, hallucination and perception, and take as constitutive that existence is dependent on the biology of the observer. Most philosophical traditions pertain to the first case assuming the independent existence of something, such as matter, energy, ideas, god, mind, spirit....or reality. I invite the reader to follow the second and to take seriously the constitutive condition of the biological condition of the observer, following all the consequences that this constitutive condition entails.

5.2 Objectivity In Parenthesis

The assumption of objectivity is not needed for the generation of a scientific explanation. Therefore, in the process of being a scientist explaining cognition as a biological phenomenon I shall proceed without using the notion of objectivity to validate what I say, that is, <u>I shall put objectivity in parenthesis</u>. In other words, I shall go on using an object language because this is the only language that we have (and can have), but although I shall use the experience of being in language as my starting point while I use language to explain cognition and language, <u>I shall not claim</u> that what I say is valid because there is an objective independent reality that validates it. I shall speak as a biologist, and as such I shall use the criterion of validation of scientific statements to validate what I say, accepting that everything that takes place is brought forth by the observer in his or her praxis of living as a primary experiential condition, and that any explanation is secondary.

5.3   The Universum Versus The Multiversa

The assumption of objectivity, objectivity without parenthesis entails the assumption that existence is independent of the observer, that there is an independent domain of existence , the __universum__, that is the ultimate reference for the validation of any explanation. With objectivity without parenthesis, things, entities, exist independently of the observer who distinguishes them, and it is this independent existence of things (entities, ideas) which specifies the truth. Objectivity without parenthesis entails unity, and , in the long run, reductionism, because it entails reality as a single ultimate domain defined by independent existence. He or she who has access to reality is necessarily right in any dispute, and those who do not have such access are necessarily wrong. In the universum coexistence demands obedience to knowledge.

Contrary to all this, objectivity with parenthesis entails accepting that existence is brought forth by the distinctions of the observer, that there are as many domains of truths as domains of existence he or she brings forth in his or her distinctions. At the same time, objectivity in parenthesis entails that different domains of existence constitutively do not intersect because they are brought forth by different kinds of operations of distinction, and therefore, it constitutively negates phenomenal reductionism. Finally, under objectivity in parenthesis each versum of the multiversa is equally valid if not equally pleasant to be part of, and disagreements between observers, when they arise not from trivial logical mistakes within the same versum, but from observers standing in different versa, will have to be solved not by claiming a privileged access to an independent reality, but through the generation of a common versum through coexistence in mutual acceptance. In the multiversa coexistence demands consensus, that is, common knowledge.

6.   BASIC NOTIONS

Everything said is said by an observer to another observer that could be him or herself. Since this condition is my experiential starting point in the praxis of living as well as my problem, I shall make explicit some of the notions that I shall use as my tools for explaining the phenomena of cognition and language, and I shall do so by revealing the actions in the praxis of living that they entail in our daily life when we do science. Indeed, by revealing what we do as observers I am making explicit the ontology of the observer as a constitutive human condition.

6.1 The Observer

An observer is, in general, any being operating in language, or in particular, any human being in the understanding that language defines humanity. In our individual experience as human beings we find ourselves in language, we do not see ourselves growing into it; we are already observers by being in language when we begin as observers to reflect upon language and the condition of being observers. In other words, whatever takes place in the praxis of living of the observer takes place as distinctions as language through languaging, and this is all that he or she can do as such. One of my tasks is to show how the observer arises.

6.2 Unities

The basic operation that an observer performs in the praxis of living is the operation of distinction. In the operation of distinction an observer brings forth a unity (an entity, a whole) as well as the medium in which it is distinguished, and entails in this latter all the operational coherences that make the distinction of the unity possible in his or her praxis of living.

6.3 Simple And Composite Unities

An observer may distinguish in the praxis of living two kinds of unities, simple and composite unities. A simple unity is a unity brought forth in an operation of distinction that constitutes it as a whole by specifying its properties as a collection of dimensions of interactions in the medium in which it is distinguished. Therefore, a simple unity is exclusively and completely characterized by the properties through which it is brought forth in the praxis of living of the observer who distinguishes it, and no further explanation is needed for the origin of these. A simple unity arises defined and characterized by a collection of properties as a matter of distinction in the praxis of living of the observer.

A composite unity is a unity distinguished as a simple unity that through further operations of distinction is decomposed by the observer in components that through their composition would constitute the original simple unity in the domain in which it is distinguished.

A composite unity, therefore, is operationally distinguished as a simple unity in a metadomain with respect to the domain in which its components are distinguished because it results as such from an operation of composition. As a result, the components of a composite unity and its correlated simple unity are in a constitutive relation of mutual specification. Thus, the properties of a composite unity distinguished as a simple one entail the properties of the components that constitute it as such, and conversely, the properties of the components of a composite unity and their manner of composition determine the properties that characterize it as a simple unity when distinguished as such. Accordingly, there is no such a thing as the distinction of a component independently of the untiy that it integrates, nor can a simple unity distinguished as a composite one be decomposed in an arbitrary set of components disposed in an arbitrary manner of composition. Indeed, there is no such thing as a free component floating around independently of the composite unity that it integrates. Therefore, whenever we say that we treat a simple unity as a composite one, and we claim that we do so by distinguishing in it elements that when put together do not regenerate the original unity, we are in fact not decomposing the unity that we believe we are decomposing, but another one, and the elements that we distinguish are not components of the composite unity that we say they compose.

6.4 Organization And Structure

A particular composite unity is characterized by the components and relations between compnents which constitute it as a composite unity that can be distinguished, in a metadomain with respect to its components, as a particular simple unity of a certain kind. As such, a particular composite unity has both organization and structure. These can be characterized as follows:

a) The relations between components in a composite unity that make it a composite unity of a particualr kind, specifying its class identity as a simple unity in metadomain with respect to its components, constitute its organization. In other words, the organization of a composite unity is the configuration of static or dynamic relations between its components that specifies its class identity as a composite that can be distinguished as a simple unity of a particular kind.Therefore, if the organization of a composite unity changes, the composite unity loses its class identity, that is, disintegrates. The organization of a composite unity is necessarily an invariant while it conserves its class identity, and <u>vice versa</u>, the class identity of a composite unity is necessarily an invariant while the composite unity conserves its organization.

b) In a composite unity, be this static or dynamic, the actual components plus the actual relations that take place between them while realizing it as a particular composite unity characterized by a particular organization, constitute its structure. In other words, the structure of a particular composite unity is the manner in which it is actually made by actual static or dynamic components and relations in a particular space, and a particular composite unity conserves its class identity only as long as its structure realizes in it the organization that defines its class identity. Therefore, in any particular composite unity the configuration of relations between components that constitutes its organization must be realized in its sturcture as a subset of all the actual relations that are held between its components as actual entities interacting in the composition.

It follows from all this, that the characterization of the organisation of a composite unity as a configuration of relations between components, says nothing about the characteristics or properties of these other than that they must satisfy the relations of the organization of the composite unity through their interaction in its composition. It also follows that the structure of a composite unity can change without it losing its class identity if the configuration of relations that constitutes its organization is conserved through such structural changes. At the same time, it also follows that if the organization of a composite unity is not conserved through its structural changes, the composite unity loses its class identity, it disintegrates and something else appears in its stead. Therefore, a dynamic composite unity is a composite unity in continuous structural change with conservation of organization.

6.5   Structure Determined Systems

Since the sturcture of a composite unity consists in its components and their relations, any change in a composite unity consists in a structural change, and arises in it at every instant necessarily determined by its structure at that instant through the operation of the properties of its components. Furthermore, the structural changes that a composite unity undergoes as a result of an interaction are also determined by the structure of the composite unity, and this is so because such structural changes take place in the interplay of the properties of the components of the composite unity as they are involved in its composition. Therefore, an external agent that interacts with a composite unity only triggers in it a structural change that it does not determine. Since this is a constitutive condition for composite unities, nothing external to them can specify what happens in them: There are no instructive interaction for composite unities.

Finally, and as a result of this latter condition, the structure of a composite unity also determines with which structural configurations of the medium it may interact. In general then, everything that happens in a composite unity is a structural change, and every structural change occurs in a composite unity determined at every instant by its structure at that instant. This is so both for static and for dynamic composite unities, and the only difference between these is that dynamic composite unities are in a continuous structural change generated as part of their structural constitution in the context of their interactions, while static ones are not. It follows from all this that composite unities are structure determined systems in the sense that everything is determined by their structure. This can be systematically expressed by saying that the structure of a composite unity determines in it at every instant:
a) the domain of all the structural changes that it may undergo with conservation of organization (class identity) and adaptation at that instant; I call this domain the instantaneous domain of the possible changes of state of the composite unity.
b) the domain of all the structural changes that it may undergo with loss of organization and adaptation at that instant; I call this domain the instantaneous domain of the possible disintegrations of the composite unity.
c) the domain of all the different structural configurations of the medium that it admits at that instant in interactions that trigger in it changes of state; I call this domain the instantaneous domain of the possible perturbations of the composite unity.
d) the domain of all the different structural configurations of the medium that it admits at that instant in interactions that trigger in it its disintegration; I call this domain the instantaneous domain of the possible destructive interactions of the composite unity.

These four domains of structural determinism that characterize every structure determined system at every instant are obviously not fixed, and they change as the structure of the structure determined system changes in the flow of its own internal structural dynamics or as a result of its interactions. These general characteristics of structure determined systems have several additional consequences of which I shall mention six. The first is that during the ontogeny of a structure determined system its four domains of structural determinism change following a course contingent to its interactions and its own internal structural dynamics. The second is that some structure determined systems have recurrent domains of structural determinism because they have recurrent structural configurations, while others do not because their structure changes in a non-recurrent way.

The third is that although the structure of a structure determined system determines the structural configurations of the medium with which it may interact, all its interactions arise as coincidences with independent systems that cannot be predicted from it. The fourth is that a composite unity exists only while it moves through the medium in interactions that are perturbations, and that it disintegrates at the first destructive interaction. The fifth is that since the medium cannot specify what happens in a structure determined system because it only triggers the structural changes that occur in this as a result of its interactions, all that can happen to a composite unity in relation to its interaction in the medium, is that the course followed by its structural changes is contingent to the sequence of these interactions. Finally, the sixth is that since mechanistic systems are structure determined systems, and since scientific explanations entail the proposition of mechanistic systems as the systems that generate the phenomena to be explained, in scientific explanations we deal and we can only deal, with structure determined systems.

6.6 Existence

By putting objectivity in parenthesis we accept that constitutively we cannot claim the independent existence of things(entities, unities, ideas, etc.), and we recognize that a unity exists only in its distinction in the praxis of living of the observer that brings it forth. But we also recognize that the distinction takes place in the praxis of living of the observer in an operation that specifies simultaneously the class identity of the unity distinguished, either as a simple unity or as a composite one, and its domain of existence as the domain of the operational coherences in which its distinction makes sense also as a feature of his or her praxis of living. Since the class identity of a composite unity is defined by its organization, and since this can be realized in a composite unity only while this interacts in a domain of perturbations, existence in a composite unity entails the conservation of its organization as well as the conservation of its operational structural correspondence in the domain of operational coherences in which it is distinguished. Similarly, since the class identity of a simple unity is defined by its properties, and since these are defined in relation to the operational domain in which it is distinguished, existence in a simple unity entails the conservetion of the properties that define it and the operational structural correspondences in which these properties are realized.

6.7 Structural Coupling Or Adaptation

I call structural coupling or adaptation the relation of dynamic structural correspondence with the medium in which a unity conserves its class identity (organization in the case of a composite unity, and operation of its properties in the case of a simple one), and which is entailed in its distinction as it is brought forth by the observer in his or her praxis of living. Therefore, conservation of class identity and conservation of adaptation are constitutive conditions of existence for any untiy (entity, system, whole, etc.) in the domain of existence in which it is brought forth by the observer in his or her praxis of living. As constitutive conditions of existence for any unity, conservation of class identity and conservation of adaptation are paired conditions of existence that entail each other so that if one is lost the other is lost, and the unity exists no more. When this happens, a composite unity disintegrates and a simple unity disappears.

6.8 Domain Of Existence

The operation of distinction that brings forth and specifies a unity, also brings forth and specifies its domain of existence as the domain of the operational coherences entailed by the operation of the peoperties through which the unity is characterized in its distinction. In other words, the domain of existence of a simple unity is the domain of operational validity of the properties that define it as such, and the domain of existence of a composite unity is the domain of operational validity of the properties of the components that constitute it. Furthermore, the constitutive operational cohenence of a domain of existence as the domain of operational validity of the properties of the entities that define it, entails all that such validity requires. Accor dingly, a simple unity exists in single domain of existence specified by its properties, and a composite unity exists in two, in the domain of existence specified by its properties as it is distinguished as a simple unity, and in the domain of existence specified by the properties of its components as it is distinguished as a composite one. The entailment in the distinction of a unity of its domain of existence as the domain of all the operational coherences int the praxis of living of the observer in which it conserves class identity and adaptation, is a constitutive condition of existence of every unity. A unity cannot exist outside its domain of existence, and if we imagine a unity outside its domain of existence, the unity that we imagine exists in a different domain than the unity that we claim that we imagine.

6.9   Determinism

To say that a system is deterministic is to say that it operates according to the operational coherences of its domain of existence. And this is so because due to our constitutive inability to experientially distinguish between what we socially call perception and illusion, we cannot make any claim about an objective reality. This we acknowledge by putting objectivity in parenthesis. In other words, to say that a system is deterministic is to say that all its changes are structural changes that arise in it through the operation of the properties of its components in the interactions that these realize in its composition, and not through instructive processes in which an external agent specifies what happens in it. Accordingly, an operation of distinction that brings forth a simple unity brings forth its domain of existnece as the domain of operational applicability of its properties, and constitutes the simple unity and its domain of existence as a deterministic system. At the same time, the operation of distinction that brings forth a composite unity brings forth its domain of existence as a domain of determinism in terms of the operational applicability of the properties that characterize its components, in the praxis of living of the observer. Accordingly, the operation of distinction that brings forth a composite unity brings forth the composite unity as well as its domain of existence and both together, as deterministic systems in the corresponding domains of operational coherences of the praxis of living of the observer.

6.10   Space

The distinction of a unity brings forth its domain of existence as a space of distinctions whose dimensions are specified by the properties of the unities whose distinction entail it as a domain of operational coherences in the praxis of living of the observer. Thus, a simple unity exists and operates in a space specified by its properties, and a composite unity exists and operates in a space specified by its properties as a simple unity if distinguished as such, and in a space specified by the properties of its components if distinguished as a composite one. Accordingly, as a simple unity exists and operates in a single space, a composite unity exists and operates in two. Finally, it follows that without the distinction of a unity there is no space, and that the notion of a unity out of space, as well as the notion of an empty space, are non-sensical. A space is a domain of distinction.

6.11 Interactions

Two simple unities interact when they, as a result of the interplay of their properties, and in a manner determined by such interplay, change their relative position in a common space or domain of distinctions. A composite unity interacts when some of its components as a result of their interactions as simple unities with other simple unities that are not its components, change their manner of composing it, and it undergoes a structural change. It follows that a simple unity interacts in a single space, in the space that its properties define, and that a composite unity interacts in two, in the space defined by its properties as a simple unity, and in the space that its components define through their properties, also as simple unities, as they constitute its structure.

6.12 Phenomenal Domains

A space is constituted in the praxis of living of the observer when he or she performs a distinction. The constitution of a space brings forth a phenomenal domain as the domain of distinctions of the relations and interactions of the unities that observer distinguishes as populating that space. A simple unity operates in a single phenomenal domain, the phenomenal domain constituted through the operation of its properties as a simple unity. A composite unity operates in two phenomenal domains, the phenomenal domain constituted through the operation of its properties as a simple unity, and the phenomenal domain constituted through the operation of the properties of its components, which is where its composition takes place. Furthermore, the two phenomenal domains in which a composite unity operates do not intersect and cannot be reduced one to the other because there is a generative relation between them. The phenomenal domain in which a composite unity operates as a simple unity is secondary to the composition of the composite unity, and constitutes a metaphenomenal domain with respect to the phenomenal domain in which the composition takes place. Due to this a composite unity cannot participate as a simple unity in its own composition.

6.13 Medium, Niche And Environment

I call the medium of a unity the containing background of distinctions, including all that is not involved in its structure if it is a composite one, with respect to which an observer distinguishes it in his or her praxis of living, and in which it realizes its domain of existence. The medium includes both that part of the background that is distinguished by the observer as surrounding the unity, and that part of the background the observer conceives as interacting with it, and which it obscures in its operation in structural coupling (in its domain of existence). I call this latter part of the medium operationally defined moment by moment in its encounter with the medium in structural coupling, the <u>niche</u> of the unity. Accordingly, a unity continuously realizes and specifies its niche by actually operating in its domain of perturbations while conserving adaptation in the medium. As a consequence, the niche of a unity is not a fixed part of the medium in which a unity is distinguished, nor does it exist with independence of the unity that specifies it, and changes as the domain of interactions of the unity changes (if it is a composite one) in its dynamics of structural change (section 6.6 c) . In these circumstances an observer can distinguish the niche of a unity regardless of whether this is a simple or composite one, only by the use of the unity as an indicator of it. Finally, I call the <u>environment</u> of a unity all that an observer distinguishes as surrounding it. In other words, while the niche is that part of the medium that a unity encounters (interacts with) in its operation in structural coupling, and obscures with its presence from the view of the observer, the environment is that part of the medium that an observer sees around a unity. Thus, a dynamic composite unity (like a living system), as it is distinguished in the praxis of living of the observer, is seen by this in an environment as an entity with a changing niche that it specifies while it slides through the medium in continuous structural change with conservation of class identity and adaptation. A composite unity in its medium is like a tight-rope walker who moves on a rope in a gravitational field and conserves his balance (adaptation) while its shape (structure) changes in a manner congruent with the visual and gravitational interactions that it undergoes as it walks (realizing its niche), and falls when this stops being the case.

7. BASIS FOR THE ANSWER: THE LIVING SYSTEM

The answer to the question of cognition requires now that we reflect upon the constitution and operation of living systems, and that we make some additional epistemological and ontological considerations about the conditions that our understanding of them must satisfy.

7.1 Science Deals Only With Structure Determined Systems

To the extent that a scientific explanation entails the proposition of a structure determined system as the mechanism that generates the phenomenon to be explained, we as scientists can only deal with sturcture determined systems, and we cannot handle systems that change in a manner specified by the external agents that impinge upon them. Accordingly, whatever I say about living systems will be said in the understanding that all the phenomena to which they give rise arise through their operation as structure determined systems in a domain of existence brought forth by the observer in their distinction also as   structure determined systems.

7.2   Regulation and Control

As was indicated in section 6.12, the distinction of a composite unity entails the distinction in the praxis of living of the observer of two phenomenal domains that do not intersect because the operation of a composite unity as a simple one is secondary to its compostion. As a result, the whole cannot operate as its own component, and a component cannot operate instead of the whole thatit is a part of. In these circumstances, notions of control or regulation do not connote actual operations in the composition of a composite unity, because this takes place only in the realization in the present of the properties of its components in their actual interactions. Notions of regulation and control only connote relations taking place in a descriptive domain as the observer relates mappings in language of his or her distinctions of a whole and its components in his or her praxis of living.

7.3   Living Systems Are Structure Determined Systems

In order to explain the phenomenon of cognition as a biological phenomenon, I must treat living systems as structure determined systems. I consider that to do so is legitimate for several reasons. I shall mention three. The first is an operational one: we know as a feature of our praxis of living that any structural change in a living system results in a change in its characteristics and properties, and that similar structural changesin different members of the same species, result in similar changes in their characteristics and properties. The second is an epistemological one: if we do not treat living systems as structure determined systems we cannot provide scientific explanations for the phenomena proper to them. The third is an ontological one: the only systems that we can explain scientifically are structure determined systems, therefore, if I provide a scientific explanation of the phenomenon of cognition in living systems, I provide a proof that living systems are structure determined systems in our praxis of living as standard observers, which is where we distinguish them.

7.4   Determinism And Prediction

The fact that a structure determined system.is deterministic does not mean that an observer should be able to predict the course of its structural changes. Determinism and predictability pertain to different operational domains in the praxis of living of the observer. Determinism is a feature that characterizes a system in terms of the operational coherences that constitute it, and its domain of existence, as it is brought forth in the operations of distinction of the observer. Accordingly, there are as many different domains of determinism as domains of different operational coherences the observer brings forth in his or her domain of experiences. Differently to this, a prediction is a computation that an observer makes of the structural changes of a structure determined system as he or she follows the consequences of the operation of the properties of the components of the system in the realization of the domain of determinism that these properties constitute. As such, a prediction can only take place after the observer has completely described the system as a structure determined one in terms of the operational coherences that constitute it in his or her domain of experiences. Therefore, the success or failure of a prediction only reflects the ability or inability of an observer not to confuse phenomenal domains in his or her praxis of living, and to indeed make the computation that constitutes the prediction in the phenomenal domain where he or she claims to make it. In these circumstances, there are two occasions in which an observer, dealing with a structure determined system that does not confuse phenomenal domains, will not be able to predict its structural changes. One occasion is when an observer knows that he or she is dealing with a structure determined by experience in the praxis of living with its components, but cannot encompass it in his or her descriptions, and, thus, cannot effectively treat it as such in its domain of existence and compute its changes of state. The other occasion is when an observer in his or her praxis of living aims at characterizing the present unknown state of a system assumed to be structure determined, by interacting with some of its components. By doing this the observer triggers in the system an unpredictable change of state that he or she then uses to characterize its initial state and predict in it a later one within the domain of determinism specified by the properties of its components. Therefore, since the domain of determinism of a structure determined system as the domain of operational coherences of its components is brought forth in its distinction in the praxis of living of the observer, and since in order to compute a change of state in a system the observer must determine its present state through an interaction with its components, any attempt to compute a change of state in a structure determined system entails a necessary uncertainty due to the manner of determination of its initial state within the constraints of the operational coherences of its domain of existence. This predictive uncertainty may vary in magnitude in different domains of distinctions, but it is always present because it is constitutive of the phenomenon of cognition as a feature of the ontology of observing and not of an objective independent reality. With this I am also saying that the uncertainty principle of physics pertains to the ontology of observing, and that it does not characterize an independent universe because, as I shall show further on, the physical domain of existence is a cognitive domain brought forth in the praxis of living of the observer by the observer as an explanation of his or her praxis of living.

7.5 Ontogenic Structural Drift

It is said that a boat is drifting when it slides floating on the sea without rudder and oars, following a course that is generated moment by moment in its encounter with the waves and wind that impinge upon it, and which lasts as long as it remains floating (conserves adaptation) and keeps the shape of a boat (conserves organization). As such, a drifting boat follows a course without alternatives that is deterministically generated moment by moment in its encounters with the waves and the wind. As a consequence of this, a drifting boat is also always, and at any moment, in the only place where it can be, in a present that is continuously emerging from the sequence of its interactions in the drift. The deterministic process that generates the course followed by a drifting boat takes place as a feature of the structural dynamics of the structure determined system constituted by the boat, the wind, and the waves, as these are brought forth by the observer in his or her praxis of living. Therefore, if an observer cannot predict the course of a drifting boat, it is not because his or her distinction of the boat, the wind, and the waves, in his or her domain of experiences, does not entail a structure determined system in which the course followed by the boat arises in a deterministic manner, but because he or she cannot encompass in his or her description of the interactions between the boat, the wind and the waves, the whole structure of the structure determined system in which the course followed by the boat is a feature of its changes of structure.

What happens with the generation of the course followed by a drifting boat, is the general case for the generation of the course followed by the structural changes of any structure determined system that the observer distinguishes in his or her praxis of living, as it interacts in the medium as if with an independent entity with conservation of class identity (organization) and adaptation (structural coupling). Since living systems are dynamic structure determined systems, this applies to them, and the ontogeny of a living system as its history of structural changes with conservation of organization and adaptation, is its ontogenic structural drift. All that applies to the course followed by a drifting boat, applies to the course followed by the structural changes that take place in the ontogeny of a living system and to the course followed by the displacement of a living system in the medium during its ontogeny. Let me make this clear. In general terms, a drift is the course followed by the structural changes in a structure determined system that arises moment after moment generated in the interactions of the system with another independent system, while its relation of correspondence (adaptation) with this other system (medium) and its organization (class identity) remain invariant. According to this the individual life history of a living system as a history of continuous structural changes which follows a course generated moment by moment in the braiding of internally generated structural dynamics with the structural changes triggered in it by its recurrent interactions with the medium as an independent entity, and which lasts as long as its organization and adaptation are conserved, takes place as a structural drift. Similarly, since the course of the displacement of a living system in the medium is generated moment by moment as a result of its interactions with the medium as an independent entity while its organization and adaptation are conserved, the displacement of a living system in the medium while it realizes its niche takes place as a drift. Living systems exist in continuous structural and positional drift (ontogenic drift) while they are alive as a matter of constitution.

As is the case with a drifting boat, at any moment a living system is where it is in the medium, and has the structure that it has, in the present of its ontogenic drift in a deterministic manner, and could not be anywhere else than where it is, nor could it have a structure different from the one that it has. The many different paths that an observer may consider possible for a drifting boat to follow at any instant, or the many different ontogenic courses that an observer may consider for a living system at any moment, are possible only as imagined alternatives in the description of what would happen in each case if the conditions were different, and not actual alternatives in the course of the boat or in the ontogeny of the living system. A drift is a process of change, and as is the case with all processes of change in structure determined systems, courses without alternatives in the domain of determinism in which it is brought forth by the distinctions of the observer. Indeed, such imagined alternatives are imaginable only from the perspective of the inability of the observer to treat the boat, the wind and the waves, or the living system and the medium, that he or she brings forth in his or her praxis of living, as structure determined systems whose changes of structure he or she computes. If we are serious about our explanations as scientists, then we must accept as an ontological feature of what we do as observers that every entity that we bring forth in our distinctions is where it is, and has the structure that it has, in the only manner that it can be given in the domain of operational coherences (domain of determinism) that we also bring forth as its domain of existence in its distinction.

Finally, let me mention several implications of all this for the entities that we bring forth as living systems in our praxis of living: a) Since for a living system a history of interactions without disintegration can only be a history of perturbations, that is, a history of interactions in the niche, a living system while living necessarily slides in ontogenic drift through the medium in the realization of its niche. This means that aim, goal, purpose or intention do not enter in the realization of a living system as a structure determined system.
b) Since the structure of a living system is continually changing, both through its internal dynamics and through the structural changes triggered in its interactions with operationally independent entities, the niche of a living system (the features of the medium that it actually encounters in its interactions) is necessarily in continuous change congruent with the continuous structural drift of the living system while this remains alive. Furthermore, this is so regardless of whether the observer considers that the environment of the living system changes or remains constant. This means that as an observer brings forth a living system in his or her praxis of living, this may appear to him or her as continuously changing in its use of a constant environment, or, conversely, as unchanging in a continuously changing environment, because the observer cannot see the encounter of a living system and its niche, which is where conservation of adaptation takes place.
c) Conservation of adaptation does not mean that the manner of living of a living system remains invariant. It means that a living system has an ontology only while it conserves its class identity and its dynamic structural correspondence with the medium as it undergoes its interactions, and that there is no constitutive restriction about the magnitude of its moment by moment structural changes other than that they should take place within the constraints of its structural determinism and its conservation of organization and adaptation. Indeed, I could speak of the laws of conservation of organization and adaptation as ontological conditions for the existence of any structure determined system in the same manner as physicists speak of the laws of conservation in physics as ontological conditions for the occurence of physical phenomena.

Every living system, including us observers, is at any moment where it is, has the structure that it has, and does what it does at that moment, always in a structural and relational situation that is the present of an ontogenic drift that starts at its inception as such in a particular place with a particular structure, and follows the only course that it can follow. Different kinds of living systems differ in the spectrum of ontogenies that an observer can consider possible for each of them in his or her discourse as a result of their different initial structures and different starting places, but each ontogeny that takes place, takes place as a unique ontogenic drift in a process without alternatives.

7.6 Structural Intersection

When an observer brings forth a composite unity in his or her praxis of living, he or she brings forth an entity in which the configuration of relations between components that constitutes its organization, is a subset of all the actual relations that take place between its components as these realize its structure and constitute it as a whole in the domain of existence in which they are brought forth (see section 6.4). As such, the organization of a composite unity does not exhaust the relations and interactions in which the components that realize it may paricipate in their domain of existence. The result of this is that in the structural realization of a composite unity, its components may participate, through other properties than those that involve them in the realization of its organization, in the realization of the organization of many other composite unities which thus, intersect structurally with it. Furthermore, when the components of a composite unity are themselves composite unities, this may participate in structural intersections that take place through the components of its components. In any case, when an observer distinguishes two or more structurally intersecting systems, he or she distinguishes two or more different composite unities realized through the same body.

Structurally intersecting systems exist and operate as simple unities in different phenomenal domains specified by their different organizations. Yet, depending on how their structural intersection takes place, structurally intersecting composite inities may exist as such in the same or different domains of existence. Thus, when two composite unities structurally intersect through their components, they share components and have as composite unities the same domain of existence. But, when two composite unities structurally intersect through the components of the components, of one or both, they do not share components and as composite unities have different domains of existence. Nevertheless, since in a structural intersection there are components or components of components, or both, that simultaneously participate in the structure of several systems, structural changes that take place in one of several structurally intersecting systems as part of its ontogenic drift, may give rise to structural changes in the other intersecting systems, and thus participate in their otherwise independent ontogenic drifts. In other words, structurally intersecting systems are structurally interdependent because, either through the intersection of their domains of structural determinism, or

through the intersection of the domains of structural determinism of
their components, or through both, they affect each other's structure
in the course of their independently generated structural changes, and
although they may exist as composite unities in different domains their
ontogenic drifts intersect forming a network of co-ontogenic drifts.
Thus, an observer may distinguish in the structural realization of a
human being as a living system the simultaneous or successive intersection of a mammal, a person, a woman, a doctor, and a mother, all
of which are different composite unities defined by different organizations that are simultaneously or successively conserved while they
are realized in their different domains of existence, with particular
characteristics that result from the continuous braiding of their
different ontogenic drifts through the continuous interplay of their
structural changes. Furthermore, these structural intersections result
in dependent domains of disintegrations as well as dependent domains
of conservations which need not be reciprocal, when the conservation
of one class identity entails the conservation of structural features
that are involved in the conservation of another. For example, in the
structural intersection of a student and a human being in a living
system, the conservation of the class identity student entails the
conservation of the class identity human being, but not the reverse:
the disintegration of the student does not entail the disintegration
of the human being, but the disintegration of the human being carries
with it the disintegration of the student. Also, a particular composite unity may disintegrate through different kinds of structural changes, like disintegrating as a student through failing an examination
or through attaining the final degree, with different consequences
in the network of structural intersections to which it belongs.

The structural intersection of systems does not mean that the
same system is viewed in different manners from different perspectives
because due to their different organizations structural intersecting
systems exist in different phenomenal domains and are realized through
different structural dynamics. It only means that the elements that
realize a particular composite unity as its components through some
of their properties as simple unities, participate through other of
their properites as simple unities as components of other unities that
exist as legitimately different ones because they have different domains of disintegrations. The interactions and relations in which the
components of a system participate through dimensions other than those
through which they constitute it, I call orthogonal interactions and
relations, and it is through these that structurally intersecting systems may exist in non-intersecting phenomenal domains and yet have unidirectional or reciprocal relations of structural dependency. Finally,
it is also through the orthogonal interaction of their components that

structurally independent systems that exist in non-intersecting phenomenal domains may also have co-ontogenic drifts.

7.7 The Living System

In 1970 I proposed that living systems were dynamic systems constituted as autonomous unities through being closed circular concatenations (closed networks) of molecular productions in which the different kinds of molecules that composed them participated in the production of each other, and in which everything could change in the manner in which they were realized except the closed circularity that constituted them as unities (see Maturana 1970, in Maturana and Varela 1980). In 1973 Francesco Varela and I expanded this characterization of living systems by saying: first, that a composite unity whose organization can be described as a closed network of productions of components that through their interactions constitute the network of productions that produce them, and specify its extension by constituting its boundaries in their domain of existence, is an autopoietic system; and second, that a living system is an autopoietic system whose components are molecules. Or, in other words, we propose that living systems are molecular autopoietic systems, and that as such they exist in molecular space as closed networks of molecular productions that specify their own limits (see Maturana and Varela 1973, in Maturana and Varela 1980 and Maturana 1975). Nothing is said in this description of the molecular constitution of living systems as autopoietic systems about thermo-dynamic constraints. This is so because the realization of living systems as molecular systems entails the satisfaction of such constraints. In fact, the statement, a composite unity exists as such in the domain of existence of its components, implies the satisfaction of the conditions of the existence of these.

The recognition that living systems are molecular autopoietic systems carries with it several implications and consequences of which I shall mention a few:

7.7.1 <u>Implications</u>: a) Living systems as autopoietic systems are structure deterrminded systems, and everything that applies to these applies to them. In particular this means that everythingthat occurs in a living system takes place in it in the actual operation of the properties of its components through relations of neighbourhood (relations of contiguity) constituted in these very same operations. According to this, notions of regulation and control do not and cannot reflect actual operations in the structural realization of a living system because they do not connote actual relations of neighbourhood in it. These notions only reveal relations that the observer establishes when he or she compares different moments in the course of transformations of the network of processes that take place in the structural realization of a particular living system. Therefore, the only peculiar thing about living systems as structure determined systems is that they are molecular autopoietic systems. b) Autopoiesis is a dynamic process that takes place in the ongoing flow of its occurrence, and cannot be grasped in a static instantaneous view of distribution of components. Due to this, a living system exists only through the continuous structural transformation entailed in its autopoiesis, and only while this is conserved in the constitution of its ontogeny. This has two basic results; one is that living systems can be realized through many different changing dynamic structures, the other is that in the generation of lineages through reproduction, living systems are constitutively open to continuous phylogenic structural change. c) A living system either exists as a dynamic structure determined system in structural coupling in the medium in which it is brought forth by the observer, that is, in a relation of conservation of adaptation through its continual structural change in the realization of its niche, or it does not exist. Or, in other words, a living system while living is necessarily in a dynamic relation of correspondence with the medium through its operation in its domain of existence, and to live is to glide through a domain of perturbations in an ontogenic drift that takes place through the realization of an ever-changing niche. d) A living system as a structure determined system operates only in the present of the structural realization of its autopoiesis in molecular space, and as such it is necessarily open to the flow of molecules through it. At the same time, a living system as an autopoietic system operates generating only states in autopoiesis, otherwise it disintegrates; due to this, living systems are closed systems with repect to their dynamics of states.

7.7.2 <u>Consequences</u>: a) To the extent that a living system is a structure determined system, and everything in it takes place through neighbourhood relations between its components in the present of the operation of their properties, notions of purpose and goal which imply that at every instant a later state of a system as a whole operates as part of its structure in the present, do not apply to living systems and cannot be used to characterize their operation. A living system may appear to operate as a purposeful or goal-directed system only to an observer who, having seen the ontogeny of other living systems of the samekind in the same circumstances in his or her praxis of living, confuses phenomenal domains by putting the consequences of its operation as a whole among the processes that constitute it. b) Due to the condition of structure determined systems for living systems there is no inside or outside in their operation as autopoietic unities; they are in autopoiesis as closed wholes in their dynamics of states or they disintegrate. At the same time, and due to this same reason, living systems do not use or misuse an environment in their operation as autopoitic unities, nor do they commit mistakes in their ontogenic drifts. In fact, a living system in its operation in a medium with conservation of organization and adaptation as befit it as a structure determined system, brings forth its ever changing niche as it realizes itself in its domain of existence as a background of operational coherences which it does not distinguish and with which it does not interact. c) Living systems necessarily form, through their recurrent interactions with each other as well as with the non-biotic medium, co-ontogenic and co-phylogenic systems of braided structural drifts that last as long as they conserve their autopoiesis through the conservation of their reciprocal structural couplings. Such is biological evolution. As a result, every living system, including us human beings as observers, is always found in its spontaneous realization in its domain of existence in congruence with a biotic and a non-biotic medium. Or, in other words, every living system is, at every instant, as it is and where it is, a node of a network of co-ontogenic drifts that necessarily involves all the entities with which it interacts in the domain in which it is brought forth by the observer in his or her praxis of living. As a consequence, an observer as a living system can only distinguish an entity as a node of the network of co-ontogenic drifts to which it belongs, and where it exists in structural coupling. d) The only thing peculiar to living systems is that they are autopoietic systems in molecular space. In these circumstances, a given phenomenon is a biological phenomenon only to the extent that its realization entails the realization of the autopoiesis of at least one autopoietic system in the molecular space.e) Modern procariotic and eucariotic cells are typical autopoietic systems in the molecular space, and because their autopoiesis is not the result of their being composed by more basic autopoietic subsystems, I call them first order autopoietic systems. I call second order autopoietic systems, systems whose autopoiesis is the result of their being composed of more basic autopoietic unities; organisms as multicellular systems are such. Yet, organisms may also "be", and I think that most of them actually are, first order autopoietic systems as closed networks of molecular productions that involve intercellular processes as much as intracellular ones. Accordingly, an organism would exist as such in the structural intersection of a first order autopoietic system with a second order one, both realized through the autopoiesis of the cells that compose the latter. The same happened originally with the eucariotic cell as this arose through the endosymbiosis of procariotic ones (Margoulis).

f) An organism as a second order autopoietic system is an ectocellular symbiont composed of cells, usually of common origin but not always so, that constitute it through their co-ontogenic drift. An organism as a first order autopoietic system, however, is not composed of cells even through its realization depends on the realization of the autopoiesis of the cells that intersect structurally with it as they constitute it in their co-ontogenic drift. The first and second order autopoietic systems that intersect structurally in the realization of an organism, exist in different non-intersecting phenomenal domains.

7.8 Phylogenic Structural Drift

Reproduction is a process in which a system gives origin through its fracture to systems characterized by the same organization (class identity) that characterized the original one, but with structures that vary with respect to it (Maturana 1980). A reproductive phylogeny or lineage, then, is a succession of systems generated through sequential reproductions that conserve a particular organization. Accordingly, each particular reproductive lineage or phylogeny is defined by the particular organization conserved through the sequential reproductions that constitute it. Therefore, a reproductive phylogeny or lineage lasts only as long as the organization that defines it is conserved, regardless of how much the structure that realizes this organization in each successive member of the lineage changes with each reproductive step (see Maturana 1980 and Maturana and Varela 1984). It follows that a reproductuve phylogeny or lineage as a succession of ontogenic drifts, constitutively occurs as a drift of the structures that realize the organization conserved along it. It also follows that each of the reproductive steps that constitute a reproductive phylogeny is the occasion that opens the possibility for a change, large or small, in the course of its structural drift. As such, a reproductive phylogeny or lineage comes to and end through the structural changes of its members. And this occurs either because autopoiesis is lost after the last of them, or because through the conservation of autopoiesis in the offspring of these, a particular set of relations of the drifting structure begins to be conserved through the following sequential reproductions as the organization that defines and starts a new lineage. This has several general implications of which I shall mention only a few:

a) A memberof a reproductive phylogeny either stays in structural coupling (conserves adaptation) in its domain of existence until its reproduction, and the phylogeny continues, or it disintegrates before then and the phylogeny ends with it.

b) A living system is a member of the reproductive phylogeny in which it arises only if it conserves through its ontogeny the organization that defines the phylogeny, and continues the phylogeny only if such organization is conserved through its reproduction.

c) Many different reproductive phylogenies can be conserved operationally embedded in each other, forming a system of nested phylogenies, if there is an intersection of the structural realization of the different organizations that define them. When this happens there is always a fundamental reproductive phylogeny whose realization is necessary for the realization of all the others. This has occurred in the evolution of living systems in the form of the phylogenic drift of a system of branching nested reproductive phylogenies in which the fundamental reproductive phylogeny is that in which autopoiesis is conserved (see Maturana 1980 and Maturana and Varela 1984). Thus, the system of branching phylogenies defined by the conservation of autopoiesis through reproductive cells in eucariotic organisms, has carried embedded in it, through the structural intersection of their realizations, many staggered nested organizations that characterize the coincident lineages conserved through it. This we recognize in the many nested taxonomic categories that we distinguish in any organism when we classify it. For example, a human being is a vertebrate, a mammal, a primate, an Homo, an Homo sapiens, all different categories corresponding to different systems of partially overlapping phylogenies that are conserved together through the conservation of its autopoiesis.

d) The ontogenic drifts of the members of a reproductive phylogeny take place in reciprocal structural coupling with many different, and also continuously changing, living and non-living systems that form part of the medium in which they realize their niches. As a result, every individual ontogeny in living systems courses embedded in a system of co-ontogenies that constitute a network of co-phylogenic structural drifts. This can be generalized by saying that evolution is constitutively a co-evolution, and that every living system is at any moment where it is, and has the structure that it has, as an expression of the present of the domain of operational coherences constituted by the network of co-phylogenic structural drifts to which it belongs. As a result, the operational coherences of every living system in its present necessarily entail the operational coherences of the whole biosphere.

e) The observer as a living system is not an exception to all that has been said above. Due to this, an observer can only make distinctions that, as operations in his or her praxis of living, take place as operations in the present of the domain of operational coherences constituted by the network of co-ontogenic and co-phylogenic structural drifts to which he or she belongs.

7.9 Ontogenic Possibilities

The ontogeny of every structure determined system starts with an initial structure that is the structure that realizes the system at the beginning of its existence in its inception. In living systems such initial structure is a cellular unity that may either originate as a single cell or as a small multicellular entity through a reproductive fracture from a cellular maternal system whose organization it conserves, or as a single cell de novo from non-cellular elements. In every living system its initial structure constitutes the structural starting point that specifies in it what an observer sees as the configuration of all the courses of ontogenic drifts that it may undergo under different circumstances of interactions in the medium. As a result, what constitutes a lineage in living systems, is the conservation through their reproduction of a particular initial structure that specifies a particular configuration of possible ontogenic drifts, and what constitutes the organization conserved through reproduction which specifies the identity of the lineage, is that configuration. Accordingly, a lineage comes to and end when the configuration of possible ontogenic drifts that defines it stops being conserved. I call the configuration of possible ontogenic drifts that specifies a lineage through its conservation, the ontogenic phenotype of the lineage. In each particular living system, however, only one of the ontogenic courses thought as possible in the ontogenic phenotype by the observer , is realized as a result of its internal dynamics under the contingencies of the particular perturbations that it undergoes in its domain of existence with conservation of organization and adaptation. Consequently, and in general, different composite unities may have different or similar ontogenic structural drifts under different or similar histories of perturbations in their domains of existence, only within the domain of possibilities set by their different or similar initial structures. Indeed, nothing can happen in the ontogeny of a living system as a composite unity that is not permitted in its initial structure. Or, in other words, and with the understanding that the initial structure of a living system is its genetic constitution, it is apparent that nothing can happen in the ontogenic structural drift of a living system that is not allowed in its genetic constitution as a feature of its possible ontogenies. At the same time, with this understanding, it is also apparent that nothing is determined in the initial structure or genetic constitution of a living system, because for anything to occur in a living system, this must undergo an actual ontogenic structural drift as an actual epigenic structural transformation that takes place in an actual history of interactions in the realization of a domain of existence. This is so even in the case of those particular ontogenic features of characters that we call genetically determined because they can be expected to appear in all the ontogenic drifts that a living system can possibly undergo up to the moment of its observation, because such a feature or character will appear only if there is an actual ontogeny. In these circumstances, a biological system of lineages, or system of phylogenies, is defined by the ontogenic phenotype conserved in the living systems that constitute it through their sequential reproductions. As a result, all the members of a system of lineages resemble each other through the ontogenic phenotype that defines the system of lineages, and not through a common genetic constitution maintained by means of a genetic flow.

7.10 Selection

An observer may claim that the actual ontogenic course followed by the structural changes of a living system is, moment by moment, selected by the medium from the many other ontogenic courses that he or she considers available to it at every instant along its life history. Yet, strictly, selection does not take place in the life history of a living system. The life history of a living system is the particular course followed by its ontogenic drift under the contingencies of a particular sequence of interactions. As such, a life history is deterministically generated instant by instant as the structure of the living system changes through its own structurally determined dynamics in its continuous encounter with the medium as an independent entity, and lasts while the living system lasts. Each ontogeny, therefore, is uniquely generated as it takes place as a process that courses without actual alternatives or decision points along it. The different ontogenic courses that an observer may describe as possible for a living system, are alternative ontogenic courses only for him or heras he or she imagines the living system in different circumstances in the attempt to compute through treating the living system and the medium as a known structure determined system. The same is valid for the phylogenic structural drift, or for the historic genêtic change in a population. What an observer in fact does when speaking of selection in relation to living systems then, is to refer to a discrepancy between an expected and an actual historical outcome, and does so by comparing the actual with the imagined in the phylogenic and the ontogenic structural drifts of living systems. Selection is not the mechanism that generates phylogenic structural change and adaptation. In fact, ontogenic and phylogenic structural changes and adaptation need not be explained, they are constitutive features of the condition of existence of living systems. All that has to be explained is the course followed by the continuous structural change that takes place in living systems, both in ontogeny and phylogeny, and this is explained by the mechanism of structural drift.

8. THE ANSWER

It follows from all that I have said about living systems that they exist only in conservation of organization and conservation of adaptation as constitutive conditions of their existence, and that this applies to the observer as a living system as well. It also follows that the present of any living system, the observer included, or, in general terms, the present of any system or entity distinguished is always that of a node in an ongoing network of co-phylogenic and co-ontogenic structural drifts. At the same time it also follows that as long as it is distinguished, any system is distinguished in conservation of organization and adaptation in its domain of existence, and that a domain of existence is a domain of structural coupling that entails all the operational coherences that make possible the system that specifies it. Or, in other words, from all that I have said so far it follows: first, that every entity distinguished is distinguished in operational correspondence with its domain of existence, and, therefore, that each living system distinguished is necessarily distinguished in adequate action in its domain of structural coupling; second, that an observer can only operate in adequate action in his or her domain of existence, and that as such he or she does so as an expression of his or her conservation of organization and adaptation in it; and third, that an observer can only distinguish that which he or she distinguishes, and that he or she does so is an expression of the operational coherences of the domain of praxis of living in which he or she makes the distinction. Let us now consider the question of cognition with all this in mind.

8.1   Cognition

Since the only criterion that we have to assess cognition is to assess adequate action in a domain that we specify with a question, I proposed, in section 2 of this article, that my task in explaining cognition as a biological phenomenon was to show how adequate action arises in any domain during the operation of a living system. This I have done through the previous sections by showing that a living system is necessarily always in adequate action in the domain in which it is distinguished as such in the praxis of living of the observer. And I have shown that this is so because it is constitutive of the phenomenon of observing, that any system distinguished should be distinguished both in conservation of organization and structural coupling, and as a node in a network of structural drifts. In the distinction of living systems this consists in bringing them forth in the praxis of living of the observer, both in conservation of autopoiesis and adaptat-

ion, and as a moment in their ontogenic drift in a medium, under conditions that constitute them in adequate action in their domains of existence. In other words, I have shown that for any particular circumstance of distinction of a living system, conservation of living (conservation of autopoiesis and of adaptation) constitutes adequate action in those circumatances, and, hence, knowledge: living systems are cognitive systems, and to live is to know. But, by showing this I have also shown that any interaction with a living system can be viewed by an observer as a question posed to it, as a challenge to its life that constitutes a domain of existence where he or she expects adequate action of it. And, at the same time, I have also shown, then, that the actual acceptance by the observer of an answer to a question posed to a living system, entails his or her recognition of adequate action by the living system in the domain specified by the question, and that it consists in the distinction of the latter in that domain under conditions of conservation and adaptation. In what follows, I present this general explanatory proposition under the guise of a particular scientific explanation:
a) The phenomenon to be explained is adequate action by a living system at any moment in which an observer distinguishes it as a living system in action in a particular domain. And I propose this in the understanding that the adequate actions of a living system are its interactions with conservation of class identity in the domain in which it is distinguished.
b) Given that structural coupling in its domain of existence (conservation of adaptation) is a condition of existence for any system distinguished by an observer, the general mechanism for adequate action in a living system as a structurally changing system, is the structural drift with conservation of adaptation through which it stays in continuous adequate action while it realizes its niche, or disintegrates. Since a system is distinguished only in structural coupling, when an observer distinguishes a living system he or she necessarily distinguishes it in adequate action in the domain of its distinction, and distinguishes it as a system that constitutively remains in structural coupling in its domain of existence regardless of how much its structure, or the structure of the medium, or both, change while it stays alive.
c) Given the generative mechanism proposed in (b), the following phenomena can be deduced to take place in the domain of experiences of an observer: i) the observer should see adequate action taking place in the form of co-ordinated behaviour in living systems that are in co-ontogenic structural drift while in recurrent interactions with conservation of reciprocal adaptation; ii) the observer should see that living systems in co-ontogeny should separate or disintegrate, or both

when their reciprocal adaptation is lost.

d) <u>The phenomena deduced in</u> (c) are apparent in the domain of experiences of an observer in the dynamics of constitution and realization of a social system, and in all circumstances of recurrent interactions between living systems during their ontogenies, in what appears to us as learning to live together. One of these cases is our human operation in language.

The satisfaction of these four conditions results: a) in the validation, as a scientific explanation, of my proposition that cognition as adequate action in living systems is a consequence of their structural drift with conservation of organization and adaptation; b) in showing that adequate action (cognition) is constitutive to living systems because it is entailed in their existence as such; c) in that different living systems differ in their domains of adequate actions (domains of cognition) to the extent that they realize different niches; and d) in that the domain of adequate actions (domain of cognition) of a living system changes as its structure, or the structure of the medium, or both, change while it conserves organization and adaptation.

At the same time it is apparent from all this, that what I say of cognition as an explanation of the praxis of living, takes place in the praxis of living, and that to the extent that what I say is effective action in the generation of the phenomenon of cognition, what I say takes place as cognition. If what I say sounds strange, it is only because we are in the habit of thinking about cognition with objectivity without parenthesis, as if the phenomenon connoted by the word cognition entailed pointing to something whose existence can be asserted to be independent of the pointing of the observer. I have shown that this is not and cannot be the case. Cognition cannot be understood as a biological phenomenon if objectivity is not put in parenthesis, nor can it be understood as such if one is not willing to follow all the consequences of such an episthemological act.

Let us now consider human operation in language as one of the phenomena which takes place as a consequence of the operation of cognition as adequate(or effective)action. This is particularly necessary because our operation in language as we are observers in the praxis of living is, at the same time, our problem and our instrument for analysis and explanation.

8.2 Language

We human beings are living systems that exist in language. This means that although we exist as human beings in language and our cognitive domains (domains of adequate actions) as such take place in the domain of languaging, our languaging takes place through our operation as living systems. Accordingly, in what follows I shall consider what takes place in language as this arises as a biological phenomenon from the operation of living systems in recurrent interactions with conservation of organization and adaptation through their co-ontogenic structural drift, and thus show it as a consequence of the same mechanism that explains the phenomenon of cognition:

a) When two or more autopoietic systems interact recurrently, and in each of them its dynamic structure follows a course of change contingent upon the history of its interactions with the others, there is a co-ontogenic structural drift that gives rise to an ontogenically established domain of recurrent interactions between them which appears to an observer as a domain of consensual co-ordinations of actions or distinctions in an environment. I call this ontogenically established domain of recurrent interactions, a domain of consensual co-ordinations, or, more generally, a consensual domain of interactions, because it arises as a particular manner of living together contingent upon the unique history of recurrent interactions of the participants during their co-ontogeny. Furthermore, because an observer can describe such a domain of recurrent interactions in semantic terms, by referring the different co-ordinations of actions (or distinctions) involved, to the different consequences that they have in the domain in which they are distinguished, I also call a consensual domain of interactions a linguistic domain. Finally, I call the behaviour through which an organism participates in an ontogenic domain of recurrent interactions, consensual or linguistic according to whether I want to emphasize the ontogenic origin of the behaviour (consensual), or its implications in the present of the ongoing interactions (linguistic). Similarly, I speak of co-ordinations of actions or co-ordinations of distinctions, according to whether I want to emphasize what takes place in the interaction in relation to the participants (co-ordinations of actions), or what takes place in the interactions in relation to an environment (co-ordinations of distinctions).

b) When one or more living system continues its co-ontogenic structural drift through its recurrent interactions in a consensual domain, it is possible that a recursion takes place in its consensual behaviour with the production of a consensual co-ordination of consensual co-ordinations of actions. What an observer sees when this happens is that the participants of a consensual domain of interactions operate in their consensual behaviour making consensual distinctions of their consensual distinctions, in a process that recursively makes a consensual action a consensual token for a consensual distinction that it

obscures. Indeed, this is indistinguishable from what takes place in
our languaging in the praxis of living. Accordingly, I claim that the
phenomenon of language takes place in the co-ontogeny of living systems when two or more organisms operate, through their recurrent ontogenic consensual interactions, in an ongoing process of recursive consensual co-ordinations of consensual co-ordinations of actions or distinctions (Maturana 1978). Or, in other words, I claim that such recursive consensual co-ordination of consensual co-ordinations of actions
or distinctions in any domain, is the phenomenon of language. Furthermore, I claim that objects arise in language as consensual co-ordinations of actions that operationally obscure for further recursive consensual co-ordinations of actions by the observers the consensual co-ordinations of actions (distinctions) that they co-ordinate. Objects
are in the process of languaging consensual co-ordinations of actions
that operate as tokens for the consensual co-ordinations of actions
that they co-ordinate. Objects do not pre-exist language. Finally, I
claim that all the phenomena that we as observers distinguish in our
operation in language arise in the living of living systems, through
their co-ontogenic structural drift when this results in an ongoing
process of consensual co-ordinations of actions, as a consequence of
the proposed mechanism for the generation of the phenomenon of cognition.

c) Languaging takes place in the praxis of living: we human beings find
ourselves as living systems immersed in it. In the explanation of language as a biological phenomenon it becomes apparent that languaging
arises, when it arises, as a manner of coexistence of living systems.
As such, languaging takes place as a consequence of a co-ontogenic
structural drift under recurrent consensual interactions. For this reason, language takes place as a system of recurrent interactions in a
domain of structural coupling. Interactions in language do not take
place in a domain of abstraction, on the contrary, they take place in
the concreteness of the "bodyhood" of the participants. Interactions
in language are structural interactions. Notions such as transmission
of information, symbolization, denotation, meaning or syntax are secondary to the constitution of the phenomenon of languaging in the living of the living systems that live it. Such notions arise as reflexions in language upon what takes place in languaging. It is for this
reason that what takes place in language has consequences in our
"bodyhood" and the descriptions and explanations that we make become
part of our domain of existence. We undergo our ontogenic and phylogenic drifts as human beings in structural coupling in our domain of existence as languaging systems. Languaging pertains to the praxis of
living of observing, and generates the praxis of living of the observer.

9. CONSEQUENCES

The answer given for the phenomenon of cognition has several fundamental consequences that I shall now consider:

9.1 Existence Entails Cognition

To the extent that cognition is the operation of a living system in its domain of structural coupling, that is, in its domain of existence, existence of living systems entails cognition as their realization as such, not as a characterization or as a representation or as a disclosure of something independent of them. Cognition as a biological phenomenon takes place in a living system as it operates in its domain of perturbations, and as such it has no content and is not about anything. Therefore, when we say that we know some-thing we are not connoting what happens in the mechanism of the phenomenon of cognition as a biological phenomenon, we are reflecting in language upon what we do.

9.2 There Are As Many Cognitive Domains As There Are Domains Of Existence

I speak of cognition only in relation to living systems. This is arbitrary since what I have said in relation to existence applies to every entity brought forth through an operation of distinction. Therefore, I make this distinction only because I am speaking of living systems and the word cognition is historically bound to them through us. Within this restriction we as observers can say that there are as many domains of cognition as there are domains of existence specified by the different identities that living systems conserve through the realization of their autopoiesis. These different cognitive domains intersect in the structural realization of a living system as this realizes the different identities that define them as different dimensions of simultaneous or successive structural couplings, orthogonal to the fundamental one in which the living system realizes its autopoiesis. As a result, these different cognitive domains may appear or disappear simultaneously or independently according to whether the different structurally intersecting unities that specify them integrate or disintegrate independently or simultaneously (see section 7.6). Thus, when a student graduates, the cognitive domain specified by the operation in the domain of structural coupling that defines the identity student, disappears together with the disintegration of the student, or, when a bachelor marries, the cognitive domain that the identity bachelor defines as a domain of operational coherences in structural coupling, disappears together with the disintegration of the bachelor. Conversely, when a student graduates and a bachelor marries, the identities graduate and husband appear with the corresponding cognitive domains specified by the operational coherences that they entail.

It follows from all this that a living system may operate in as many different cognitive domains as different identities the different dimensions of its structural coupling allow it to realize. It also follows from all this that the different identities that a living system may realize are necessarily fluid, and change as the dimensions of its structural coupling change with its structural drift in the happening of its living. To have an identity, to operate in a domain of cognition, is to operate in a domain of structural coupling.

9.3   Language Is The Human Cognitive Domain

Human beings as living systems operating in language operate in a domain of recursive reciprocal consensual perturbations that constitute their domain of existence as such. Therefore, language as a domain of recursive consensual co-ordinations of actions is a domain of existence, and as such a cognitive domain defined by the recursion of consensual distinctions in a domain of consensual distinctions. Furtermore, human beings as living systems operating in language constitute observing, and become observers, by bringing forth objects as primary consensual co-ordinations of actions distinguished through secondary consensual co-ordinations of actions in a process that obscures the actions that they co-ordinate. Human beings, therefore, exist in the domain of objects that they bring forth through languaging. At the same time, human beings by existing as observers in the domain of objects brought forth through languaging, exist in a domain that allows them to explain the happening of their living in language through reference to their operation in a domain of dynamic reciprocal structural coupling.

9.4   Objectivity

Objects arise in language as consensual co-ordinations of actions that in a domain of consensual distinctions are tokens for more basic co-ordinations of actions that they obscure. Without language and outside language there are no objects because objects are only constituted as consensual co-ordinations of actions in the recursion of consensual co-ordinations of actions that languaging is. There are no objects for living systems that do not operate in language, or in other words, for them objects are not part of their cognitive domains. Since we human beings are objects in a domain of objects that we bring forth and handle in language, language is our peculiar domain of existence and peculiar cognitive domain. In these circumstances, objectivity arises in language as a manner of operating with objects without distinguishing the actions that they obscure. In this operation, descriptions arise as concatenations of consensual co-ordinations of actions that result in further consensual co-ordinations of actions which if done without distinction of how objects arise, can be distinguished as manners of languaging that take place as if objects existed outside of language. Objects are operational relations in languaging.

9.5 Languaging: Operation In A Domain Of Structural Coupling

To the extent that language arises as a consensual domain in the co-ontogenic structural drift of living systems involved in recurrent interactions, the organisms that operate in language operate in a domain of reciprocal co-ontogenic structural coupling through reciprocal structural perturbations. Therefore, to operate in language is not an <u>abstract</u> activity as we usually think. To language is to interact structurally. Language takes place in the domain of relations between organisms in the recursion of consensual co-ordinations of consensual actions, but at the same time language takes place through structural interactions in the domain of the "bodyhood" of the languaging organisms. In other words, although languaging takes place in the social domain as a dance of recursive relations of co-ordinations of actions, interactions in language as structural interactions are orthogonal to that domain, and as such trigger in the "bodyhood"of the participants structural changes that change as much the physiological background (emotional standing) on which they continue their languaging, as the course that this takes. The result is that the social co-ordinations of actions of languaging, as elements of a domain of recursive operation in structural coupling, become part of the medium in which the participant living systems conserve organization and adaptation through the structural changes that they undergo contingent to their participation in that domain. Thus, although the domain of co-ordinations of actions and the domain of structural change of the participants in language do not intersect, their changes are coupled orthogonally through the structural interactions that take place in language. As the body changes languaging changes, and as languaging changes the body changes. Here resides the power of words. Words are abstract entities in languaging and structural interactions in language, and it is through this that the world that we bring forth in languaging becomes part of the domain in which our ontogenic and phylogenic drifts take place.

9.6 Language Is A Domain Of Descriptions

Language is a system of recursive consensual co-ordinations of actions in which every consensual co-ordination of actions becomes an object through a recursion in the consensual co-ordinations of actions in a process that becomes the operation of distinction that distinguishes it and constitutes the observer. In these circumstances, all participants in a language domain can be ovservers with respect to the sequences of co-ordinations of actions in which they participate, constituting a system of recursive distinctions in which systems of distinctions become objects of distinctions. Such recursive distinctions of distinctions in the happening of living in language that bring forth systems of objects, constitute the phenomenon of description. As a result, all that there is in the human domain are descriptions in the happening of living in language which, as happenings of living in language become objects of descriptions in language. Descriptions, however, do not replace the happening of living that they constitute as descriptions, they only expand it in recursions that follow its operational coherences. In agreement with this, scientific explanations, as systems of descriptions, do not replace the phenomena that they explain in the domain of happening of living of the observer, but bring forth operational coherences in that domain that allow for further descriptions in it.

9.7 Self-consciousness Arises With Language

For a living system in its operation as a closed system there is no inside or outside, it has no way of making the distinction. Yet, in language such a distinction arises as a particular consensual co-ordination of actions in which the participants are recursively brought forth as distinctions of systems of distinctions. When this happens self-consciousness arises as a domain of distinctions in which the observers participate in the consensual distinctions of their participations in language through languaging. It follows from this that the individual exists only in language, and that the self exists only in language, and that self-consciousness as a phenomenon of self-distinction takes place only in language. Furthermore, it also follows that since language as a domain of consensual co-ordinations of actions is a social phenomenon, and as such it does not take place within the anatomical confines of the "bodyhood" of the living systems that generate it, on the contrary, it is external to them and pertains to their domain of interactions as a manner of coexistence.

9.8   History

The significance or meaning of any given behaviour resides in the circumstances of its enaction, not in the characteristics of the dynamics of states of the behaving living system or in any particular feature of the behaviour itself. In other words, it is not the complexity of the inner states of a living system or of its nervous system nor any aspect of the behaviour itself, that determines the nature, meaning, relevence or content of any given behaviour, but its placement in the ongoing historical process in which it arises. The higher human functions do not take place in the brain; language, abstract thinking, love, devotion, reflection, rationality, altruism etc., are not features of the dynamics of states of the human being as a living system, nor of its nervous system as a neuronal network, they are socio-historical phenomena. At the same time, history is not part of the dynamics of states of a living system because this takes place only in the present, instant by instant in the operation of its structure in changes that occur out of time. History, time, future, past or space, exist in language as forms of explanation of the happening of living of the observer, and partake of the involvement of language in this. Therefore, it is in the explanation of the happening of living through the coherences of language that an observer can claim that the structure of a living system that determines its changes of state in the present, always embodies its history of interactions because it continuously arises in the present in a structural drift contingent to such history.

9.9   The Nervous System Expands The Domain Of States Of The Living System

For living systems to operate in language the diversity and plasticity of their internal states must match the diversity of the changing circumstances generated in their recursive consensual coordinations of actions. In other words, although language does not take place within the "bodyhood" of the living system, the structure of the living system must provide the diversity and plasticity of states required for it to take place. The nervous system participates in this by expanding the domain of states of the organism through the richness and recursiveness of its dynamics as a closed network of changing relations of neuronal activities (see Maturana 1983), and by expanding in the organism the domain of its changes of states that follow in it a course contingent upon both its own changes of states and its interactions in the medium. And, this the nervous system does: a) by admitting the interactions of the organism as orthogonal perturbations from the medium, a condition that makes its structural drift as a cellular network, as well as the structural drift of the organism and its participation in the generation of behaviour, contingent upon the history of those interaction; and b) by admitting orthogonal interactions from the components of the organism, a condition that makes its structural drift as a cellualr network, as well as the structural drift of the organism and its participation in the generation of behaviour, recursively contingent upon the dynamics of structural changes of the organism. The result of all this for the organism (including its nervous system) is the possibility of the recursive involvement of its dynamics of states with the ongoing flow of its own dynamics of states through its behaviour, if it has sufficient plasticity in the nervous system and participates in a sufficiently large domain of recurrent interactions with other organisms. Indeed, this is what permits the production of language as this arises when the internal recursiveness of the dynamics of states of the nervous system couples with the recurrence of social consensual co-ordinations of actions, giving rise to the recursion of consensual co-ordinations as an ongoing process in the generation of social behaviour.

The ongoing recursive coupling of behavioural and structural changes that give origin to language, is possible because a structure determined system exists in two non-intersecting phenomenal domains realized through orthogonally dependent structures, namely, its domain of states and its domain of interactions. It is our basic double existence as structure determined systems in two non-intersecting but orthogonally coupled phenomenal domains, that permits us in our operation in language to generate endless orthogonally interdependent and yet non-intersecting phenomenal domains in the happening of our living.

9.10 Observing Takes Place In Languaging

The nervous system is a closed network of interacting active neuronal elements (neurons, effectors and receptors) that are structurally realized as cellular components of the organism. As such it operates as a closed network of changing relations of activity between its components; that is, it is constitutive to the organization of the nervous system that any change of relations of activity between its components should lead to further changes of relations of activity between them, and that in that sense it should operate without inputs or outputs. Therefore, any action upon an environment that an observer sees as a result of the operation of the nervous system, is a feature of the structural changes that take place in it as a cellular network (muscular contraction, glandular secretion), and not a feature of its operation as such. Indeed, the operation of the nervous system and the actions of the organism take place in non-intersecting phenomenal domains realized by orthogonally related structures. Similarly, any perturbation of the medium impinging upon the organism is a perturbation in the structure of the nervous system, not an input in its dynamics of states, and if this changes it does so because the structure of the nervous system changes in a manner contingent to the perturbation, not because it admits an input to its operation. As a result of all this, all that takes place in the nervous system is a dance of changing relations of neuronal activities that, in the domain of structural coupling where the observer beholds the organism, appears as a dance of changing configurations of effector/sensor correlations. An observer that sees an effector/sensor correlation as an adequate behaviour does so because he or she beholds the organism in the domain of structural coupling in which the distinguished behaviour takes place in the flow of its conservation of adaptation. The organism in its operation does not act upon an environment, nor does the nervous system operate with a representation of one in the generation of the adequate behaviour of the organism; the environment exists only for an observer (see section 6.13), and as such it is a phenomenon of languaging.

That the nervous system should operate as a closed network of changing relations of activity between its components, and not with representations of an environment, has two fundamental consequences:
a) For the operation of the nervous system everything is the same. Or in other words, all that takes place in the operation of the nervous system are changes of relations of activity between its components, and it does not distinguish in its operation whether its changes of state arise through its internal dynamics or as a result of structural changes triggered in it through what an observer sees as external structural perturbations.
b) For the observer the organism operates in many different domains of structural coupling which intersect operationally in the domain of states of the nervous system through the structural perturbations generated in it by the interactions of the organism in these different domains. As a result of this, several things happen that are relevant to understand   the domains of reality that the observer brings forth (see next sections). Firstly, an observer can always treat a state of activity of the nervous system (a configuration of changes of relations of activity) that arises as a result of a particular interaction of the organism, as a representation of that interaction, and to do so by constituting the domain of descriptions as a metaphenomenal domain in which both are distinguished together. Secondly, different states of activity of the nervous system that for an observer

represent interactions of the organism in non-intersecting phenomenal domains (different domains of structural coupling), can affect each other and give rise to behaviours of the organism that constitute meta domains of relations between the phenomena that take place in those non-intersecting phenomenal domains. Thirdly, the meta domains of relations established through their operational intersection in the domain of states of the nervous system by the operation of the organism in its different domains of structural coupling, constitute, through the behaviours that these intersections generate, new domains of structural coupling of the organism that do not intersect with the others. And fourthly, the operational intersection of the differernt domains of interaction (different domains of structural coupling) of an organism in the operation of its nervous system, allows it to operate in recurrent interactions with other organisms in the continuous recursive generation of meta domains of relations which become phenomenal domains in their own right in the ongoing flow of those recurrent interactions. The result of all this is the possibility of the constitution of the observer when two or more organisms generate observing in their recursive consensual co-ordinations of actions, as they operate in reciprocal references for the constitution of meta domains of relations in their ongoing recurrent interactions. Also as a result of this, observing is to operate in language with operational coherences brought forth in language. Or, in other words, since the operation of the nervous system appears in the domain of operation of the organism as sensory/effector correlations, observing is co-ordinations of the "bodyhood" of observers through their generation of a coreography of interlaced sensory/effector correlations. In fact, all that there is for the operation of the nervous system of the observer is observing, is its closed dynamics of changing relations between its neuronal components. It is only for an observer that sees two or more interacting organisms in his or her praxis of living, that the sensory/effector correlations of these appear recursively involved with each other in a network of recursive sensory/effector correlations constituted through the orthogonal interactions of their nervous system. And, finally, it is only for an observer that such a network of recursive sensory/effector correlations becomes language, and a meta domain with resect to the operation of the nervous system where explanations and observing takes place, when it becomes a recursive system of consensual co-ordinations of consensual actions.

10. THE DOMAIN OF PHYSICAL EXISTENCE

A domain of existence is a domain of operational coherences entailed by the distinction of a unity by an observer in his or her praxis of living. As such, a domain of existence arises as the domain of the operational validity of the properties of the unity distinguished if it is a simple unity, or as the domain of validity of the properties of the components of the unity distinguished if this a composite one. As a consequence, the distinction of a unity entails its domain of existence as a composite unity that includes it as a component. Therefore, there are as many domains of existence as kinds of unities an observer many bring forth in his or her operations of distinction. In these circumstances, since the notion of determinism applies to the operation of the properties of the components of a unity in its composition (see section 6.10 and 7.4), all domains of existence, as composite entities that include the unities that specify them, conform deterministic systems in the sense indicated above. This has certain consequences for us living systems existing in language, and for the explanations that we generate as such. The following are some of them.

i) Our domain of existence as the composite unities that we are as molecular autopoietic systems, is the domain of existence of our component molecules, and entails all the operational coherences proper to the molecular existence. Therefore, our existence as autopoietic systems implies the satisfaction of all the constraints that the distinction of molecules entails, and our operation as molecular systems implies the determinism entailed in the distinction of molecules.

ii) If we distinguish molecules as composite entities, they exist in the domain of existence of their components, and as such their existence implies the satisfaction of the determinism that the distinction of the latter entails. The same applies to the decomposition of the components of molecules, and so on recursively. Since unities and their domains of existence are brought forth and specified in their distinctions in the happening of living of the observer, the only limit to the recursion in distinctions is the limit of the diversity of experiences of the observer in his or her happening of living (praxis).

iii) Since the observer as a living system is a composite entity, the observer makes distinctions in his or her interactions as a living system through the operation of the properties of his or her components. If the observer uses an instrument, then his or her distinctions take place through the operation of the properties of the instrument as if this were one of its components. The result of this is that an observer cannot make distinctions outside its domain of existence as a composite entity.

iv) Descriptions are series of consensual distinctions subject to recursive consensual distinctions in a community of observers. Observers operate in language only through their recursive interactions in the domain of structural coupling in which they recursively co-ordinate consensual actions as operations in their domains of experiences through the praxis of their living. Therefore, all interactions in language between observers take place through the operations of the properties of their components as living systems in the domain of their reciprocal structural coupling. Or, in other words, we as human beings operate in language only through our interactions in our domain of exestence as living systems, and we cannot make descriptions that entail interactions outside this domain. As a consequence, although language as a domain of recursive consensual distinctions is open to unending recursions, language is a closed operational domain in the same sense that it is not possible to step outside language through language, and descriptions cannot be characterizations of independent entities.

v) Since everything said is said by an observer to another observer, and since objects (entities, things) arise in language, we cannot operate with objects (entities or things) as if they existed outside the distinctions of distinctions that constitute them. Furthermore, as entities in language, objects are brought forth as explanatory elements in the explanation of the operational coherences of the happening of living in which languaging takes place. Without observers nothing exists, and with observers everything that exists, exists in explanations.

vi) As we put objectivity in parenthesis because we recognize that we cannot experientially distinguish between what we socially call perception and illusion, we accept that existence is specified by an operation of distinction: <u>nothing pre-exists its distinction</u>. In this sense houses, people, atoms or elementary particles are not different. Also in this sense, existence as an explanation of the praxis of living of the observer, is a cognitive phenomenon that reflects the ontology of observing in such praxis of living, and not a claim about objectivity. Therefore, with objectivity in parenthesis, an entity has no continuity beyond or outside that specified by the coherences that constitute its domain of existence as this is brought forth in its distinction. The claim that the house to which I return every evening from work is the same that I left in the morning, or that whenever I see my mother I see the same person that gave birth to me, or that all the points of the path of an electron in a bubble chamber are traces left by the same electron, are claims that constitute cognitive statements that define sameness in the distinction of the unity (house, mother or electron) as this is specified in the operation of distinction that brings it forth together with its domain of existence. Since according to all that I have said , cognitive statements are not, and cannot be statements about the properties of independent objects, sameness is necessarily always a reflection of the observer in observing in the domain of existence that he or she brings forth in his or her distinctions. Furthermore, since no entity can be distinguished outside its domain of existence as the domain of operational coherences in which it is possible, every distinction specifies a domain of existence as a domain of possible distinctions; that is, every distinction specifies a domain of existence as a versum in the multiversa, or colloquially, every distinction specifies a domain of reality.

vii) A scientific explanation entails the proposition of a mechanism (or composite entity) that, if realized, would generate the phenomenon to be explained in the domain of experiences (praxis or happening of living) of the observer (see section 4). The generative character of the scientific explanation is constitutive to it. Indeed, this ontological condition in science carries with it the legitimacy of the fundamental character of the phenomenal domain in which the generative explanatory mechanism takes place, as well as the legitimacy of treating every entity distinguished as a composite unity, asking for the origin of its properties in its organization and structure. And because this is also the case for our common sense explanations in our effective operation in our daily life, it seems natural to us to ask for a substratum independent of the observer as the ultimate medium in which everything takes place. Yet, although it is an episthemological necessity to expect such a substratum, we constitutively cannot assert its existence distinguishing it as a composite entity and characterize it in terms of components, and relations between components. In order to do so we would have to describe it, that is, we would have to bring it forth in language and give it form in the domain of recursive consensual co-ordinations of actions in which we exist as human beings. However, to do so would be tantamount to characterizing the substratum in terms of entities (things, properties) that arise through languaging, and which, as consensual distinctions of consensual co-ordinations of actions, are constitutively not the substratum. Through language we remain in language, and we lose the substratum as soon as we attempt to language it. We need the substratum for epistemoligical reasons, but in the substratum there are no objects, entities or properties; in the substratum there is nothing (no-thing) because things belong to language. In other words, nothing exists in the substratum.

viii) Distinctions take place in the domain of experiences, in the happening or praxis of living of the observer as a human being. For this reason, the domain of operational coherences that an observer brings forth in the distinction of a unity as its domain of existence, also occurs in his or her domain of experiences as a human being as part of his or her praxis of living. Therefore, since language is operation in a domain of recursive consensual co-ordinationsof consensual actions in the domain of experiences of the observers as human beings, all dimensions of the domains of experiences of the observers exist in language as co-ordinations of actions between observers. As such, all descriptions constitute configurations of co-ordinations of actions in some dimensions of the domains of experiences of the members of a community of observers in co-ontogenic structural drift. Physics, biology, mathematics, philosophy, cooking, politics etc., are all different domains of languaging, and as such are all different domains of recursive consensual co-ordinations of consensual actions in the praxis or happening of living of the members of a community of observers. In other words, it is only as different domains of languaging that physics, biology, philosophy, cooking, politics, or any cognitive domain exists. Yet, this does not mean that all cognitive domains are the same, it only means that different cognitive domains exist only as they are brought forth in language, and that languaging constitutes them. We talk as if things existed in the absence of observers, as if the domain of operational coherences that we bring forth in a distinction would operate as it operates in our distinctions regardless of them. We now know that this is constitutively not the case. We talk for example, as if time and matter were independent dimensions of physical space. Yet, it is apparent from my explanation of the phenomenon of cognition that they are not and cannot be. Indeed, time and matter are explanations of some of the operational coherences of the domains of existence brought forth in the distinctions that constitute the ongoing languaging of the praxis of living of the members of a community of observers. Thus, time, with past, present and future, arises as a feature of an explanatory mechanism that would generate what the observer experiences as successive non-simultaneous phenomena and matter arises as a feature of an explanatory mechanism that would generate what he or she experiences as mutually impenetrable simultaneous distinctions. Without observers nothing can be said, nothing can be explained, nothing can be claimed,... in fact, without observers nothing exists because existence is specified in the operation of distinction of the observer. For epistemological reasons we ask for a substratum that could provide an independent ultimate justification or validation of distinguishability, but, for ontological reasons, such a substratum remains beyond our reach as observers. All that we can say ontologically about the substratum that we need for epistemological reasons, is that it permits what it permits, and that it permits all the operational coherences that we bring forth in the happening of living as we exist in language.

ix) As we operate in language we operate in a domain of reciprocal structural coupling in our domain of existence as composite unities (molecular autopoietic systems), that is, we operate in the domain of existence of our components. Therefore, anything that we say, any explanation that we propose, can only entail distinctions that involve the operation of our components in their domain of existence as we operate as observers in language. Accordingly, it is in the domain where we exist as composite entities where we distinguish molecules, atoms or elementary particles, as entities that we bring forth in language through operations of distinction that specify them and the operational coherences of their domains of existence. If what we call the physical domain of existence is the domain where physicists distinguish molecules, atoms or elementary particles, then we as living systems specify the domain of physical existence as our limiting cognitive domain as we operate as observers in language, interacting in the domain of existence of our components as we bring it forth as an explanation of the happening of our living. We do not exist in a pre-existing domain of physical existence, we bring it forth and specify it as we exist as observers. The experience of the physicist, be this in classic, relativistic or quantum physics, does not reflect the nature of the universe, it reflects the ontology of the observer as a living system as he or she operates in language bringing forth the physical entities and operational coherences of their domains of existence. Einstein made the assertion that "scientific theories (explanations) are free creations of the human mind", and then, in what seemed to reveal a paradox, he asked the question, "how is it, if that is the case, that the universe is intelligible through them?". In this article I have shown that there is no paradox if one reveals the ontology of observing and the ontology of scientific explanations through putting objectivity in parenthesis. Indeed, I have shown that a scientific explanation entails: a) the proposition of a phenomenon to be explained, brought forth as such a priori in the praxis of living (domain of experiences) of the observer; b) the proposition of an ad hoc generative mechanism, also brought forth a priori in the praxis of living of the observer, that if allowed to operate would generate the phenomenon being explained as a consequence to be witnessed by the observer in his or her praxis of living; c) the operational coherence of the four operational conditions that constitute its criterion of validation, as they are realized in the praxis of living of the observer; and d) the superfluity and impertinence of the assumption of objectivity. From all this it follows that the explanatory mechanism proposed in a scientific explanation is constitutively "a free creation of the human mind" because it is brought forth constitutively a priori in the praxis of living of the observer, that is without any other justification than its ad hoc generative character of the phenomenon explained. It also follows from this, that a scientific explanation constitutively explains the universe (versum) in which it takes place because both the explanatory mechanism and the phenomenon being explained, occur, in a generative relation, as non-intersecting phenomena of the same operational domain of the praxis of living of the observer. Or, in other words, it also follows from all this that since the operation of distinction specifies the entity distinguished as well as its domain of existence, a scientific explanation constitutively explains the universe (versum) in which it takes place because it brings with it the domain of operational coherences (the versum of the multiversa) of the praxis of living of the observer that it makes intelligible. Strictly, then, there is no paradox, scientific explanations do not explain an independent world or universe, they explain the praxis of living (the domain of experiences) of the observer making use of the same operational coherences that constitute it in languaging. It is there where science is poetry.

11. REALITY

The word reality comes from the latin noun *res* that means object (thing), and is commonly used signifying objectivity without parenthesis. The real, and sometimes the really real, is meant to be that which exists independently of the observer. Now we know that the concepts entailed in this way of speaking cannot be sustained. Objects, things, arise in language when a consensual co-ordination of actions by being consensually distinguished in a recursion of consensual co-ordinations of actions, obscures the actions that it co-ordinates in the praxis of living in a consensual domain. Since according to this an object, a unity, is brought forth in language in an operation of distinction that is a configuration of consensual co-ordinations of actions, when an object is distinguished in language its domain of existence as a coherent domain of consensual co-ordinations of actions becomes a domain of objects, a domain of reality, a versum of the multiversa such that all that there is in it is all that is entailed in the consensual co-ordinations of actions that constitute it. Every domain of existence is a domain of reality, and all domains of reality are equally valid domains of existence brought forth by an observer as domains of coherent consensual actions that specify all that there is in them. Once a domain of reality is brought forth, the observer can treat the objects or entities that constitute it both as if they were all that there is and as if they existed independently of the operations of distinctions that bring them forth. And this is so because a domain of reality is brought forth in the praxis of living of the observer as a domain of operational coherences that require no internal justification.

It follows from all this, that an observer operating in a domain of reality necessarily operates in a domain of effective actions, and that another observer claims that the first one makes a mistake or has an illusion only when this observer begins to operate in a domain of reality different from the one that he or she expected. Thus, if we specify the operation of distinction ghost, then ghosts exist, are real in the domain of existence brought forth in their distinctions and we can do effective actions with them in that domain, but they are not real in any other domain. Indeed, everything is an illusion outside its domain of existence. In other words, every domain of reality as a domain of operational coherences brought forth in the happening of living of the observer in language, is a closed domain of effective consensual actions, that is, a cognitive domain; and conversely, every cognitive domain as a domain of operational coherences is a domain of reality. What is uncanny, perhaps, is that although different domains of reality are seen by an observer as different domains of co-ordinations of actions in an environment, they are lived by the observer as different domains of languaging which differ only through their ongoing transformation in the different circumstances of recursion in which they arise. We as observers can explain this now by saying that, as we operate in language through our consensual interactions in the happening of living of a community of observers, our structural drift in the happening of our living becomes contingent upon the course of those consensual interactions, and that this takes place in a manner that keeps the transformation of the happening of our living congruent with the domain of reality that we bring forth in that community of observers, or we disintegrate as members of it. It is this which makes us observing systems capable, through language, of an endless recursive generation of new cognitive domains (new domains of reality) as new domains of praxes of observing in our continuous structural drifts as living systems.

12. SELF-CONSCIOUSNESS AND REALITY

The self arises in language in the linguistic recursion that brings forth the observer as an entity in the explanation of his or her operation in a domain of consensual distinctions. Self-consciousness arises in language in the linguistic recursion that brings forth the distinction of the self as an entity in the explanation of the operation of the observer in the distinction of the self from other entities in a consensual domain of distinctions. As a result, reality arises with self-consciousness in language as an explanation of the distinction between self and non-self in the praxis of living of the observer. Self, self-consciousness and reality exist in language as explanations of the happening of living of the observer. Indeed, the observer as human being in language is primary with respect to self and self-consciousness, and these arise as he or she operates in language explaining his or her experience, his or her praxis of living as such. That the entities brought forth in our explanations should have an unavoidable presence in our domain of existence, is because we are realized as observers as we distinguish them in the domain of operational coherences that they define as we distinguish them. We do not go through a wall in the praxis of living becuase we exist as living systems in the same domain of operational coherences in which a wall exists as a molecular unity, and a wall is distinguished as a composite entity in molecular space as that entity through which we cannot go as molecular entities.

The observer is primary, not the object. Better, observing is given in the praxis of living in language, and we are already in it when we begin to reflect upon it. Matter, energy, ideas, notions, mind spirit, god,... are explanatory propositions of the praxis of living of the observer. Furthermore, matter, energy, ideas, notions, mind, spirit or god, as explanatory propositions entail different manners of living of the observer in recursive conservation of adaptation in the domains of operational coherences brought forth in their different distinctions. Thus, when the observer operates with objectivity without parenthesis, he or she operates in an explanatory avenue that entails neglecting the experiential indistinguishability between what we call perception and illusion, and when he or she operates with objectivity in parenthesis he or she operates in an explanatory avenue that entails accepting this indistinguishability as a starting point. In the explanatory path of objectivity without parenthesis, the observer, language and perception, cannot be explained scientifically because a contradiction arises with the structural determinism of the living system, while with objectivity in parenthesis there is not such a contradiction. As one operates within any given domain of reality one can operate with objectivity without contradictions, but when a disagreement arises with another observer, and one thinks that it is not a matter of a simple logical mistake, one is forced to claim a privileged access to an objective reality to resolve it, and to deal with errors as if they were <u>mistakes</u> of what it is. If in similar circumstances one is operating with objectivity in parenthesis, one finds that the disagreeing parties operate in different domains of reality and that the disagreement disappears only when they begin to operate in the same one. Furthermore, one also finds that errors are changes of domain of reality in the operation of an observer that he or she notices only <u>a posteriori</u>. Finally, by operating with objectivity without parenthesis we cannot explain how an observer operates in the generation of a scientific explanation because we take for granted the abilities of the observer. Contrary to this, as we operate with objectivity in parenthesis, scientific explanations and the observer appear as components in a single closed generative explanatory mechanism, in which the properties or abilities of the observer are shown to arise in a different phenomenal domain that the one in which its components operate.

We human beings exist only as we exist as self-conscious entities in language. It is only as we exist as self-conscious entities that the domain of physical existence exists as our limiting cognitive domain in the ultimate explanation of the human observer's happening of living.The physical domain of existence is secondary to the happening of living of the human observer, even though in the explanation of observing the human observer arises from the physical domain of existence. Indeed, the understanding of the ontological primacy of observing is basic for the understanding of the phenomenon of cognition. Human existence is a cognitive existence and takes place through languaging, yet, cognition has no content and does not exist outside the effective actions that constitute it. This is why nothing exists outside the distinctions of the observer. That the physical domain of existence should be our limiting cognitive domain, does not alter this. Nature, the world, society, science, religion, the physical space, atoms, molecules, trees..., indeed all things, are cognitive entities, explanations of the praxis or happening of living of the observer, and as such, as this very explanation shows, they only exist as a bubble of human actions floating on nothing. Everything is cognitive, and the bubble of human cognition changes in the continuous happening of the human recursive involvement in co-ontogenic and co-phylogenic drifts with the domains of existence that he or she brings forth in the praxis of living. Everything is human responsibility.

The atom and the hydrogen bombs are cognitive entities. The big bang, or whatever we claim from our present praxis of living gave origin to the physical versum, is a cognitive entity, an explanation of the praxis of living of the observer bound to the ontology of observing. That is their reality. Our happening of living takes place regardless of our explanations, but its course becomes contingent upon our explanations as they become part of the domain of existence in which we conserve organization and adaptation through our structural drifts. Our living takes place in structural coupling with the world that we bring forth, and the world that we bring forth is our doing as observers in language as we operate in structural coupling in it in the praxis of living.We cannot do anything outside our domains of structural coupling; we cannot do anything outside our domains of cognition.

378

This is why nothing that we do as human beings is trivial, and everything that we do becomes part of the world that we live in as we bring it forth as social entities in language. Human responsibility in the multiversa is total.

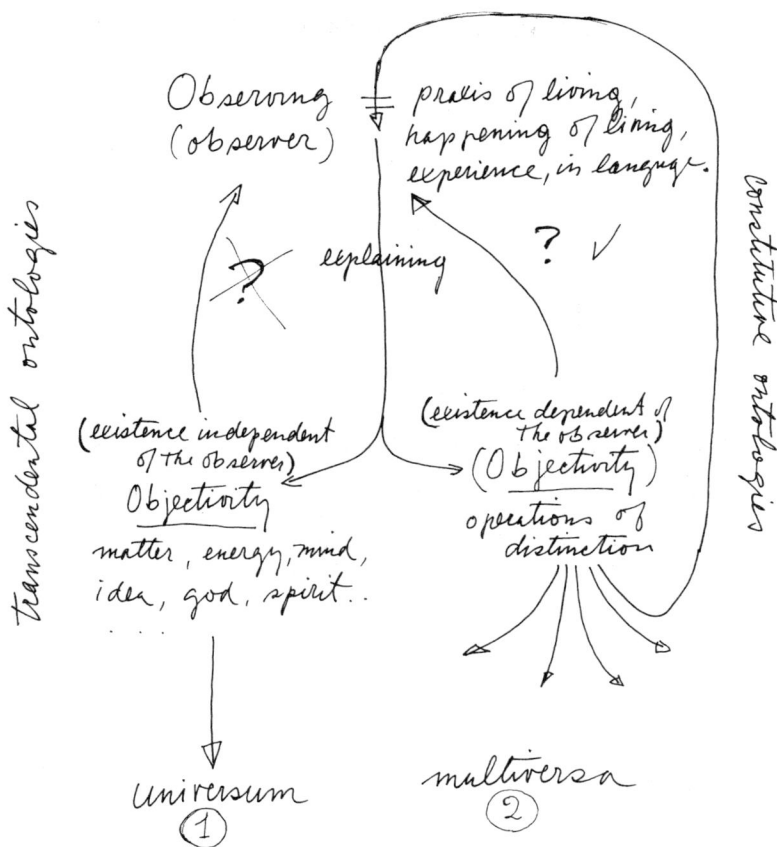

Fig. 1. - The observer is in the praxis of living in language prior to any explaining. Path ① in explaining does not take in consideration that we human beings cannot exponentially distinguish between what on reflexion we call perception or illusion, path ② does. As a result path ① entails the acceptance of the existence of entities (matter, energy, mind, consciousness, god...) that are independent of the observer, and that in the ultimate end validate what he or she says. I call this path the path of the <u>transcendental ontologies</u>. Path ② entails accepting that existence depends on the operations of distinction of the observer, and that it is only this that validates whatever he or she says. Path ① entails a single domain of ultimate reality, path ② entails many domains of equally legitimate realities, each brought forth by the operations of distinction of the observer. I call path ②, the path of the constitutive ontologies. Whether we as observers operate through path ① or through path ② is the result of our emotioning, but it is only through path ② that observing, languaging and cognition can be explained as biological phenomena.

ON ASSOCIATIVE MEMORIES

G. Palm

Max-Planck-Institut für Biologische Kybernetik

7400 Tübingen, FRG

ABSTRACT

The paper provides an algorithmic analysis of various methods of storage and retrieval of information. It turns out that associative memory methods should be preferred in general, if the number of items to be stored is large and their individual information content is small. To work in an economic way associative memory methods require an unusual, highly redundant sparse coding.

1. INTRODUCTION

The word "associative memory" provokes various associations, of which I want to distinguish three. Firstly, one may think of associations in the psychological sense as some kind of mental activity that a given word or concept makes another word or concept come to mind, which is then said to be "associated to" the first one. Secondly, one may think of a mathematical problem (the matching problem) which can be understood as a mathematical formulation of "association" in the first sense and which is treated extensively in Kohonen's book "Associative Memory" [11]. Thirdly, one may think of computer storage devices that are claimed to work in an associative way and are therefore called associative memories.

In this paper, I want to deal mainly with the third connotation, i.e. with associative computer memories. I shall try to clarify what it means to store information in an associative way and I shall compare the different procedures that result from slightly different interpretations of the word "associative": For this purpose it will be useful to relate the third connotation of "associative memory" with the other two.

Associations, in the psychological sense, can be described mathematically by a mapping $x \to m(x)$, where x stands for the input word and $m(x)$ for the word associated to x. By letting someone "associate freely", one can try to investigate his state of mind. The idea is very broadly speaking that the associations (i.e. the mapping m) are dependent on the state of mind, and since this state of mind may change there may be corresponding changes in the associations (i.e. in the mapping m). This means that the mapping m cannot be regarded as fixed, but that it should be flexible. Furthermore, changes of the mapping m should depend on previous experiences of the corresponding individual. Therefore, the present state of the mapping m can be said to contain (or store) some information about these previous experiences: m can be regarded as some kind of memory. There is more that could be said about the connection between mappings and memories, but I have already expanded on this in an earlier paper [22] and I do not want to repeat myself here.

The mathematical problem that arises here is the following: How can one construct a flexible mapping? More specifically: Given a set S of pairs (x,y) of "messages", how can one construct a mapping m that maps x into y for every pair (x,y) in S?

Furthermore: Given such a set S, a mapping m that does the job for S, and another pair (x'y'), how can one modify m to a mapping m' that does it also for (x',y'), i.e. that fulfills m'(x)=y for every (x,y) in S and m'(x')=y'?

This problem is also called the matching problem. If we take the messages x and y to be vectors of numbers ($x \in \mathbb{R}^m$, $y \in \mathbb{R}^n$) and try to perform the mapping by a linear transformation (i.e. an mxn matrix), then there is an optimal approximate solution to this matching problem. This solution can be found by means of linear regression, it can be written down in terms of the pseudoinverse of a matrix, and all this is discussed extensively by Kohonen [11] and very briefly in my earlier paper [22]. This solution of the matching problem by means of the pseudoinverse can also be regarded as a refinement of the idea to store the "crosscorrelations" $x \otimes y$, $(x \otimes y)_{ij} = x_i \cdot y_j$, of the messages $(x,y) \in S$ in a matrix.

In summary, a solution to the matching problem is provided in terms of a matrix $A=(a_{ij})$ of real numbers.

In the same context there have been some attempts to restrict the range of possible values for the coefficients a_{ij} from $\mathbb{R}$ for example to $\{-1,0,1\}$. In an attempt to estimate the information storage capacity of a matrix $A=(a_{ij})$ of coefficients when used in this way [22], I even restricted the range to $a_{ij} \in \{0,1\}$. Such a matrix of zeros and ones (or of -1,0, and 1) and the way it is used for storage and recall bares a strong resemblance to the old core-memories in computers, where individual 1-bit storage units were placed in a matrix arrangement.

This observation is the basic link between associative matrix memories that arise from a solution to the matching problem and associative information storing devices. In the remainder of this section I want to formulate two types of operational problems for which associative storing devices should be useful and which are derived from the above considerations. In the following sections I shall introduce some concrete storage and recall procedures and discuss their relative advantages when used for these problems.

First problem: Pattern mapping. This is mainly the above mentioned problem. The messages that have to be associated with each other are called "patterns", and are thought of as sequences of symbols from some (finite) alphabet. Of course, they could also be sequences of numbers or of numerals (i.e. vectors). In many cases it will be most convenient to think of sequences of zeros and ones (since arbitrary messages can be coded into this form). Many paris of patterns (x,y) are read into the storage device somehow, and later it automatically performs the mapping from x to y. This means that when x is given as an input to the machine, it responds with y. A typical application for such a machine would be an automatic telephone-book, i.e. a machine that works on names and addresses as inputs and gives the corresponding telephone numbers as outputs.

Second problem: Pattern completion. Here again the patterns are thought of as sequences of symbols. Many patterns x are read into the storage device, and later it automatically performs the pattern completion, i.e. given a sufficiently large part of the pattern (sequence of symbols) x, it responds with the whole pattern x.

Note that the second problem requires more flexibility from the machine than the first one. Indeed any sufficiently large part of the sequence of symbols shall provoke the whole sequence, it need not always be the first half of the sequence, for example, or the second half, but it can be any part. In a way the first problem can be regarded as special case of the second problem. We simply consider a pair (x,y) of patterns as one pattern again: (x,y)=z. Then we have a very restricted problem of pattern completion, namely to associate the whole pattern z=(x,y) to the first part x of it.

If we had a more flexible machine for the first problem, namely one that responds with y not only to the complete input x but also to sufficiently large parts of it, then we could in turn regard the second problem as a special case of the first one, where x=y. It should also be clear that it remains a point for the subsequent discussion, how large a part of a pattern has to be, in order to be "sufficiently large" to serve as an address for the whole pattern.

For the two problems mentioned above we shall use one basic storage device that can be viewed as a general equivalent to any kind of computer memory. For this one device we shall present different methods of storage and read-out of the data that represent basically the various contemporary proposals of using existing computer memories or non-existing similar structures for the two problems of pattern mapping and pattern completion. I shall try to carry out the description and comparison of the methods on this (not too high) level of abstraction, because a fair comparison is much easier when all the methods have to make use of the same "hardware device". I do not want to discuss the detailed implementation of the methods on existing computer memory hardware for the following reason:

Some methods are used on existing hardware and any expert in hardware will recognize them (most likely by the fact that for these methods the translation into existing hardware is quite obvious), whereas others cannot be so easily implemented on existing hardware. In this general discussion I do not want to get biased by the existing hardware.

2. THE BASIC DEVICE

The basic device consists of a two dimensional array of switches which can be addressed by two sets of wires running through the array in the two orthogonal directions (see Fig. 1). In the following these wires will be referred to as the horizontal and the vertical wires, respectively.

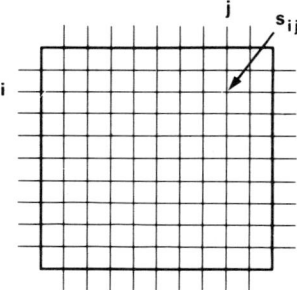

Fig. 1

The switches can be switched on and off by means of these wires, which can carry different values of activity (e.g. voltage). For the switching three different values of activity are sufficient, e.g. $\{-1,0,1\}$. If the two wires passing through a switch carry the value -1, it is switched off; if they both carry the value +1 it is switched on. In all other cases the switch remains as it was. Note that by this mechanism each switch can be controlled individually: we can control the switch s_{ij} (see Fig. 1) by means of i horizontal and the j vertical wire without any effect on all the other switches, provided that all other wires carry the value 0 of activity.

The state of the switches in the matrix can be inquired by the following additional mechanism. Let us denote the state of the switch s_{ij} by $s_{ij}=0$, when it is switched off and $s_{ij}=1$, when it is switched on. Then an input (activity) $a_1,...,a_m$ on the horizontal wires is related to the vertical wires through the switches s which are switched on and the incoming activity on each vertical wire is added. This means that the input $a=(a_1,...,a_m)$ on the horizontal wires will yield the output $b=(b_1,...b_n)$ on the vertical wires, where $b_j=\sum_i s_{ij} a_i$, in matrix notation this can be written as $b=S^*a$, where $S=(s_{ij})$ and $S^* = (s_{ji})$. Similarly an input $b=(b_1,...,b_n)$ on the vertical wires will yield on output $a=(a_1,...,a_m)$ on the horizontal wires, which is given by $a_i=\sum_j s_{ij}b_j$, i.e. $a=Sb$.

Note that by this mechanism the state of each individual switch in the matrix can be inquired. If we give the input 1 to the ith horizontal wire (0 on all other wires) and record the output of the jth vertical wire, it will be $b_j=s_{ij}$. Of course, we can also find s_{ij} the other way round, i.e. by putting in a 1 only on the jth vertical wire and recording the output from the ith horizontal wire.

This is our basic device and I will call it the switching matrix. Similar devices are well known in the literature [30,29,15]. Sometimes we shall need a variation of this device where the values of the s_{ij} are -1 if the corresponding switch is switched off and +1, if it is switched on. In this case the same formulas as above shall hold for the retrieval, i.e. an input a on the vertical wires yields the output $b=S^*a$ on the vertical wires and an input b on the vertical wires yields an output $a=Sb$ on the horizontal wires. Note, however, that this relation now can no longer be interpreted as a switching between the wires (-1•-1=1, but at -1 the switch is switched off), rather, we can say that in this case the formation of $b_j=\sum_i s_{ij} a_i$ corresponds to a correlation between the input a and the jth column of the matrix and the formation of $a_i = \sum_j s_{ij}b_j$ corresponds to a correlation between the input b and the ith row of the matrix, especially if the input patterns also consist of -1's and 1's.

Now we can describe the various ways of using this device as a storage device in the two set-ups of pattern mapping and pattern completion, as defined in the introduction.

3. METHODS FOR PATTERN MAPPING

We want to use the switching matrix to implement a flexible mapping $m: x \to y$ which works for various sets S of pairs (x,y). The patterns (x,y) in these pairs are sequences of symbols of a certain (maximal) length, this implies that they are taken from a limited set of possible patterns: $x \in X$ and $y \in Y$, where X denotes the set of all possible input patterns and Y denotes the set of all possible output patterns. Of course, this implies that $|S| \leq |X|$. In many cases the length $|S|$ of the list is not known exactly in advance, but one usually has an estimate (or an upper estimate) z for it.

3.1. Addressing by All Possible Inputs

By a fixed code c the input patterns $x \in X$ are coded into numbers. This means that $c: X \to \{1,...,m\}$ is an injective mapping. Of course, this implies that $m \geq |X|$. By another fixed code d the output patterns are coded into 0-1-sequences. This means that $d: Y \to \{0,1\}^n$ is another injective mapping. Of course this implies that $2^n \geq |Y|$. The association (x,y) is then stored in the switching matrix by giving the $c(X)^{th}$ horizontal wire the activity 1 and the pattern d(y) of activity to the vertical wires. This causes the corresponding switches in the $c(X)^{th}$ horizontal row to be switched on.

In the recall, given the input x we simply compute c(X) (by the fixed code c), input activity 1 to the corresponding horizontal wire, record d(y) on the vertical wires, and compute y from this (by the inverse of the fixed code d).

The dimensions of the mxn matrix needed for this procedure: Obviously, we need m=|X| and n≥ld|Y|. Therefore, this method is good if $|S| \approx |X|$, but not if $|S| << |X|$.

Assume that we know that $|S| \leq z$, where z is much smaller than |X|, That means that we know that by far not all possible input patterns $x \in X$ will occur in the set S. Then one would like to try m=z or $z \leq m < |X|$ instead of m=|X|. There are several methods for this.

3.2. Serial Search Through List of Actual (x,y) Pairs

We choose m=|S| (or we take for m the maximal value z that we expect for |S|. In the rows of the matrix we simply store the pairs (x,y) of messages. For this we need a code d: $X \times Y \rightarrow \{0,1\}^n$.

Now the first pair (x,y) is stored in the first row by giving the first horizontal wire activity 1 and applying the pattern d(x,y) to the vertical wires. Then the second pattern is stored in the second row in the same way, and so on. In the recall of $x \rightarrow y$ one has to put a 1 into the first horizontal wire, record d(x',y') on the vertical wires, compute (x',y') and check whether x' corresponds to x. In this case y'=y is the correct output; if not, one has to repeat this procedure with the second horizontal wire, and so on.

Obviously the main problem of this procedure is the long time (the large number of comparisons) needed for the recall. Even, if one codes x and y separately, i.e. uses a code $d(x,y) = (d_1(x), d_2(y))$, and instead of comparing x and x' compares $d_1(x)$ and $d_1(x')$ bit by bit, one needs on the average $m+ld|X|$ one bit comparisons, before the right address is found and the right output can be computed. Therefore we have to look for other methods that reduce the look-up time. The next method reduces the look-up time at the cost of some of the storage capacity. The 4th and 5th methods save look-up time by comparing $d_1(x)$ with all the stored $d_1(x')$ in parallel, instead of serially.

3.3. Hash Coding of (x,y) Pairs

We choose m between $|S|$ and $|X|$ and use a variety $c_1, c_2, c_3,...$ of codes $c_i: X \to \{1,...,m\}$. For the storage, we give the $c_i(x)^{th}$ horizontal wire activity 1. But before setting the switches we first check the output on the vertical wires.

1) If this output $(b_1,...,b_n)$ is zero, we know that nothing has been stored yet in the $c_1(x)^{th}$ row, and proceed to set the switches. This time we need a code $d: X \times Y \to \{0,1\}^n$ and set the activity in the vertical wires according to $d(x,y)$, i.e. we store not only y but also x in the $c_1(x)^{th}$ row of the matrix.

2) If the output $(b_1,...,b_n)$ is not zero, this means that some pattern has already been stored in the $c_1(X)^{th}$ row, and we repeat the procedure with the $c_2(X)^{th}$ row, and so on until we find an empty space where the pattern (x,y) can be finally stored (in a coded form as $d(x,y)$.

For the recall we first compute $c_1(x)$, then input activity 1 to the $c_1(x)^{th}$ row of the matrix, record $d(x',y')$ on the vertical wires, compute (x',y') from that (by the inverse of the fixed code d), and check whether $x'=x$.

1) If $x'=x$ we are done and the correct output is $y'=y$.

2) If $x'\neq x$ we have to do the same with $c_2(x)$ and so on, until finally we find the correct pattern (x',y') with $x'=x$.

In this procedure, obviously the main question is: how many repetitions of the storage (and recall) procedure are needed on the average until an empty space has been found. Here usually it is assumed that the codes c are 'quasi random' and probabilistic arguments are used to estimate this number of repetitions, i.e. the number up to which the codes c_1, c_2, c_3,... have to be used on the average. It turns out that this number is rather small (1,2 or 3).

The argument is roughly the following: If k patterns have been stored in the matrix, then the probability of hitting an occupied place with $c_1(x)$ is k/m, the probability of hitting an occupied place twice (with $c_1(x)$ and $c_2(x)$) is $(k/m)^2$, and so on. Therefore we expect to find an empty space in

$$1+k/m+(k/m)^2 +(k/m)^3 + ... = 1/(1-k/m) \tag{1}$$

steps. Thus, on the average, for a list S with $|S|= z$ we need

$$z^{-1}\sum_{k=0}^{z-1}1/(1-k/m) \approx (m/z)\int_0^{z/m} 1/(1-x)dx = -(m/z)\ln(1-z/m) \tag{2}$$

steps.

Table 1 relates this number of steps to various values for m/z. The dimensions of the mxn matrix needed for this procedure are specified by $|S|<m<|X|$ and $n \geq ld|X|+ld|Y|$. The average number of one-bit comparisons needed in the recall is now reduced to

$[(m/z) \ln(m/(m-z))-1] \cdot 2 + ld\ |X|$.

$\frac{m}{z}$	nr. of steps needed
1.1	2.64
1.3	1.91
1.5	1.65
1.7	1.51
1.9	1.42
2.1	1.36

Table 1

3.4. Parallel Search Using Complex Switches

Here we need two matrices: one to store the inputs x occurring in the set S, and one for the corresponding outputs y. For the storage of the inputs we take the variation of the matrix that uses $s_{ij} \in \{-1,1\}$. We take an input code c: $X \rightarrow \{-1, 1\}^{n_1}$ (this implies that $|X| \leq 2^{n_1}$) and use an mxn_1 matrix with m=|S|.

For the storage of the output we take an output code d:$Y \rightarrow \{0,1\}^{n_2}$ (implying $|Y| \leq 2^{n_2}$) (we could as well take $\{-1,1\}^{n_2}$ and use the same variation of the matrix also for the output) and use an mxn_2 matrix.

In the storage of the kth pair (x,y) the pattern c(x) is applied to the vertical wires of the first matrix, and a 1 is put into the kth horizontal wire. This switches some switches in the kth row from -1 to 1, thus reproducing the pattern c(x) in this row. At the same time the pattern d(y) is applied to the vertical wires of the second matrix.

In the recall the input pattern x is coded into c(x)=b, this is then applied to the vertical wires of the first matrix, yielding the output a=S·b on the horizontal wires.

The output on each of these wires is then passed through a threshold device θ with threshold n_1 ($\theta(t) = 0$ if $t < n_1$ and $\theta(t) = 1$ if $t \geq n_1$).

This threshold of n_1 will be reached in exactly one of the horizontal wires (say the ith), since $a_i = \sum_{j=1}^{n_1} s_{ij} b_j = n_1$ ($s_{ij}, b_j \in \{-1,1\}$) means that $b_j = s_{ij}$ for every j, i.e. that the i_{th} row contains exactly the input pattern b. This means that the m threshold devices in the horizontal wires perform in parallel a comparison of the input pattern b with each of the patterns stored in the horizontal lines. The outputs of these m threshold devices will then be used as input to the second matrix yielding as output the pattern stored in the ith row of the second matrix i.e. d(y). By the inverse of the coding d this pattern is finally transformed into the output y.

This method needs an $n_1 \times m$ and an $n_2 \times m$ matrix, i.e. $m \cdot (ld|X| + ld|Y|)$ storage elements and m = z threshold devices. It needs roughly $2 \cdot m \cdot ld|X|$ operations in the recall. We should also note that in this method the input matrix needs the more complicated $\{-1,1\}$ storage element.

3.5. Parallel Search Using Simple Switches

This method is only a variation of the fourth method. Again we need two matrices: one to store the inputs x in S, and one for the corresponding outputs y. We take an input code $c: X \rightarrow \{0,1\}_k^n{}^1 = \{(a_i)_i \in \{0,1\}: \Sigma a_i = k\}$ (thus $B(n_1,k)^* \geq |X|$). For the storage of the inputs we use an $m \times n_1$ matrix with $m=|S|$. For the storage of the outputs we take an output code $d: Y \rightarrow \{0,1\}^{n_2}$ (thus $|Y| \leq 2^{n_2}$) and an $m \times n_2$ matrix.

---

* We use this abbreviation for the binomial coefficients $B(n,k) = n!/(k! (n-k)!)$.

The procedure for storage and recall is nearly the same as in the fourth method. Only in the recall we have to use a different threshold to detect the row of the first matrix in which the input x has been stored. Since every input pattern c(x) contains exactly k ones, we have to set the threshold at k.

This method needs m threshold elements an an $n_1 \times m$ and an $n_2 \times m$ matrix. For $k=n_1/2$ we have $|X| \leq B(n_1,n_1/2)$, and therefore, given $\varepsilon > 0$ we can get $n_1 \leq (1+\varepsilon) \cdot ld|X|$ for sufficiently large n_1. Realistic values for $\alpha = 1+\varepsilon$ that can be obtained by concrete simple coding schemes are given in Table 2.

$\binom{k}{i}$	$\dfrac{k}{ld\binom{k}{i}} = \alpha$
$\binom{2}{1}$	2.00
$\binom{3}{1}$	1.89
$\binom{4}{2}$	1.55
$\binom{5}{2}$	1.51
$\binom{6}{3}$	1.39
$\binom{7}{3}$	1.36
$\binom{8}{4}$	1.31
$\binom{10}{5}$	1.25
$\binom{16}{8}$	1.17
$\binom{20}{10}$	1.14
$\binom{100}{50}$	1.04
$\binom{1000}{500}$	1.005

Table 2: A 0-1-sequence of length n is built from n/k groups of k elements each. If each group contains exactly i ones, we obtain $B(k,i)^{n/k}$ possible sequences of length n containing exactly $i \cdot n/k = j$ ones. Now $B(k,i)^{n/k} < B(n,j) < B(n,n/2) < 2^n = |\{0,1\}^n|$. Still $\log(2^n)/\log(B(k,i)^{n/k}) = k/ld(B(k,i))$ can be close to 1. These values are shown in the Table.

In summary, we need $m \cdot (\alpha \cdot ld|X| + ld|Y|)$ storage elements, where typical values for are given in Table 2, and $m=|S|$ threshold devices. The total number of elementary operations performed in a recall is roughly $(\alpha/2) \cdot z \cdot ld|X|$.

3.6. Association Between x and y for All (x,y) Pairs

For j<m and k<n, we use an mxn matrix, an input coding of X into the set of all 0-1-vectors of length m with exactly j ones $c: X \rightarrow \{(a_i)_i \in \{0,1\}^m : \Sigma_i a_i = j\} = \{0,1\}_j^m$, and an output coding of Y into the set of all 0-1-vectors of length n with exactly k ones $d: Y \rightarrow \{0,1\}_k^n$. This implies that $|X| \leq B(m,j)$ and $|Y| \leq B(n,k)$.

For the storage we apply the pattern $c(x)$ to the horizontal wires and $d(y)$ to the vertical wires of the matrix. In the recall the input x is coded into $c(x)=a$, this is applied to the horizontal wires, yielding $b=S*a$ as output on the vertical wires. Each vertical wire is then passed through a threshold device with threshold j, yielding a 0,1-pattern which is then coded by the inverse of the output code d into the output y.

Obviously, there is a problem in this procedure, namely that the output $b=S*a$ might contain more than k numbers that reach the threshold j, and therefore the inverse coding of d may not be possible. The chance that this kind of error will occur increases with the number z of pairs in the set S. (Note, however, that this error will always make the decoding impossible, but will never lead to a possible decoding with a wrong output.) I have calculated the probability of such an error and related quantities[22]. Some results of these calculations are shown in Table 3. It turns out that for large matrices the error probabilities can be kept small enough to ensure an effective storage capacity of about 2/3 of a bit per switch in such a matrix[22].

This method sometimes does not work correctly (see Table 3), it needs n threshold devices an an mxn matrix. If |S| is not too small, n can be chosen considerably smaller than $|S|$ (see Table 3) and usually we get $m \cdot n \leq 2 \, (ld|Y|) \cdot |S|$. The Number of operations in one recall is roughly $j \cdot n$, where j will be considerably smaller than m.

m × n	k	l	z	I	p_0	p_1	F
100 × 100	2	6	420	4 631	0.52	0.17	0.73
100 × 100	2	7	280	3 352	0.14	0.010	0.15
1000 × 100	2	9	3 600	43 057	0.14	0.010	0.15
1000 × 100	2	7	4 800	53 042	0.51	0.16	0.72
1000 × 1000	2	9	34 800	592 600	0.72	0.37	1.28
1000 × 1000	2	10	25 000	466 200	0.13	0.009	0.14

Table 3: m × n = dimensions of the matrix, z = (estimation for) $|S|$, I = information stored (in bits), p_0 = probability of more than 0 errors, p_1 = probability of more than 1 error, F = expected number of errors (wrong additional ones in an output pattern).

4. METHODS FOR PATTERN COMPLETION

Now we use the switching matrix to store a set S of patterns in such a way that from a (sufficiently large) part x' of a pattern $x \in S$ that has been stored we can retrieve the whole pattern x. There are basically two ways of understanding this task:

i) Given a part x of a pattern we want to obtain a complete list of all patterns x in S that contain x' (or that match x' best).

ii) Given a part x' of a pattern we want to find one pattern in S that best matches x'. (In this case we assume that x' contains enough information to uniquely determine a best match in most cases.)

In both (i) and (ii) we have to specify what we mean by a "part" of a pattern. To this end we assume that the set X of possible inputs consist of strings of subelements (like sentences consist of words or letters or like different characteristics of a person x=([name], [address], [phone no.]), or like the pairs x=(q,a) of questions and answers considers above), i.e. $X = X_1 \times X_2 \times ... \times X_r$. A 'part' of a pattern x= $(x_1, x_2, ..., x_r)$ is simply a pattern like $x' = (x_1, x_2, ?, ?, x_5, x_6, ?, x_8, ..., x_r)$. In (ii) we also want an output when x' is not part of any pattern x in S; in this case we need a criterion to find the 'best match' i.e. the pattern in S that comes closest to x'. For this we may use something like the 'Hamming distance', i.e. the number of differing subelements, but we may also use a refined distance that incorporates distances d_i inside the sets $X_1, ..., X_r$ (especially if these sets are large this may become important), i.e. we use the distance $d(x,y) = \sum_{i=1}^{r} d_i(x_i, y_i)$. The Hamming distance is a special case of this formula, where $d_i(x_i, y_i) = 0$ if $x_i = y_i$ and $=1$ otherwise.

4.1. Serial Search Through List of All Patterns

We simply store the nth pattern in the nth row of the matrix. For this we need input codes $c_i: X \rightarrow \{0,1\}^n$ and an mxn matrix with m=|S| and n= $\sum_i n_i$.

The storage of the nth pattern $c(x) = (c_1(x_1), c_2(x_2), ...)$ is done by applying this pattern of activity to the vertical wires and activity 1 to the nth horizontal wire. In the retrieval we have to compare the input x' to all the pattern stored in the matrix. Again we can do this bit by bit and for each pattern x we compute d(x,x'). Then we have to remember the row (or rows) with the lowest value of the distance d (or at least the numbers of the corresponding wires), and we have to go through the whole matrix either to get a complete list of best matches, or to be sure to have found the

best match. For this we need $m \times n'$ one-bit comparisons (which takes a long time: $n' = \sum_{i \in x'} n_i$).
This method needs $m \cdot (\sum_{i=1}^{r} \mathrm{ld}|X_i|)$ storage elements.

4.2. Parallel Matching Using Complex Switches

Here we use the variation of the matrix working with $\{-1,1\}$ instead of $\{0,1\}$. We need input codes $c_i: X_i \to \{-1,1\}^n$ and store the ith pattern $c(x)=(c_1(x_1), c_2(x_2), \ldots c_r(x_r))$ in the ith horizontal wire.

In the recall, a part x of a pattern is coded by the codes c_i corresponding to the subpatterns x_i that are known, and the unknown entries are coded as zeros. E.g. for $x'=(x_1, x_2, x_3, ?, ?, \ldots)$ we get $b = c(x') = (c_1(x_1) c_2(x_2), c_3(x_3), 0, 0, \ldots)$. This pattern $c(x')$ is applied to the vertical wires of the matrix, and on the horizontal wires we get the 'scores' $a_i = \sum_j s_{ij} b_j$ indicating how well the pattern in the ith row matches the input b. Then we have to find (and remember for a short time) those numbers i that yielded the highest score. This can again be done by threshold detection as in method 3.4. For example one can start with high threshold and lower it gradually until at least one output appears at the horizontal wires, and after that, by giving activity 1 successively to those wires one can find the complete patterns x (via $c(x)$) that best match x'.
This method needs an $m \times n$ matrix, $n = \sum_{i=1}^{r} n_i$, $m = |S|$ and therefore $z \cdot \sum_i \mathrm{ld}|X_i|$ storage elements, and z threshold devices, and about $2 \cdot z \cdot n$ operations in the recall.

4.3. Parallel Matching Using Simple Switches

We use the same idea again but a different input coding. We take codes $c_i: X_i \rightarrow \{0,1\}_{k_i}^{n_i}$ implying $|X_i| \leq B(n_i, k_i)$.

Again we store the patterns $c(x)=(c_1(x_1),...,c_r(x_r))$ successively by applying them to the vertical wires and an input of 1 successively to one of the horizontal wires. In the recall we again apply x' to the vertical wires and interpret the output on the horizontal wires as a 'score', note the wires with the highest score and retrieve the corresponding patterns x successively by putting input +1 to those horizontal wires and recording the output $c(x)$ from the vertical wires.

This method needs more storage elements than the second method, namely $\alpha \cdot z \cdot \sum_i \mathrm{ld}|X_i|$. Table 2 shows the values of for different codes c. It needs m=z threshold devices and about $z \cdot (\alpha/2) \sum_i \mathrm{ld}|X_i|$ operations in the recall.

4.4. Auto-Association of All Patterns

Again we take codes $c_i: X_i \rightarrow \{0,1\}_{k_i}^{n_i}$ (implying $|X_i| \leq B(n_i, k_i)$) and define $c(x)=(c_1(x_1),..,c_r(x_r))$, $c: X \rightarrow \{0,1\}_k^n$, where $n = \sum_i n_i$ and $k = \sum_i k_i$.

For the storage, again the pattern $c(x)$ is applied to the vertical wires of the matrix, but now the same pattern is also applied to the horizontal wires. Thus we need an nxn-matrix. This procedure is repeated for every pattern x in S. In the recall, the part $c(x')$ is applied to the vertical wires and every horizontal output wire is passed through a threshold device. All thresholds always take the same value, and this value is adjusted in such a way that the output a_1 has more 1's than the input $c(x')$. Then a_1 is taken as an input, yielding a_2 and so the number of 1's is successively increased, until the final a_f has k 1's. Then a_f is recoded by the inverse of c, yielding the complete output pattern x.

This procedure needs n threshold devices and nxn storage elements. In the recall it needs about n·k operations. But you will notice that this method may not always work, since it may end up with a pattern a_f that does not have exactly k ones, or that cannot be recoded by the inverse of c. Again I have estimated the probability of such errors[22] and some results are given in Table 4. The method usually works reasonably well with about 3 switches per bit (i.e. $n^2 \approx 3 \cdot I$ in Table 4) and if |S| is not too small we can achieve n < |S| (see also Table 4).

n	k	z	I	F_1	F_2	F_3	F_4	F_5	F_6	F_7	F_8	F_9	F_{10}	F_{11}
100	5	294	3 085	4.3	9.1	19.5	42.5							
100	7	140	2 720	1.2	2.4	4.6	9.3	19.3	41.5					
1 000	8	13 020	335 423	10.7	20.2	38.4	73.0	139.4	267.2	513.9				
1 000	12	5 200	301 373	0.6	1.2	2.3	4.4	8.4	16.4	32.0	62.8	123.8	245.7	490.9

Table 4: n x n = dimensions of the matrix, B(n,k) possible patterns, z = |S|, I = information stored (in bits), F_i = expected number of errors (wrong additional ones), when in the input pattern i ones are missing (i.e. the input contains k-i ones).

Note that the procedure of recall introduced here needs a repeated use of the matrix which can be carried out most easily if the matrix is provided with feedback connections (as in Fig. 2): these can obviously also be used during storage, when a pattern x has to be associated with itself.

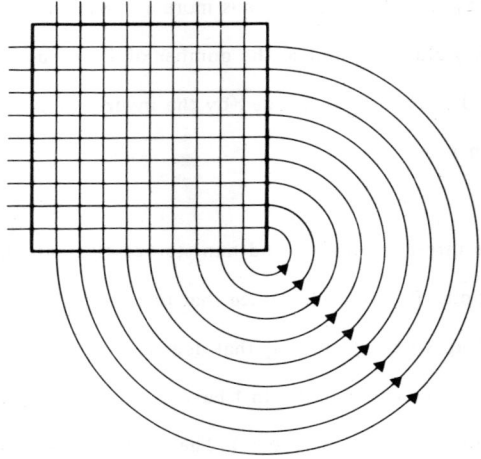

Fig. 2

In this case the recall procedure with its 'playing with the threshold' to obtain further 'hints' for the right pattern x (and the obvious criterion that the final a is 'right', if it 'holds' at the maximal threshold k), has a certain introspective appeal as a description of the way we try to remember something. I have elaborated on this in my book 'Neural Assemblies' [23]. But in this paper my aim is to compare the various algorithms from a technological and economical point of view, i.e. by means of criteria like the ratio of bits of stored information and number of storage elements, or the amount of additional hardware needed for coding, threshold detection and similar supplementary tasks.

5. COMPARISON OF THE VARIOUS METHODS

Let me begin by characterizing briefly the various methods of sections 3 and 4. Method 3.1. seems to be the most straightforward one, but it is usually not adequate because $|S|<<|X|$. Method 3.2. (and 4.1.) is the conventional storage method. The other methods introduced here become interesting only when the conventional method takes too much look-up time, i.e. when the list S becomes large (say $|S|>1000$).

The most interesting storage procedures which really make use of our basic device as presented in Section 2, are Methods 3.4, 3.5 and 3.6 and Methods 4.2, 4.3 and 4.4. These methods employ the possibility of highly parallel processing in such a matrix. Methods 3.4, 3.5 and 4.2, and 4.3 do the matching of patterns in parallel, and differ only in the coding they use. They can be referred to as <u>parallel comparison methods</u>. Methods 3.6 and 4.4 also work in parallel, but they differ from the other methods in one important respect. They really store information in a distributed way. This means that the storage of a new pattern interacts with all the previously stored patterns and implies a certain 'fuzzyness' of the working of the memory. This has disadvantages: Error probabilities have to be taken into account, erasing a stored pattern in the sense that storage and erasure of a pattern have exactly zero effect on the matrix, is not possible. But it also has advantages: If part of the matrix is destroyed or does not work properly, there is not really one whole pattern lost, but the error probabilities in all patterns are slightly increased. The memory is never really 'full', i.e. one can always put in one more pattern, thereby increasing the error probability for all patterns, but one does not have to erase any specific pattern in order to put in a new one. It allows for, in fact it requires redundant coding of the patterns, whereas in all other methods redundant coding is a waste of memory space. From these characteristics it is clear that only methods 3.6 and 4.4 are truly 'associative'. They will be referred to as <u>associative methods</u>.

5.1. Pattern Mapping

The task is to install a mapping between all pairs of patterns (x,y) in a list S. Here it seems reasonable to distinguish three cases.

1) |Y| is small compared to |X|. A typical example for this case is the association of the correct answers 'yes' or 'no' to a number of questions (|Y|=2). This case is in favour of those methods that do not need storage space for the questions, i.e. methods 1 and 6. Method 1 is only useful, however, when |S| is very large.

2) |Y|≈|X|. This is probably the most common case in practice. Examples are abundant - a telephone book is one. In this case all methods introduced here work reasonably well except the conventional method 2 which takes too much look-up time, and method 1, which is only economical if the list S is very long (|S|≈|X|).
Larger |X|/|Y| and |S| are in favour of the associative method 6, also in view of the total number of elementary operations needed in the recall.

3) |Y| is large compared to |X|. A typical example for this case is a library, where you ask a title x and get the book y. For this extreme case digital data processing may not be a reasonable solution at all. Moreover, one would usually use keys or addresses y' for the elements y Y, store the pairs (y',y) in a conventional way (i.e. by method 2) or by method 1 or simply by filing the books, and then use a faster method to store the mapping x→y', where now the set Y' is much smaller than Y. This procedure reduces

403

the problem to case (1) or (2). In principle we can again use all six methods in this case. Method 1 again will take too much look-up time. Methods 3 and 6 have the disadvantage against methods 4 and 5, that they need more storage elements, but on the other hand the associative method 6 needs less threshold elements and also less elementary operations in the recall than the parallel comparison methods 4 and 5 (compare Table 3), and thus becomes preferable for larger S.

Tables 5 and 6 show the relative advantages and disadvantages for the methods 1-6 in the various cases. Table 7 gives concrete parameter values for method 6 for concrete storage problems, when 10^4, 10^6 or 10^8 bit have to be stored. Finally one should note that I did not explicitly consider the problem of widely varying length of the list S. All methods need additional techniques and on the average more storage space to handle this problem [13]. The associative method 6, however, will work reasonably well without modification, over a considerable range of S (cf. Fig. 3 and Fig. 4).

method	switches/bit	thresholds	additional hardware	increase $z \rightarrow$	coding	retrieval time
1	$\frac{x}{z}$	0	–	no problem	very short	0
2	$\frac{ld\ x + ld\ y}{ld\ y}$	0	$ld\ x$ – bit comparator	additional data cannot be stored	very short	long
3	$1.5 \frac{ld\ x + ld\ y}{ld\ y}$	0	$ld\ x$ – bit comparator	retrieval time takes longer	very short	short
4	$\frac{ld\ x + ld\ y}{ld\ y}$	z	complex switches	additional data cannot be stored	very short	0
5	$\alpha \cdot \frac{ld\ x}{ld\ y} + 1$ *)	z	–	additional data cannot be stored	some redundancy	0
6	2.4	$\leq \frac{z}{10}$	–	higher error probability in all data	considerable redundancy	0

*) where α is taken from table 0

Table 5

	x >> y	x ~ y	x << y
$300 \le z \le 10^4$	6 > 3 > 4 ~ 5	3 ~ 4 ~ 5 ~ 6	2 ~ 3 ~ 4 ~ 5 ~ 6
$10^4 \le z \le \frac{x}{10}$	6 > 3 > 4 ~ 5	6 ~ 3 ~ 4 ~ 5	3 ~ 4 ~ 5 ~ 6
$\frac{x}{10} \le z \le x$	1 > 6 > 3	1 > 6	1 >

Table 6: Preferences among the six methods in various cases. $x = |X|$, $y = |Y|$, $z = |S|$.

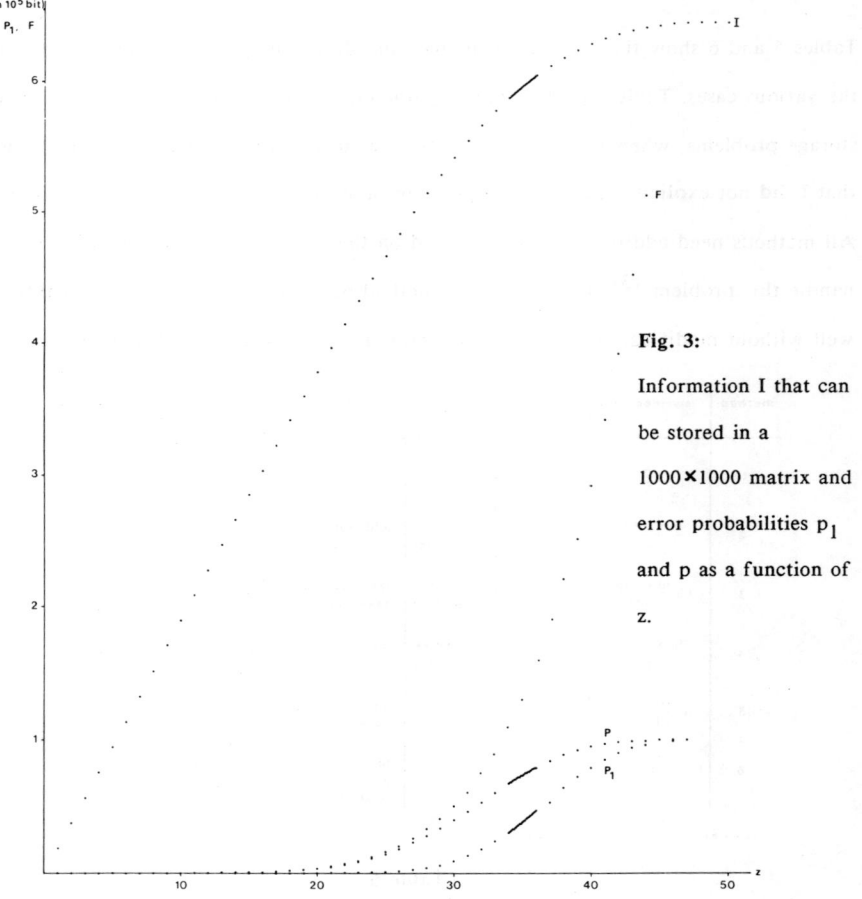

Fig. 3: Information I that can be stored in a 1000×1000 matrix and error probabilities p_1 and p as a function of z.

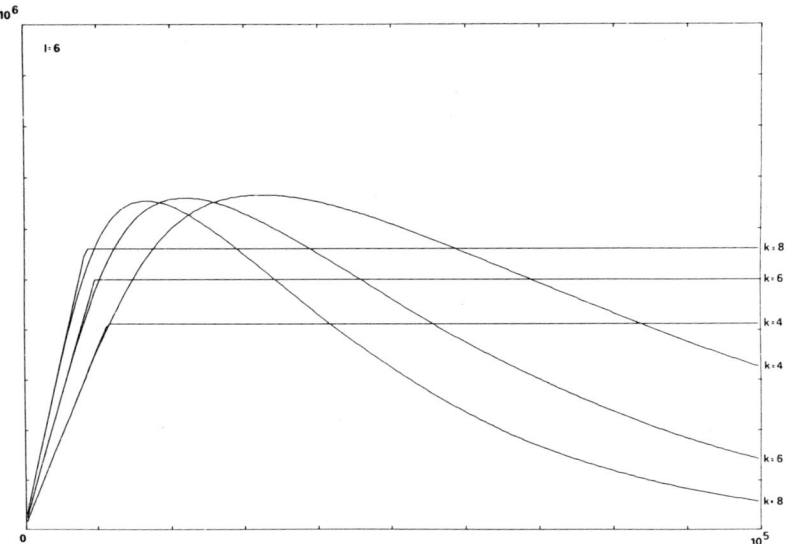

Fig. 4: Information I that can be stored in a 1000 1000 matrix as a function of z. Comparison of associative and conventional methods.

B	x	y	z	m	I	n	k	bit/sw	Dens(1)
10^4	$2^{10} - 2^{15}$	2^{16}	625	318	2	56	9	0.5633	0.4685
10^4	$2^{10} - 2^{22}$	2^{16}	625	318	3	56	7	0.5631	0.5217
10^4	$2^{10} - 2^{15}$	2^{16}	625	332	2	54	9	0.5583	0.4662
10^4	$2^{10} - 2^{23}$	2^{16}	625	384	3	49	7	0.5420	0.5024
10^4	$2^{15} - 2^{24}$	2	10 000	5 806	2	4	1	0.4306	0.5771
10^4	$2^8 - 2^{11}$	2^{50}	200	3 266	1	6	1	0.5107	0.3996
10^4	$2^8 - 2^{20}$	2^{50}	200	210	3	100	25	0.4779	0.5111
10^6	$2^{16} - 2^{27}$	2^{16}	62 500	20 670	2	80	9	0.6049	0.4924
10^6	$2^{16} - 2^{26}$	2^{16}	62 500	16 050	2	100	9	0.6231	0.5036
10^6	$2^{16} - 2^{40}$	2^{16}	62 500	26 900	3	64	7	0.5812	0.5314
10^6	$2^{18} - 2^{26}$	2^8	125 000	15 480	2	100	5	0.6495	0.5528
10^6	$2^{18} - 2^{26}$	2^8	125 000	15 480	2	100	4	0.6538	0.4731
10^6	$2^{14} - 2^{150}$	2^{200}	5 000	2 600	22	1000	38	0.3853	0.7996
10^6	$2^{14} - 2^{164}$	2^{200}	5 000	6 000	20	500	49	0.3367	0.8047
10^8	$2^{24} - 2^{40}$	2^{32}	3 125 000	29 000	3	5 000	11	0.6903	0.3723
10^8	$2^{24} - 2^{28}$	2^{16}	6 250 000	29 000	2	5 000	8	0.6918	0.1699
10^8	$2^{24} - 2^{54}$	2^{16}	6 250 000	29 000	4	5 000	4	0.6911	0.1699
10^8	$2^{20} - 2^{66}$	2^{500}	200 000	7 500	6	20 000	94	0.6727	0.5253
10^8	$2^{20} - 2^{73}$	2^{500}	200 000	15 300	6	10 000	100	0.6571	0.5365

Table 7: B = (estimate for the) number of bits to be stored, m,l,n,k = parameters of matrix as in Table 3, bit/sw = stored bits per switch, Dens(1) = density of ones in the matrix after the list S has been stored.

5.2. How to Determine the Dimensions of the Matrix for the Associative Method

If one wants to use the associative method 6, how should one go about finding the parameters m,n,k and j required in this method? Typically the parameters $x=|X|$, $y=|Y|$, and $z=|S|$ will be given. The number b of bits to be stored can then be calculated as $b=z \cdot ldy$. From the asymptotic equations given in [22], one derives that $m \cdot n \approx b/\ln 2$ (from eq. 4.12) and that

$$j \cdot k = (m \cdot n/z) \ln 2 \approx b/z = ldy \text{ (from eq. 4.2 with p=1/2)}. \tag{3}$$

But how shall one proceed? Obviously n should not be chosen too large because the method requires n threshold elements, but on the other hand the coding of the output requires that $B(n,k) \geq |Y|$.

Once n has been chosen, equation (4) gives a good idea about the size of m, and one can optimize the information content of the matrix (i.e. formula (3.1.1) of [22]) by varying k and j. Table 7 shows the results of this procedure for different values for x,y,z and n. As a rule of thumb for the choice of n, I have found that $n \geq \sqrt{(ldy)/2}$ should always be fulfilled, and that a reasonable choice for n that usually gives $m \, n \leq 2b$ is about $n = \exp(\sqrt{ldy})$ (or $ld \, n = \sqrt{2 \cdot ld \, y}$). But then one still has to find the maximal information content varying j and k, which in my examples (almost) always showed $k \geq j$.

Usually we find parameter values for n and k such that $B(n,k)$ is considerably larger than $|Y|$. This means that the output patterns in Y are coded by $d: Y \rightarrow \{0,1\}^n_k$ as a subset of all the $B(n,k)$ possible 0,1-sequences of length n with k 1's. This redundancy can be used for an error-correcting code $d:\{0,1\}^n_k \rightarrow Y$, and such a code will be needed because the matrix stores just enough information to determine the output pattern $y \in Y$ but not enough information to determine correctly the codeword $d(y)$. Therefore it will indeed produce errors (i.e. additional 1's) in the output 0,1-sequence $d(y)$, but it should be possible to determine the final output y correctly in spite of these errors.

5.3. Pattern Completion

In the case (i) of section 4 where the task is to find a complete list of best matches, the associative method 4 is not adequate, since it yields always only one output pattern, and although one could derive a strategy of obtaining several matches with method 4, one could never be sure of getting a complete list. Method 1 is only reasonable for small $|S|$. Thus in most cases one should use the parallel comparison methods 2 or 3. Since both methods, 2 and 3 require summations and threshold decisions performed in parallel, it is conceivable that for large $|S|$ one may split the list S into a few sublists and work on these serially. This would be a combination of the purely serial first method and the purely parallel second (or third) method.

In the case (ii) where the task is just to find one match (or only a few matches), the associative method becomes interesting, and we shall discuss this case for the rest of this section. I also think it is the more common case, since usually one is not really interested in a complete list - even if one can get it. A typical example for this situation is the search for literature references, where one usually wants just a few good references and is frustrated by the awfully long (but in a sense 'complete') list, the computer spits out. Here there are basically three reasons that suggest the employment of the truly associative method 4.

(1) $|S|$ becomes too large and method 4 needs significantly less summations and threshold elements than $|S|$ - as in method 2 or 3. As a rule of thumb the dividing point is when $b \approx z(ldz)^2$. Of course, a more direct comparison between the associative method 4 and the parallel comparison methods 2 and 3 would require a numerical relation between the dimensions of the matrix in method 4, and in the other methods. This will be discussed in more detail in 5.4.

(2) If one looks for best matches, the metric on X defining the optimality should be preserved by the coding. This is much easier to achieve with a redundant code - as required in the associative method 4. To put it the other way round: if the metric on X is too fine, it may turn out to be very hard, if not impossible to find a short code for the elements of S that preserves the metric.

(3) In many cases one wants an interactive procedure by which the user can arrive at a 'nice' list of best matches, instead of a complete but much too long list. Method 4 gives out only one match at a time and there are several reasonable strategies to use it in an interactive way.

Table 8 summarizes some advantages and disadvantages of the four methods.

method	switches/bit	thresholds	additional hardware	increase z →	coding	retrieval time
1	1	0	ld x − bit comparator z score-memories	additional data cannot be stored	very short	long
2	1	z	z score-memories complex switches	additional data cannot be stored	very short	short
3	$(z\cdot)$	z	z score-memories	additional data cannot be stored	some redundancy	short
4	3.4	$\leq \frac{z}{10}$	threshold control	higher error probability in all data	considerable redundancy	short

5.4. How to Determine the Dimensions of the Matrix for the Associative Method

If one wants to use an nxn matrix with method 4 in a concrete case, how should one choose the parameter values for n and k? Again we assume that the parameters $x=|X|$ and $z=|S|$ are given. Moreover, we need to know the total number b of bits to be stored. This number has to be less than $z \cdot ldx$, since in this method we need to know a part of a pattern to address it. Indeed, I shall assume that the information stored about a simple pattern on the list is less than $(1/2) \cdot ld\ x$. Thus I assume that $b \leq (z/2) \cdot ld\ x$. Now the procedure is quite simple: From the asymptotic formulae in [22] one gets $n \approx \sqrt{2b/\ln 2}$ (eq. 5.12) and

$$k \approx n\sqrt{(\ln 2)/z} \approx \sqrt{2b/z} \quad \text{(eq. 5.2 with p=1/2).} \tag{4}$$

For a concrete value of n one can simply find the best value of k by optimizing the information content of the matrix (i.e. formula 3.17 of [22]) varying k. Finally, one should check whether $B(n,k) \geq |X|$. The results of this procedure for some concrete cases are shown in Table 9.

B	x			z	n	k	bit/sw
10^4	2^{11}	—	2^{30}	1 300	176	5	0.3186
10^4	2^{12}	—	2^{25}	2 000	174	4	0.3337
10^6	2^{18}	—	2^{55}	65 000	1 708	6	0.3431
10^6	2^{20}	—	2^{38}	200 000	1 710	4	0.3423
10^6	2^{18}	—	2^{46}	100 000	1 704	5	0.3445
10^8	2^{24}	—	2^{74}	$6.5 \cdot 10^6$	17 000	6	0.3462
10^8	2^{25}	—	2^{62}	10^7	17 000	5	0.3464

Table 9

6. ASSOCIATIVE MEMORY AND SPARSE CODING

From the discussion in the preceding section it becomes clear that a very unusual coding of patterns into 0-1-sequences is needed for the associative methods 3.6. and 4.4. What we need is a <u>sparse coding</u> with a very low fixed number k of ones distributed over a much larger number n of places. How can this be done in practice?

One practical way of constructing a redundant code $d: Y \rightarrow \{0,1\}^n_k$ for $Y=\{0,1\}^r$, starts by dividing a sequence z in $\{0,1\}^n$ (and y in $\{0,1\}^r$) into k sections of roughly equal length $l \approx n/k$ (and $l' \approx r/k$ respectively). Then an input section of length l' from y is interpreted as a binary expression of a number x between 0 and $2^{l'}-1$ and this number is coded into a 0-1 sequence of length $l=2^{l'}$ consisting of only 0's except one 1 in the $(x+1)^{st}$ digit. This is then the corresponding section of the codeword $z=d(y)$. This way one can easily and efficiently address a large matrix, since instead of n parallel wires one needs only an address of length $r \approx k \cdot l' = k \cdot ld\, l = k \cdot ld(n/k)$. (Similar coding schemes are considered in Table 2).

If such a coding is used, an error consisting of one additional 1 in the sequence d(y) is not very serious, since the additional 1 will fall into one of the subsections, and only one additional bit would be needed to determine which of the two 1's is the wrong one. If we rely on this method we get interested in those parameters k,l,m,n of the matrix, which yield a low probability for more than one additional one to occur. For the practical cases considered in Table 7 these parameter values have been calculated in Table 10.

B	x	y	z	m	l	n	k	bit/sw	p_1	F	Dens(1)
10^6	$2^{16} - 2^{105}$	2^{16}	62 500	31 300	9	80	3	0.3995	0.0073	0.1269	0.4895
10^6	$2^{16} - 2^{102}$	2^{16}	62 500	25 400	9	100	3	0.4167	0.0096	0.1463	0.4847
10^6	$2^{14} - 2^{96}$	2^{200}	5 000	3 100	12	1 000	33	0.3300	0.0090	0.1406	0.4720
10^8	$2^{24} - 2^{166}$	2^{32}	3 125 000	21 900	16	17 000	3	0.5268	0.0091	0.1416	0.3723
10^8	$2^{20} - 2^{170}$	2^{500}	200 000	26 500	16	10 000	57	0.3787	0.0098	0.1471	0.4901

Table 10

In concrete applications there will often occur more natural ways of devising such a sparse coding. Using a newly developed associative memory device [27], we have recently investigated a few such cases, namely coding of written words in terms of the pairs of consecutive letters they contained (compare Fig. 5) or coding of checkerboard-positions in terms of features that are relevant for this game. These investigations will be discussed in more detail in a forthcoming paper [26].

Input	Threshold	Output
KROKODIL	11	KROKODIL
KROKO	7	KROKODIL
ODIL	5	KROKODIL
CROCODIL	10	KROKODIL
KRODIL	8	KROKODIL
KROCKODIEL	11	KROKODIL
KRO?DIL	7	KROKODIL
KRO?IL	4	S???????OETE
S???????OETE	8	SCHILDKROETE
SCHULTKROETE	11	SCHILDKROETE
KROETENSCHILD	12	SCHILDKROETE
KROETE	6	SCHILDKROETE

Fig. 5: Pattern completion in a 448×448 associative memory in which about hundred names of animals had been stored.

Two general principles of such coding schemes can, however, be mentioned already at this point.

i) <u>The principle of similarity-preserving coding</u>: The code c should be such that original patterns that are "similar" in some natural sense are coded into 0-1-sequences that are similar in the Hamming-distance sense.

ii) <u>Coding into equally improbable features</u>: One can try to design a code into 0,1-sequences with very few ones by interpreting each place in the 0,1-sequence as signalling the presence of a particular feature in the original pattern. In this case all the n features would be about equally improbable (their probability should be $p=k/n$ in a code using sequences in $\{0,1\}_k^n$).

7. VARIATIONS OF THE BASIC DEVICE

Up to now we have discussed several strategies of information storing which work with one "basic device" - the "switching matrix". In practical situations there may be variations of this device, and some variations have been discussed in the literature. This section is devoted to the question, to which extent the results of the previous section still hold for such variations. The characteristic features of the device are

1) The fact that the memory elements are switches, i.e. have two states, "switched on", i.e. connectivity 1 between the two wires, and "switched off", i.e. connectivity 0,

2) the matrix arrangement of the memory elements and the particular wiring,

3) the switching rule.

In principle, the switching rule that we use in our basic device has been described in section 1. But we should notice that in every single method we have worked with only two activity values (usually 0 and 1, but also -1 and 1 in methods 3.4 and 4.2). Moreover in all methods we did not use the possibility to switch something off (it could for example be used to erase the whole matrix, if one wants to start afresh). This reflects the fact that in the two basic problems posed in section 1 we did not explicitly consider the problem of erasing something from the memory. In fact, this problem typically occurs in all methods except the associative methods 3.6 and 4.4, when the memory is "full" and one has to erase something in order to make space for a new pattern to be stored. In all these methods, however, erasure of one particular pattern is not at all difficult. In methods 3.6 and 4.4, where it would indeed be difficult to delete one pattern from the matrix without interfering with all the other patterns, this erasure is usually not necessary, since here one can store one additional pattern in the matrix, without deleting any previously stored pattern, only at the risk of somewhat higher error probabilities.

If we take 0 and 1 as the two activity values, then the simple switching rule that has actually been used in all methods can be described as in Table 11. This way of writing the switching rule already gives a framework for discussing other possible rules: any insertion of 'on', 'off' and ' - ' into the third row 'rule', will give a specific switching rule. I have discussed these variations elsewhere [23,24], although in a slightly different setting.

horizontal activity	1	1	0	0
vertical activity	1	0	1	0
rule	on	–	–	–

horizontal activity	1	1	0	0
vertical activity	1	0	1	0
rule	off	–	–	–

The exactly opposite rule, given in Table 12 or abbreviated as (off,-,-,-), could be used in all methods in an exactly complementary way, if all switches are in the 'on'-state initially. In this case the methods using threshold detection in the retrieval of course have to use an inverse criterion: when sufficiently high activity (=ϑ) showed that all these switches addressed by the inputs were switched on, now sufficiently low activity (=0) shows that all the switches addressed by the input are switched off.

When we consider switching rules which use both switching on and off in different situations (e.g. (on, off,-,-)), then the error detection becomes much more complicated in the associative methods when we may apply the switching rule many times in succession on the same switch. In this case we cannot distinguish a switch that has been switched on and off again from one that has not been switched at all. With our original simple rule an error in the recall would always show up as an additional 1 in the output pattern and could thus be easily detected (but not always be corrected). Now an error may also result in a missing 1 or, what is worse, in a missing 1 in one place and an additional 1 in another place, which cannot even be detected (in general). On the other hand, such a variation can provide the possibility also of erasing something from the matrix (although this is in general not possible without interfering with previously stored patterns).

Now I want to discuss in more detail the sort of variation that is most commonly mentioned for associative storage methods in particular in the context of "natural" associative memories, see also App. 2 of my book [23]. One could use a continuous version of the basic device instead of our discrete switching version - or at least one should allow for more than two states of the switches. The basic idea now is that instead of switches we have something like variable resistors that can vary the connectivity between the horizontal and vertical wires in a more or less continuous way between 0 and 1. A further variation allows also negative connectivity values.

Every presentation of a pair of patterns (or a pattern in case 4.4) to the matrix in the learning phase gives a change c in the connectivity values (i.e. the new values are obtained by adding the c-values to the old values). We may say that the switching rule is now given in a differential form. Denoting the (now possibly also continuous) values of activity in the horizontal and the vertical wire by x and y, we can specify such a differential switching rule as a function c(x,y) which gives the resulting change in activity. Thus we obtain the final connectivity between two wires (i and j) as

$$s_{ij} = c_o + \sum_{t=1}^{z} c(x_i(t), y_j(t)) \qquad (5)$$

where t=1,..,z denotes the successive presentation of the z input-output pairs, or in continuous time

$$s_{ij} = c_o + \int_0^T c(x_i(t), y_j(t))dt \qquad (6)$$

where T denotes the total presentation time. Rules of this type have been discussed in the literature, but they are still rather unrealistic, since, in principle, the sums or integrals may add up to arbitrary high values, whereas for example a resistor can only vary its connectivity between 0 and 1. Thus one should more realistically take $c_{ij} = f(s_{ij})$, where the function f is depicted in Fig. 6 - generally, one should take a sigmoid shaped function f, i.e. an increasing function f which becomes constant for very high ($s_{ij} \to \infty$) and very low ($s_{ij} \to -\infty$) values of the sum s_{ij}. The exact form of the function f may indeed turn out to be not very crucial [9,10]. For example, replacing f by a.f with a constant factor a does not matter since it only means a change of scale. Also a total shift of f in the vertical direction is not important (for example in the associative methods it can be compensated by a corresponding shift in the horizontal direction as well (it can be compensated by the initial value c_o of the connectivities). In the following discussion we shall therefore simply assume that f looks as in Fig. 6.

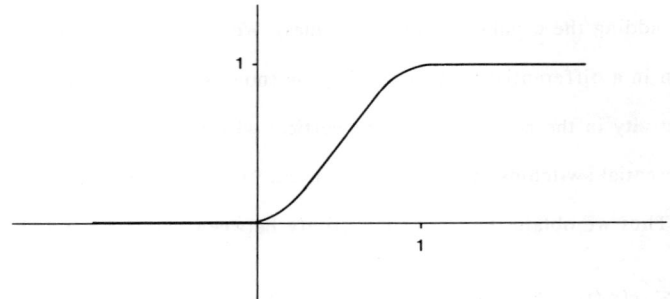

Fig. 6

If we now take only two activity values, 0 and 1, for x and y and let c(x,y)=x·y, then we obviously retain our "basic device" in this more general framework. Of course, in order to use the working range of f we should take the starting value $c_o=0$, since $c(x,y) \geq 0$. To obtain the opposite case (switching off) we simply take c(x,y)=-x·y and $c_o=1$.

What other storage rules seem to be reasonable in this framework? For the subsequent discussion we assume that the activity variables have positive values, but are not necessarily restricted (e.g. to {0,1}).

1) One could take c(x,y) = α·x·y with 0<α<1. Then the "switches" are not switched on in one step, but in several (1/α) smaller steps. This makes a better use of the working range of f - for example one could repeat a presentation of a pair of patterns if there occur errors in the recall after only one presentation -, but it also takes more bits to determine the state of each single switch. For example, if α=1/3, there are 4 possible states (0,1/3,2/3,1), but it is not at all clear whether one can use the whole device in such a way that it yields a 2-fold storage capacity, compared to our basic device. If α

becomes smaller one needs more and more bits to determine the state of a single switch in the matrix, and the operation of the device comes closer to a continuous variation of the connectivity (as in eq. 6). In this limiting case one cannot discuss the storage capacity in terms of bits anymore, but rather in terms of dimensions (or structural information-content [16,17]). In this case a good strategy is to use orthogonal vectors $(x_i)_i$ (and also $(y_j)_j$) as inputs to the matrix - or at least to store them in a kind of orthogonal way (this is only a brief side remark, remote from the main theme of this paper, it is discussed more explicitly in references [11] and [22]).

2) One could use $c(x,y) = - \alpha \cdot x \cdot y$. This is the analogous modification to the opposite of our basic device.

3) One could use correlation instead of just multiplication. If we have a set S of pairs (x,y) of vectors $x=(x_1,...,x_m)$ and $y=(y_1,...y_n)$ to store, we define the average vectors $\bar{x} = \sum_{t=1}^{z} x(t)/z$ and $\bar{y}$ similarly. Then taking the products leads to the matrix $x \otimes y = (x_i \cdot y_j)_{ij}$, whereas the correlation of x and y yields $(x-\bar{x}) \otimes (y-\bar{y})$. This strategy seems reasonable because the function $c(x,y)=(x-\bar{x})(y-\bar{y})$ is neither all positive as in our basic device, or in (1), nor all negative, as in (2), but gives on the average an equal amount of positive and negative contributions. In this case the initial value c_0 should be about 1/2. One should also notice that this type of rules also have the advantage that one can erase a pattern from the matrix by using anticorrelation, and this can be done without interfering with previously stored pattern as long as the switches are in the linear working range of the function f. This interesting modification of the simple multiplication, will not give a dramatically different behaviour (see 5 below), but it may in some actual simulations turn out to be slightly more effective, although this remains to be seen. It has been used for example in the Perceptron [28,21]. As I said above, the calculation or estimation of the storage capacity becomes more difficult for such methods, since the connectivities are now

not only changed in one direction, and I know of no theoretical results (the keen reader may probably be able to distill something out of reference [4]).

4) One could use addition instead of multiplication, e.g. $c(x,y)=\beta \cdot x + \gamma \cdot y$ ($\beta, \gamma \in \mathbb{R}$), but this is not a good idea: even if we add arbitrary functions $\beta(x)$ and $\gamma(y)$, we can obtain

$$s_{ij}=c_o + \int c(x_i(t),y_i(t))dt = c_o + \int \beta(x_i(t))dt + \int \gamma(y_i(t))dt \qquad (7)$$

and thus the whole storage matrix $c_{ij}=f(s_{ij})$ can be determined from the n+m numbers $\beta_i = \int \beta(x_i(t))dt$ (i=1,...,m) and $\gamma_j = \int \gamma(y_j(t))dt$ (j=1,...,n). Therefore the storage capacity can at most be proportional to n+m instead of n m, as in our basic device.

5) Combining our remarks in (1), (2) and (4) we obtain a qualitative picture of the information storage capacities of all rules of the type

$$c(x,y)= \alpha \cdot x \cdot y + \beta(x) + \gamma(y), \qquad (8)$$

where α may be positive or negative and β and γ may be arbitrary functions of x and y, respectively: The two opposite cases for α have been discussed in (1) and (2), and (4) shows that $\beta(x)$ and $\gamma(y)$ do not contribute much to the information storing capacities (although they may have a stabilizing effect). The correlation method (3) is also of the type (5), since $(x-\bar{x})\cdot(y-\bar{y})=x \cdot y - \bar{x} \cdot y + \bar{x} \cdot \bar{y}$, where $\bar{x}$ and $\bar{y}$ are constants.

8. PHYSICAL REALIZATION OF THE BASIC DEVICE

It is quite clear that our "basic device" is a storing device that works highly parallel, and therefore it should be considered as an important building block for parallel computers that have to be built in the future. There is already some commercial interest in building this kind of device [6,12], and already in the fifties, similar devices have been built [5]. Apparently the main problem with these devices was their size. From the preceding tables it should be clear that methods 3.6 and 4.4 only work well with large matrices (see also Fig. 7), and only now do we have the technology to build these efficiently.

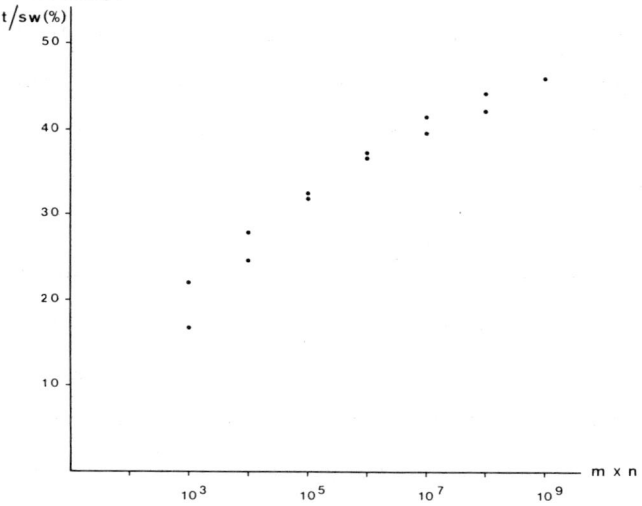

Fig. 7: storage efficiency (in bits per switch) against the size m×n of an associative matrix ($p_1 \leq 1$ per mille).

From the engineering point of view, our basic device clearly can easily be constructed, and there are actually several possible designs [6,27]. The addition of activity along the wires, for example, could be performed in a digital or analogue way, but I think that in the long run analog devices are to be preferred. The main reason is that it makes "fuzzy switching" possible. Let me explain this in a little more detail: In the discussion of our storage methods 3.6 and 4.4 we have mentioned that the method needs error correcting codes, which has the additional effect that these methods can tolerate ill-functioning of some of the switches. In the last section we have seen that a continuous change of connectivity instead of proper switching from 0 to 1 will also work. If switches are realized by means of transistors it may actually happen that some or even many of the switches do work in a somewhat fuzzy way: they do not switch from "off" to "on" but just increase their connectivity in a less well defined way. For the engineer it is good to know that this does not deteriorate the function of the whole device altogether (provided that the connectivity values are not interpreted digitally but as analog numbers). Therefore, the physical realization of our basic device - even as a very large integrated circuit - should be no serious problem for the engineer.

A completely different motive for work on associative memories is the idea that our brain may work by this principle [30,31,18,19,20,1]. Especially when we consider auto-association, i.e. our method 4.4, we can imagine that in some part of the brain the neurons are used as the threshold detectors and the synapses operate as some kind of fuzzy switches. This idea requires that the synapses increase their connectivity upon coincident activity in the pre- and the postsynaptic neuron they connect between. This law for synaptic variation has already been postulated in 1949 by Donald Hebb [7,8], and is today well-known among neuroscientists as Hebb's Law. Many experimental results in the neurosciences (recent ones are for example [14] and [2],[3]) do fit quite well with the idea that our "basic device" may be realized in the cortex of the brain in terms of neurons and synapses, and that it is used for auto-association (as in method 4.4). I have recently collected some of these results [23,25].

REFERENCES

1. Anderson, J.A., A simple neural network generating an interactive memory. Math Biosci, 14, 197 (1972).
2. Baranyi, A. and Feher, O., Intracellular studies on cortical synaptic plasticity. Exp. Brain Res., 41, 124-134 (1981).
3. Baranyi, A. and Feher, O., Synaptic facilitation requires paired activation of convergent pathways in the neocortex. Nature, 290, 413-415 (1981).
4. Borisyuk, F.N., Borisyuk, R.M., Dunin-Barkovsky, V.L., Kovalenko, V.H. and Kovalenko, E.I., Estimation of information capacity of Purkinje cells. In: Dobrushin, Krynkov, and Toom (eds.), Locally Interacting Systems and their Application in Biology. Springer-Verlag, Berlin, Heidelberg, New York (1978).
5. Frank, H., Pavlovs bedingte Reflexe und Steinbuchs Lernmatrizen. In: H. Frank (ed.), Kybernetik. Umschau Verlag, Frankfurt (1962).
6. Goser, K., Foelster, C. and Rueckert, U., Intelligent memories in VLSI. Information Sci., 34, 81-82 (1984).
7. Hebb, D.O., The Organization of Behavior. John Wiley, New York (1949).
8. Hebb, D.O., Textbook of Psychology. Sanders, Toronto, Philadelphia, London (1958).
9. Hopfield, J.J., Neurons with graded responses have collective computational properties like those of twostate neurons. Proc Natl Acad Sci USA, 81, 3088 (1984).
10. Hopfield, J.J. and Tank, D.W., Computing with neural circuits. Science, 233, 625-633 (1986).
11. Kohonen, T., Associative Memory. Springer-Verlag, Berlin, Heidelberg, New York (1977).
12. Lamb, S., An add-in recognition memory for S-100 bus microcomputers. Computer Design, 17, 140-186 (1978).
13. Larson, P., Dynamische Hashverfahren. Spektrum, 6, 7-19 (1983).
14. Levy, W.B. and Steward, O., Synapses as associative memory elements in the hippocampal formation. Brain Res., 175, 233-245 (1979).
15. Longuett-Higgins, H.C., Willshaw, D.J. and Buneman, O.P., Theories of associative recall. Q.Rev.Biophys., 3, 223 (1970).
16. MacKay, D.M., Complementary measures of scientific information content. Methodos, 7, 63 (1955).
17. MacKay, D.M., The structural information capacity of optical instruments. Information and Control, 1, 148-152 (1958).
18. Marr, D., A theory of cerebellar cortex. J. Physiol., 202, 437 (1969).
19. Marr, D., A theory of cerebral neocortex. Proc. R. Soc. London, B262, 23 (1970).
20. Marr, D., Simple memory. Philos. Trans. R. Soc. London, B176, 161 (1971).
21. Minsky, M. and Papert, S., Perceptrons. MIT Press, Cambridge, Mass. (1969).
22. Palm, G., On associative memory. Biol. Cybern., 36, 19-31 (1980).
23. Palm, G., Neural Assemblies: An Alternative Approach to Artificial Intelligence. Springer-Verlag, Berlin, Heidelberg, New York (1982).
24. Palm, G., Rules for synaptic changes and their relevance for the storage of information in the brain. In: R. Trappl (ed.), Cybernetics and Systems Research. North-Holland Publishing Company, Amsterdam (1982).
25. Palm, G., Local synaptic modification can lead to organized connectivity patterns in associative memory. In: E. Frehland (ed.), Synergetics, from Microscopic to Macroscopic Order. Springer-Verlag, Berlin, Heidelberg, New York, Tokyo (1984).
26. Palm, G., On associative coding. In preparation.

27 Palm, G. and Bonhoeffer, T., Parallel processing for associative and neural networks. Biol Cybern, **51**, 201-204 (1984).
28 Rosenblatt, F., Principles of Neurodynamics MacMillan, New York (1962).
29 Steinbuch, K., Die Lernmatrix. Kybernetik, 1, 36 (1961).
30 Uttley, A.M., Conditional probability machines and conditioned reflexes.
31 Uttley, A.M., Temporal and spatial patterns in a conditional probability machine. In: Shannon and McCarthy (eds.), Automata Studies. Princeton University Press, (1956).

On the Function of the Cat's Visual Cortex

W. von Seelen, H. P. Mallot, F. Giannakopoulos

Arbeitsgruppe III (Biophysik), Institut für Zoologie,
Johannes Gutenberg-Universität, Saarstr. 21, D-6500 Mainz,
FRG

1 Introduction

The analysis of neuronal systems is essentially determined by the kind and scope of obtainable data. As a rule it is impossible to observe or control these systems completely at the single cell level which sets fundamental limits on analysis. A viable alternative is estimating models dimensioned according to avalaible data, which is in our opinion the only possible way of analysing problems of the functions of neuronal layers and areas in the cortex. Although neurophysiology has been very successful in analysing local functions, no definition of integrated elementary systems has yet been achieved. This paper attempts to develop a basic component for modelling the visual cortex which can be adapted to various problems and coupled to form complex networks. It combines neuroanatomical and neurophysiological findings and takes into account the following factors:

1.) There is a limited neuroanatomical uniformity for the areas representing a retinotopic map, i.e. the visual areas, which number between 15 and 20 in the cat. These areas can be interpreted as a functional unit.

2.) As a rule the areas in the visual cortex consist of six vertically intercoupled, horizontal layers in which feedback plays a decisive role and the layers 2 - 5 generate output.

3.) If several inputs compete for a cortical site the individual images are segmented and stored beside each other in a defined order. This leads to the concept of columns which, however, are not visible in the intrinsic anatomy of the fibers.

4.) All areas and layers are connected in networks either directly or by feedback.

5.) The geometrical imaging of the retina onto the cortex areas is usually heavily distorted and it is possible to define limited regions of homogeneity only in area 17.

Figure 1 illustrates the fundamental difficulties for the description of the system resulting from points 2 - 5 above. It is extremely difficult and possibly not very meaningful to separate functions in a system that is based on cooperation of subsystems. In view of these problems, system simulation acquires a key role and the definition of elementary systems and the strategy of their coupling becomes decisive. If a viable definition is achieved, then the analysis can be supplemented by synthesis. Computer simulation can be used to observe certain coupling principles separately and generate possibly more adequate models accessible to experimental testing.

We will now define a basic functional unit consisting of a spatially two-dimensional, time-independent network which is initially homogenous and linear. Feedback, mapping, "cortical" cooperation, and definite nonlinearity are integrated successively.

2 Dynamic Two-dimensional Spatial Networks

The following considerations refer to neuronal networks in the visual cortex. A two-dimensional spatial network is defined as the basic system that is coupled according to the principle of lateral inhibition and has, in the first instance, the following properties, which will be revised from case to case later:

1.) the system is linear and time invariant

2.) the distance between two neurons is small in comparison to the size of the stimuli

3.) the coupling of the neurons depends only on their distance from each other.

If $x(r,s,t)$ and $y(r,s,t)$ are the time- and space-dependent input and output values and $H(r,s,t)$ describes the coupling of the system, then

$$y(r,s,t) = \sum_{j=1}^{m} \tilde{h}_j(r,s,t) \ast x_j(r,s,t) \qquad (1)$$

holds, leaving feedback out of account for the first. The term j is an index for the various input values, to each of which a coupling function $\tilde{h}$ is assigned. The time dependence of the system is generated by the synaptic events which are assumed to be the same at all positions in one layer so that

$$\tilde{h}_j(r,s,t) = g_j(t)\, h_j(r,s), \qquad j = 1, 2, \ldots, m \qquad (2)$$

and consequently space and time dependence are separable. In many cases $h_j(r,s)$ is interpreted by

$$h_j(r,s) = \sum_{i=1}^{n} h_{ij}(r,s) = \sum_{i=1}^{n} m_{ij} \exp(-(r^2+s^2)/B_{ij}^2) \qquad (3)$$

as a sum of partial couplings in accordance with neurophysiological findings. Fig. 2 depicts the elementary system defined by (1 - 3) schematically. It is characterized by parallel processing of a complete "picture".

Using the usual description for linear systems in the

spatiotemporal frequency range (ξ,η,ω), (1 - 3) lead to

$$Y(\xi,\eta,\omega) = \sum_{j=1}^{m} (\sum_{i=1}^{n} G_{ij}(\omega) \; H_{ij}(\xi,\eta)) \; X_j(\xi,\eta,\omega) \qquad (4)$$

where the uppercase letters are the Fourier transformations. So such networks are to be seen as filters that perform important subfunctions in pattern identification, visually aided orientation in space, and separation of object and background (Reichardt and MacGinitie 1962; v. Seelen 1968; Marko 1969). If $H_{ij}(\xi,\eta)$ is complex conjugated to a stimulus, then such networks are to be interpreted as correlators that have an essential role in problems of optimal receivers and associative memory structures. The formally relatively simple treatment of such filters should not obscure the fact that they are very efficient and versatile tools especially when X_j represents different signal parameters that are imaged onto the spatial coordinates such as time frequency in the acoustic domain or disparity in the visual system.

3 Neuronal Feedback Systems

3.1 Specific Properties

As far as is known at present, all cortical systems are parts of multiple feedback overall networks (cf. Fig. 1). Feedback must be considered an essential structural principle of brains. The characteristics of such systems, especially under the conditions of control theory, were treated in detail concentrating on stability and dynamic behavior in particular. In the following, some aspects of neuronal feedback systems will be discussed from the point of view of information processing. The aims will be to interpret a series of repeatedly measured results that serve as a basis for complicated structural concepts as direct and simple consequences of feedback. We also want to show that feedback is in general essential for the generation of adequate models of neuronal systems.

The basic system defined in Section 2 is expanded to include the feedback $h_R(r,s,t)$ which can have a positive or a negative effect. According to Fig. 2 and with $i,j = 1$, one then obtains

$$Y(\xi,\eta,\omega) = X(\xi,\eta,\omega) \, G(\omega) \, H(\xi,\eta) \, 1/(1 \pm G_R(\omega) \, H_R(\xi,\eta)) \qquad (5)$$

for the transmission behavior. Using the Tailor expansion, one obtains

$$Y = X \, GH(1 \mp G_R H_R + G_R^2 H_R^2 \mp G_R^3 H_R^3 + \ldots) \qquad (6)$$

from (5) for $|G_R(\omega) \, H_R(\xi,\eta)| < 1$. The upper sign in (5) and (6) stands for the negative feedback and the lower one for the positive feedback. The arguments of the functions in (6) correspond to those in (5). For a large number of cases, it can be shown that $h(r,s)$ can be approximated by (3) in neuronal systems (Krone et al. 1986). Consequently, in the simplest configuration ($i = j = 1$), the following features can be inferred from the structure of (6):

1.) Spatial and temporal behaviour in feedback networks

can no longer be factorized. This space-independent temporal behavior is the reason for apparently complicated space-dependent transient characteristics in receptive fields. This feature is frequently misinterpreted as nonlinearity.

2.) If (3) holds, the power function of H_R leads to a decrease in bandwidth and so to an increasing width of the measurable coupling function. In the case of feedback, the receptive fields, interpreted as spatial distributions of the signs of the effects in a plane, do not reflect the direct neuroanatomical connection, but rather extend over the whole plane. Many field classifications should be revised bearing this in mind. If H_R characterizes a bandpass the different terms in (6) show an increased selectivity in the frequency domain.

3.) Coupling spatial and temporal behavior makes a high variability in the system features possible. If $G_R(\omega) = e^{-i\omega t_0}$ with t_0 defining the time delay of the system then (6) is the explicit representation of a recursive algorithm and so an interesting mathematical instrument. This interpretation further shows that the characteristics described by the power of $H_R(\xi,\eta)$ (e.g. selectivity) occur with time lag as long as a low pass predominates in G_R. Therefore coupling functions and consequently receptive fields have to be interpreted as time-dependent in principle (cf. Section 5, point 1). With suitably chosen G, G_R, H and H_R spatio-temporal oscillations are possible, which can, amongst other things, scale the temporal axis for co-operation with other subsystems (cf. Section 5).

4.) A high selective amplification can be achieved with positive feedback if values of $H_R G_R$ are close to 1 or, in complicated networks, the system determinant has very small values. The selectivity of the amplification can affect parameters (e.g. intensity), picture features, and parts of scenes (e.g. background).

The following specializations (for simplicity's sake described for the one-di...sional case) are intended as examples and refer to characteristics frequently measured.

For the case of moving stimuli and direction specific neuronal responses, e.g. in area 18 (Dinse, v. Seelen 1981), let

$$h_R(r) = r \exp(-r^2/B^2) \; , \; G_R(\omega) = 1/(1 + T_1 i\omega)$$

and $x(r,t) = x(r + \tilde{v}t)$ where $\tilde{v}$ is the velocity. With (6) one obtains

$$Y(\xi,\omega) = X(\xi) \, \delta(\omega - \xi\tilde{v}) \, \widetilde{H}(\xi,\omega) \left[1 \mp \frac{1}{1 + T_1 i\omega} \xi i \exp(-\xi^2 B^2/4) \right. \\ \left. + \left(\xi i/(1+T_1 i\omega) \right)^2 \exp(-2\xi^2 B^2/4) \mp \cdots \right] \qquad (7)$$

If $T_1 u\tilde{v} \gg 1$ then

$$Y(\xi,t) \approx X(\xi) \widetilde{H}(\xi,\xi\tilde{v}) \left[1 \mp \frac{1}{T_1 \tilde{v}} \exp(-\xi^2 B^2/4) + \frac{1}{T_1^2 \tilde{v}^2} \exp(-2\xi^2 B^2/4) \mp \cdots \right] \exp(i\xi\tilde{v}t) \qquad (8)$$

For sufficiently high velocities, the "long range" terms (n B) are suppressed, i.e. the feedback becomes less effective and the "receptive fields" shrink. Moreover, due to the spatial asymmetry all feedback terms in (8) become real so that a moving pattern can be correctly estimated with respect to position and without dispersion even though there are time lags in the system. If the order of the spatial filter is higher than the order of the time filter the prediction of a moving object's position is possible. The combination of temporal and spatial asymmetry applied to the feedback system here can also be applied to $\widetilde{H}(\xi,\xi\breve{v})$ in (8).

If $h(r)$ is antisymmetric and h_R symmetric and negative then, by combining terms that follow each other in the Tailor expansion, it can be shown that the "long range" terms in (8) cause reversed direction specificity, i.e. the movement of small objects and the shift in retinal images due to eye movements can occur in opposite directions. This receptive field characterisic, known as "double opponency" frequently occurs in the cortex and is connected with the relationships involved in eye pursuit movements.

When feedback is used in neuronal systems the sign is decisive. If h_R is negative and has a low pass characteristic, the sequence of the terms in (6) shifts the lower filter flanks to higher spatial frequencies. If G_R characterizes dead time or a low pass the coupling function becomes spatially narrower as time increases and the system can become more selective with respect to spatial frequency (cf. Section 4). Positive feedback results in suppression of the higher spatial frequencies as time increases and the effect of a stimulus seems to spread "dispersively".

Coupling two systems $(x_1, h_1; h_{R12}; y_1; x_2, h_2; h_{R21}; y_2)$ with the "lateral" coupling (h_{R12}, h_{R21}) being negative leads to an amplification of the difference (Reichardt and MacGinitie 1962). Invariance can be implemented if h_{R12} is negative and $h_{R21} \gg -h_{R12}$ which makes y_1 dependent on x_2 and invariant against x_1. If positive feedback (h_{R21}) is replaced by weighted summations of x_1 and y_2, invariance is generated on the basis of the reafference principle. Coupling several systems increases the diversity of possibilities so that discussing the characteristics has to be related to the task being performed.

Moreover, with the aid of feedback slight synaptical changes, as long as they do not take place in H_1 but rather in H_R, could considerably alter the network. If a filter is described by $F = H_1/(1 - H_R)$ then the change following a variation in H_R is

$$\frac{\partial F}{\partial H_R} = \frac{\partial F}{\partial H_1} \quad (H_1/(1 - H_R)) \tag{9}$$

So changes in feedback are considerably more efficient than changes in the direct coupling as long as such changes occur at locations where H_R is close to 1. Consequently, it is very advantageous to implement learning processes as feedback loops.

The aspects of feedback dealt with so far concern special problems in information processing. A further characteristic of this kind of coupling renders its use generally meaningful if one takes the fundamental structure of neuronal data into account.

3.2 Regularization by Feedback

In the discussion of (6), we already mentioned the interpretation of feedback as a recursive or iterative algorithm. Suppose that the feedback loop includes a temporal delay which may be due to synaptic transmission. Let y denote an initial excitation of neurons and $\Phi(y)$ the excitation which is brought about by intrinsic coupling. The feed forward term, which transforms an external stimulus x into a distribution of excitation will be denoted by Ψ. Feedback with a temporal delay then results in the iterative algorithm:

$$y_o = 0$$
$$y_t = \Phi(y_{t-1}) + \Psi(x), \qquad (10)$$

The properties of that highly parallel form of information processing will now be studied in one dimension, treating x and y as discrete vectors. It can be shown in general that (10) with linear Φ and Ψ can be used to minimize any quadratic functional of x and y by a gradient method. This interpretation is inspired by the work of Poggio et al. (1985), who showed that vision involves the usually ill-posed inversion problems of optics. These cannot be solved by an explicit inversion formula. Rather, additional constraints have to be imposed on the solutions in order to make them well-defined. Such constraints are the optimization of certain quadratic functionals (cf. Poggio et al. 1985). In our discrete description, we have to minimize functions $R^n \to R$ instead of functionals. Consider as an example:

$$F(y) := \|y-x\|^2 + \lambda \|Dy\|^2, \quad \lambda > 0 \tag{11}$$

For given data x and regularization parameter λ, the minimization of that function results in an excitation y, which fits or interpolates the data and satisfies a certain smoothness-constraint, which is specified by the linear operator D. For example, D may perform a spatial derivation, i.e. $(Dy)_i = \frac{1}{2}(y_{i+1} - y_{i-1})$. Some easy computation yields:

$$\text{grad } F(y) = 2(y-x) + 2\lambda D^T Dy, \tag{12}$$

where D^T denotes the transposed matrix of D. Note that $D^T D$ is a symmetric matrix for all D.

To formulate a recursive algorithm for the minimization of (11), let q denote the amount by which y is altered in each step, $q \in R^+$:

$$\begin{aligned} y_o &= 0 \\ y_t &= y_{t-1} - q \text{ grad } F(y_{t-1}). \end{aligned} \tag{13}$$

We substitute for grad F from (3) and obtain by comparison with (10):

$$\begin{aligned} \Phi &= (1-2q)E - 2\lambda q D^T D \\ \Psi &= 2qE, \end{aligned} \tag{14}$$

where E is the identity.

Equation (14) has the following interesting properties: 1. the constant q can always be chosen in such a way that the algorithm converges. Furthermore, since F is a quadratic functional, the convergence approaches the global minimum. 2. The matrix of Φ in (14) is always symmetric. Therefore, feedback with a nonsymmetric coupling does not minimize a functional of type (11). 3. If D is chosen to be the spatial derivative mentioned above, $\bar{\Phi}$ has non-negative components throughout with maximal values along the diagonal. That is to say, Φ is a lowpass coupling in a positive feedback loop.

Positive feedback with a lowpass filter is thus able to perform a regularization operation. Poggio et al. (1985) showed that regularizations of the type (11) are extremely useful in computer vision applications. As a possible neural implementation, they proposed analog circuitry on the level of single dendrites and spines. The abundance of feedback coupling in the visual cortex, however, leads us to the conclusion that the iterative algorithm may be directly implemented in the cortical network. In general, optimization problems more complicated than interpolation and smoothing may be solved by the feedback algorithm, if couplings other than lowpasses are considered.

4 Receptotopic mapping

In order to describe cortical layers as elements of an integrated network, we have to consider the receptotopic mappings characterizing these layers. Typically, sensoric surfaces are mapped to the corresponding cortical areas continuously. It is usually supposed that the relevance of this mapping for information processing is to preserve neighborhoods, which is a consequence of continuity. However, we want to stress that it is the systematic alteration of neighborhoods rather than their strict preservation, which is important. This alteration is a kind of preprocessing performed on the stimulus, which allows the cortex to use just one space-invariant or uniform operation for a variety of different tasks subsequently.

An important support for this view is the existence in the cortex of more than one representation of the visual field each with a different mapping. This fact, which holds for all investigated mammals, clearly shows that mapping is an information processing strategy rather than a mere consequence of an economical principle in connectivity and representation.

A retinotopic map may be formalized as a piecewise continous, one-to-one function R: $R^2 \rightarrow R^2$. The areal magnification factor $M_a(r,s)$ is given by the Jacobian of that function, the linear magnification along a direction φ, $M_l(r,s,\varphi)$, by the norm of the Gateaux-derivative along that direction.

An overall description of different mappings can be obtained by decomposing them into simple steps. It turns out that apparently totally different mappings can be constructed out of very similar parts (Mallot 1985). In a model of the maps of the visual areas 17, 18, and 19 of the cat, the most important step is the complex power function which produces the decrease of the areal magnification with excentricity in all three areas. This function maps the right hemiplane continously onto a sector. The mapping of the left hemiplane, however, includes a discontinuity like the one termed field discontinuity by van Essen et al. (1981): the horizontal meridian is split and its peripheral part is represented twice. Thus, both the first order and the second order transformations can be modelled by one single function, namely a power function with an exponent of 0.43.

Fig. 3 compares the model of the mappings in the cat visual cortex with the neurophysiological results of Tusa et al. (1979). The model reproduces the general form of the mappings, the areal magnification, the branching of the horizontal meridian in the areas 18 and 19, the continuous connections at the boundaries, and the differences between the upper and lower part of the visual field. The gap between the representations of the periphery in the areas 18 and 19, which does not occur in reality, may be closed by folding the plane model three-dimensionally. The resulting fold corresponds to the lateral sulcus in the cat brain.

We now combine the retinotopic mapping with the space-invariant, linear systems theory discussed in section 2. Let e denote a distribution of excitation in a neural layer, due to a two-dimensional stimulus (picture) x. Intracortical processing is characterized by a kernel k. In the case of a retinotopic mapping, R: $(r,s) \mapsto (u,v)$, the stimulus at position (u,v) in the cortex is determined by the intensity of the picture at position $R^{-1}(u,v)$ in the visual field, i.e.:

$$\begin{aligned} e(u,v) &= \int\int x(R^{-1}(u',v')) \, k(u-u',v-v') \, du'dv' \\ &= \int\int x(r,s) \, k((u,v)-R(r,s)) \, |J_R(r,s)| \, drds \quad (15) \\ &=: \int\int x(r,s) \, g_{u,v}(r,s) \, drds. \end{aligned}$$

With respect to the retinal coordinates (r,s), the sequence of mapping and convolution yields a new, space-variant operator with the kernel g. This kernel g may be interpreted as the receptive field, whereas k is proportional to the point-spread-function of the operator (15).

The biological interpretation of (15) is that the magnification of a given region of the visual field is not followed by a "dilution" of the excitation in that region. If the magnification is brought about by the different densities of retinal ganglion cells at different excentricities, the "dilution" is prevented by the sensitivity and amplification of these ganglion cells, which is proportional to the cell density (Fischer & May 1970). If areal magnification includes an increase in cell numbers ("cellular magnification"), the interpretation is even simpler: the influence which a point in the visual field has on the cortical excitation depends on the size of its representation. Therefore, (15) is a model of image processing via retinotopic mapping in both the cat, where no cellular magnification occurs, and the monkey, where areal magnification is partly due to an increase in cell numbers.

According to (15), the space-variance induced by the retinotopic mapping is twofold: first, the domain of the kernel is distorted, and second, the kernel is multiplied by an asymmetric function, namely the magnification factor. The combination of isotropic kernels with a retinotopic mapping may thus lead to orientation-specific receptive fields which are distributed in the visual field in a way suitable for analysing optical flow, for instance.

5 Linear Theory of the Visual Cortex

In this section we want to combine the structural principles of the cortex already treated, i.e. **two-dimensional filtering, feedback, and retinotopic mapping**, into a comprehensive theory of linear image processing in an area of the visual cortex. Of course, there is a variety of further structural principles which could have been included, such as discreteness of external input, interareal feedback, thalamic processing or nonlinearities. We think, however, that a model of two-dimensional layers, feedback, and mapping is already sufficient for a number of important properties of the visual cortex.

We first consider the intracortical processing, i.e. the kernel k in (15), which now additionally depends on time. A model deriving this spatio-temporal kernel from cortical anatomy combines six layers of the type described in section 2 into a stack with strong vertical connectivity (Krone et al. 1986, von Seelen et al. 1986). These vertical connections constitute a global feedback loop, which is shown in Fig. 4. The equations of the model implemented in the computer simulations are:

$$y(t) = \sum_{z=1}^{6} \left(D^+(z) \, f^+(t-z) - D^-(z) \, f^-(t-z) \right)$$

$$f^+(t) = e(t-10)/T^+ + f^+(t-1)(1-1/T^+)$$

$$f^-(t) = f^+(t-10)/T^- + f^-(t-1)(1-1/T^-)$$

$$e(t) = x(t) + \sum_{z=1}^{6} A(z) \, y(t-z)$$

The interpretation of the various terms can be seen from Fig. 4.

The most important outcome of the simulations is the spatio-temporal behavior of the response. The feedback loops produce an oscillation in time, whose period, however, is not simply the sum of propagation times and synaptic delays. Rather, the period is much longer (some 50 ms) and results from the interference of all the feed forward and feedback connections depicted in Fig. 4. More specifically, the results important for our discussion are the following:

(i) Excitation reaches its maximum with a certain delay after the onset of stimulation. The delay is lamina-dependent and corresponds well to neurophysiological latency times.

(ii) The iteration of the processing in the feedback loop results in a smoothing of the spatial distribution of excitation which occurs independently of the details of fiber anatomy. Therefore, the response of an element of the network ("neuron"), depends primarily on its position in the network, and not on the details of its connectivity.

(iii) Temporal and spatial parts of the responses are not separable into independent factors. (Cf. Fig. 5) This effect has already been discussed in section 3, since it is a general consequence of feedback. In the multi-input multi-output system, non-separability leads to the interesting phenomenon that the spatial filter applied to a given stimulus depends essentially on the temporal structure of the presentation. As a rule, transient stimuli are treated less specifically than sustained ones. This behavior is again in good agreement with experimental results.

This leads us to an important conclusion concerning the strategy of cortical image processing. The simple spatial filtering approach to image processing assumes that the role of a (neural) layer is the detection of a certain feature in the image. Multilayered systems are usually supposed to deal successively with more complicated features. In contrast to this view, the feedback model favors the idea of spatio-temporal processing rather than that of a hierarchy of increasingly complicated feature detectors. The brain's analysis of a moving scene need not be based on a sequence of instantaneous pictures but may be spatio-temporal from the very beginning, i.e. the analysis does not evaluate features but "elementary events". Accordingly, the receptive field properties in the different layers of the cortex need not represent a hierarchy either.

We can now combine the above model of intracortical coupling with the formalization of retinotopic mapping. This comprehensive model is suited to dealing with the differences between areas 17, 18, and 19 in both neuroanatomy (via the parameters of the feedback model) and mapping. It can therefore be used to study the significance of the multiple representations of the visual field in the cortex. With respect to a single area, the model includes the space-variance of neural image processing and can be used to study the different roles of fovea and periphery.

Fig. 6 shows simulated tuning curves for the velocity of a moving light spot (Brittinger 1986). The stimulus is mapped to the area 18 according to Fig. 3b and then processed by the intracortical feedback system, Fig. 4. The results are in good agreement with electrophysiological recordings in the distinction between "lowpass" and "bandpass" behaviour (Orban 1984) as well as in the different peak velocity in the two areas. Fig. 7 shows simulated direction tuning for a moving spot in area 18. Units respond best to the direction towards the center of the visual field. This specificity is due to the retinotopic mapping since the intracortical processing is isotropic (Brittinger 1986).

An important conclusion from these simulations is that a variety of receptive field properties are rather direct consequences of simple principles of cortical anatomy. This holds for example for many of the intensively studied tuning curves. On the other hand, properties which have not found much interest up to now are in the same way consequences of the cortical structure. In particular, the temporal behavior and the dependence of receptive field properties on the position in the visual field are deeply built into the anatomy of the cortex. If structure is to serve as a hint to function, one might indeed suppose that these properties are even more important than the classically investigated ones. In this sense, we wish to include a critique of the classical concept of the receptive field: the studied properties mainly fit into the perceptron concept of cortical information processing but not into cortical anatomy.

6 Nonlinear Cortex Couplings

The preliminary basic system proposed here interprets very many findings when one takes the feedback, the layering and the mapping into account from case to case. In a number of processes, however, the nonlinearity is essential and the system behavior cannot be approximated by linearization. Therefore, a nonlinearily coupled structure is proposed in the following which we consider can be used for the interpretation of cortical characteristics and which completes the basic system discussed here.

The system considered consists of two cortical layers (Fig. 8). Layer L_2 contains excitatory and inhibitory cells. In contrast to the excitatory cells, inhibitory ones cannot feedback to themselves directly. Their fibers do not leave layer L_2. Layer L_1 does not contain any somata but only processes from excitatory cells. The system has one input in each layer.

The temporal behavior of the system is defined by low passes and time delay at the synapses and latency times in the dendrites and axons. The spatial behavior is characterized by axonal and dendritic coupling functions of the form

$$b \exp(-(r^2 + s^2)/B^2).$$

The cell excitation is represented as a function of the total potential y: $y = f(y_2)$, where f is a nonlinear (logistic) function. The total membrane potential y is defined as:

$$y = d_1^+ \otimes f_1^+ + d_2^+ \otimes (f_2^+ - a^- \otimes d^- \otimes f^-).$$

where d represents the dendritic and a the axonal coupling functions. f^+ and f^- the postsynaptic excitation at the excitatory and inhibitory synapses and $\otimes$ is the symbol for the spatial convolution.

To describe y it is sufficient to consider the following system of integro-differential equations:

$$\tau \frac{\partial y}{\partial t} = -y + d_1^+ \otimes a_1^+ \otimes f(y(t - T_1)) + d_2^+ \otimes a_2^+ \otimes f(y(t - T_2))$$
$$- d_2^+ \otimes a^- \otimes d^- \otimes v(t - T_2') + d_1^+ \otimes E_1 + d_2^+ \otimes E_2 \quad (16)$$

$$\tau \frac{\partial v}{\partial t} = -v + a_2^+ \otimes f(y(t - T_2'')) + E_2$$

where τ designates the synaptic constant and T the latency times. $v = f_2^+$ is the postsynaptic excitation at the synapses in layer L_2 and E_1, E_2 are external stimuli.

The anatomical structure on which (16) is based essentially corresponds to that discussed in Section 5 (Krone et al. 1986; v. Seelen et al. 1986).

In the following, for computational reasons, the latency times T are equated with zero and only one input (E_1) is allowed, E_2 being equated with zero. We first consider only such solutions to (16) as do not depend on the spatial variables (r,s). Such solutions describe the dynamics of cell populations with only a small spatial area, e.g. in hypercolumns (cf. Wilson and Cowan 1972, 1973). Equations (16) then reduce to the following system of nonlinear differential equations:

$$\tau \dot{y} = -y + Q_{11}f(y) - Q_{12}v + P_1 E_1 \quad (17)$$

$$\tau \dot{v} = -v + Q_{21}f(y)$$

Q and P are non-negative constants composed of the dendritic and axonal parameters.

The nonlinearity of the neurons is approximated with the following function:

$$f(y) = 1/(1 + \exp(-c(y - \bar{y})))$$

with $c > 0$ and $\bar{y} = (Q_{11} - Q_{12}Q_{21} + P_1)/2$.

6.1 System Characteristics

The number and the stability behavior of the equilibria play a central part in the examination of the system (17). The equilibria are solutions to the following system of equations:

$$-y + Q_{11}f(y) - Q_{12}v + P_1E_1 = 0$$
$$-v + Q_{21}f(y) = 0$$

where E_1 is constant in time.

Let E_1 be chosen initially so that

$$|E_1 - 1/2| < \varepsilon \qquad (18)$$

where ε is a sufficiently small, positive constant.

The stability behavior of the equilibria is determined with the aid of linear stability analysis (Arrowsmith and Place 1982).

Case A: If the positive backfeed Q_{11} is sufficiently small that

$$-4 + (Q_{11} - Q_{12}Q_{21})c < 0 \qquad (19)$$
$$-8 + Q_{11}c < 0 \qquad (20)$$

then there is only one equilibrium which is asymptotically stable. Qualitatively the system behaves like the corresponding linear system.

Case B: If the positive feedback Q_{11} is so large that
$$-8 + Q_{11}c > 0 \tag{21}$$
and at the same time the negative feedback is chosen so that (19) is the case, then there is a single equilibrium, which is an unstable focus or node. In this case the existence of a periodic solution (more precisely of a limit cycle) can be demonstrated using the Poincare-Bendixson theorem (Arrowsmith and Place 1982), i.e. the system is able to carry out "self-excited" oscillations (cf. Fig. 9a).

Case C: If the positive feedback Q_{11} is so much larger than the negative $Q_{12}Q_{21}$ that
$$-4 + (Q_{11} - Q_{12}Q_{21})c > 0 \tag{22}$$
then three equilibria result. The "upper" and "lower" equilibria are asymptotically stable, while the "middle" one is instable (a saddle point).

If the system is situated in the proximity of one of the two stable equilibria it acts as a linear filter for small signals. The characteristics of this filter depend on the relevant equilibrium, e.g. it can be a low pass but with a different bandwidth in both cases.

If the values of E_1 lie outside the interval defined by (18) bifurcation phenomena occur in case B and hysteresis in case C (Wilson and Cowan 1972; Cowan and Ermentrout 1978; Giannakopoulos, publication in preparation).

6.2 Behavior with Time-Dependent Stimuli

To interpret the system characteristics we chose two

different sets of parameters Q and special stimuli of the form

$$E_1 = f_0(x_1(t)), \quad t \geqslant 0,$$

where $f_0(x_1) = 1/(1 + \exp(-x_1))$.

a) Let Q be chosen so that (19) and (21) (case B in Section 6.1) are fulfilled and let

$$x_1(t) = A_1 \sin \omega_1 t, \quad A_1 \geqslant 0, \omega_1 > 0$$

and then consider the responses of the system (17) for various amplitudes of A_1.

If $A_1 = 0$ and so $x_1 \equiv 0$ then the system is able to perform self-excited oscillations with the frequency ω_0 (cf. Fig. 9a). If A_1 is sufficiently small, the system performs "oscillations" that are still dependent on the eigen-oscillation (cf. Fig. 9b). When A_1 is sufficiently large the system oscillates at the frequency ω_1 of the external stimulus (cf. Fig. 9c).

b) Let (22) be fulfilled (case C, Section 6.1) and

$$x_1(t) = A_1 \exp(-D_1 t) \sin \omega_1 t, \quad A_1 \geqslant 0, D_1, \omega_1 > 0.$$

Let the system be passive at the "lower" equilibrium before the external stimulus is applied and become active only when the external stimulus begins to take effect. If the amplitude A_1 is sufficiently small the system returns to the "lower" equilibrium after some time (Fig. 10a). If A_1 is sufficiently large the system reaches the "upper" equilibrium and remains there after the stimulus has subsided (Fig. 10b).

Using nonlinearity in the basic system discussed here

thus makes it possible to implement qualitatively very different kinds of system behavior such as hysteresis, oscillations and various controllable filter characteristics by using simple signal and parameter variations.

The basic system discussed in this paper was developed for the estimation of cortical subsystems. Depending on the kind of data and the problem posed, the level of complexity of the model can be varied. The extensive use of feedback in the cortex and the nonseparability of spatio-temporal operations narrowly limits the significance of receptive field measurements. Consequently, estimations based on model systems seem to us to be indispensable. Anatomy plays an important role in the model since the spatial operators are frequently homogenous and so the geometrical arrangement of the cells determines the function. Nevertheless the decisive problem remains unsolved: how can the elementary systems be combined and flexibility dependent on the signals be implemented? The solution to this problem could be facilitated if the elementary neuronal systems could be linked to "elementary situations" since biotopes permit sharper distinctions than the formal restrictions in signal processing. At the moment, the model does not take into account the discrete form of projection into the cortex caused by binocularity and orientation specificity. However, very efficient smoothing of the discrete cortex images can be carried out with the positive feedback, since the power function in (6) leads to a "dense" spectrum in the transfer characteristic. In addition, the discrete representation which is probably used to combine functions - possibly in the sense of geometric modulation - will have to be included in the model.

Acknowledgements

This work was supported by the Stiftung Volkswagenwerk grant number I/60 511.

We are also grateful to Neil Beckhaus for translating parts of the paper and improving the English of the rest.

Literature

Arrowsmith DK, Place CM (1982) Ordinary differential equations. Charma and Hall Mathematics Series, London, New York

Best J, Dinse HRO (1984) Laminar dependent visual information processing in the cat's area 17. Neurosci Lett Suppl. 18: 76

Brittinger R (1986) Kybernetische Simulation bewegter Reize im visuellen Cortex. Diplomarbeit, Universität Mainz

Cowan JD, Ermentrout GB (1978) Some aspects of the "eigenbehavior" of neural nets. In: Levin SA (ed) Studies in mathematical biology part I. Mathematical Association of America, Washington DC, pp 67-117

Dinse HRO, Seelen W.von (1981) On the function of cell systems in area 18, Parts I and II. Biol Cybern. 41: 47-69

Fischer B, May HU (1970) Invarianzen in der Katzenretina: Gesetzmäßige Beziehungen zwischen Empfindlichkeit, Größe und Lage receptiver Felder von Ganglienzellen. Exp. Brain Res. 11: 448-464

Krone G, Mallot HA, Palm G, Schüz A (1986) Spatio-temporal receptive fields: a dynamical model derived from cortical architectonics. Proc Roy Soc London B 226: 421-444

Mallot HA (1985) An overall description of retinotopic mapping in the cat's visual cortex areas 17, 18, and 19. Biol Cybern 52: 45-51

Marko H (1969) Die Systemtheorie der homogenen Schichten. I. Mathematische Grundlagen. Kybernetik 5: 221-

Orban GA (1984) Neural operations in the visual cortex. Springer, Berlin Heidelberg New York Tokyo

Poggio T, Torre V, Koch C (1985) Computational vision and regularization theory. Nature 317: 314-319

Reichardt W, MacGinitie G (1962) Zur Theorie der lateralen Inhibition. Kybernetik 1: 155-165

Seelen W von (1968) Informationsverarbeitung in homogenen Netzen von Neuronenmodellen. Kybernetik 5: 133-148

Seelen W von, Mallot HA, Krone G, Dinse HRO (1986) On information processing in the cat's visual cortex. In: Palm G, Aertsen A (eds) Brain theory. Springer, Berlin Heidelberg New York Tokyo, pp 49-79

Segraves MA, Rosenquist AC (1982) The afferent and efferent callosal connections of retinotopically defined areas in cat cortex. J Neurosci. 8: 1090-1107

Tusa RJ, Rosenquist AC, Palmer LA (1979) Retinotopic organization of areas 18 and 19 in the cat. J Comp Neurol 185: 657-678

Van Essen DC, Maunsell JHR, Bixby JL (1981) The middle temporal visual area in the macaque: myeloarchitecture, connections, functional properties, and topographic organization. J Comp Neurol 199: 293-326

Wilson HR, Cowan, JD (1972) Excitatory and inhibitory interaction in localized populations of model neurons. Biophys J 12: 1-24

Wilson HR, Cowan, JD (1973) A mathematical theory of the

functional dynamics of cortical and thalamic nervous tissue.
Kybernetik 13: 55-80

Figure Captions

Fig. 1: (a) Connections in the visual system with emphasis on the primary cortical visual areas (17, 18, 19, and suprasylvian). (b) Scheme of the callosal connections (from Segraves and Rosenquist 1982).

Fig. 2: Coupling scheme of a two-dimensional neuronal network.

Fig. 3: The model of the retinotopic mapping in the areas 17, 18 and 19 compared with the results of measurements by Tusa et al. (1979).

Fig. 4: Computational structure of the cortex model. The matrix D denotes the density of excitatory (D^+) and inhibitory (D^-) postsynaptic sites of the neurons. A is the corresponding density of the presynaptic sites. The terms x and y indicate the system's input and output respectively, e denotes the total axonal excitation, $f^{+(-)}$ is the excitation (inhibition) after having passed the synapses with the time constant $T^{+(-)}$ and the time delay T^D.

Fig. 5: (a) Responses to a stimulus $x(r,t) = \delta(r)\sigma(t)$ applied to layer 4. $\sigma(t)$ indicates the temporal step function, $\delta(r)$ is a spatial impulse. (b) Two typical simulations compared with PSTHs from averaged multi-unit recordings obtained by on-off stimulations of the excitatory part of the receptive fields in area 18 of the cat (Best and Dinse 1984).

Fig. 6: Velocity tuning for each layer's maximum response to a spatial impulse moving from the fovea to the periphery. The position of the "registered neuron" is given in Fig. 7

(+45).

Fig. 7: Polar diagrams of the maxima of the cortical layers' neuronal responses to a spatial impulse moving in different directions (right). The neurons are located in area 18 at the points indicated in the visual field (left).

Fig. 8: The structure of the nonlinear model (explanation in text).

Fig. 9: Responses of the system (17), (18), and (21) to stimuli of the form $x_1(t) = A_1 \sin \omega_1 t$ with $\omega_1 = 1.0$. (a) $A_1 = 0$, (b) $A_1 = 1.25$, (c) $A_1 = 3.0$ (details in text).

Fig. 10: Responses of the system (17) and (22) to stimuli of the form $x_1(t) = A_1 \exp(-D_1 t) \sin \omega_1 t$ with $D_1 = 0.1$, $\omega_1 = 3.0$. (a) $A_1 = 2.0$, (b) $A_1 = 3.0$ (details in text).

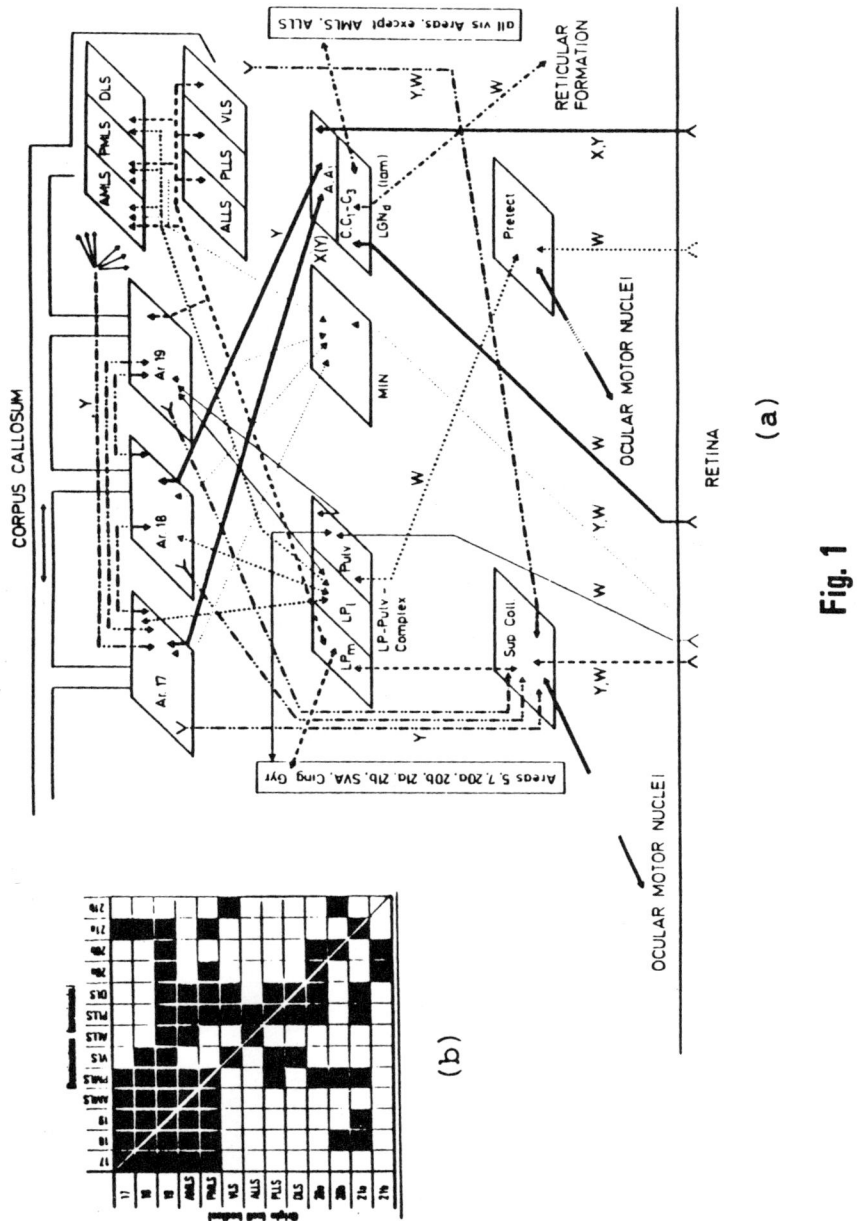

Fig. 1

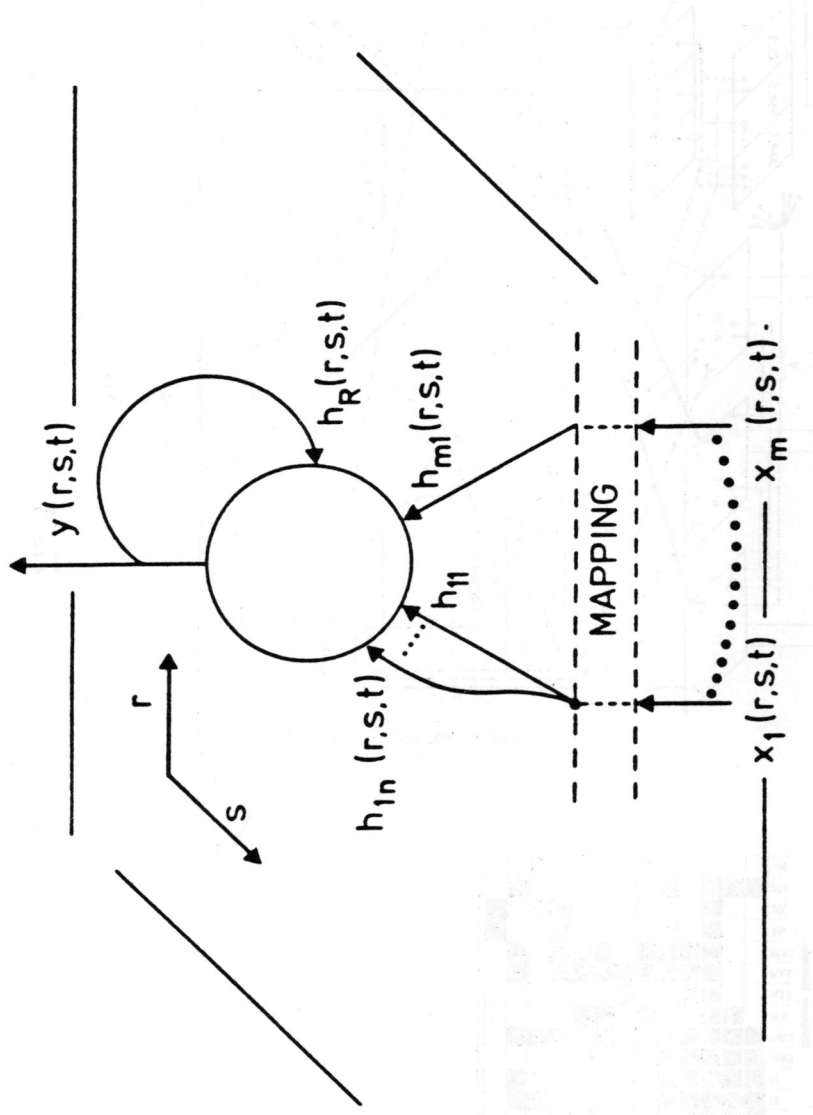

Fig. 2

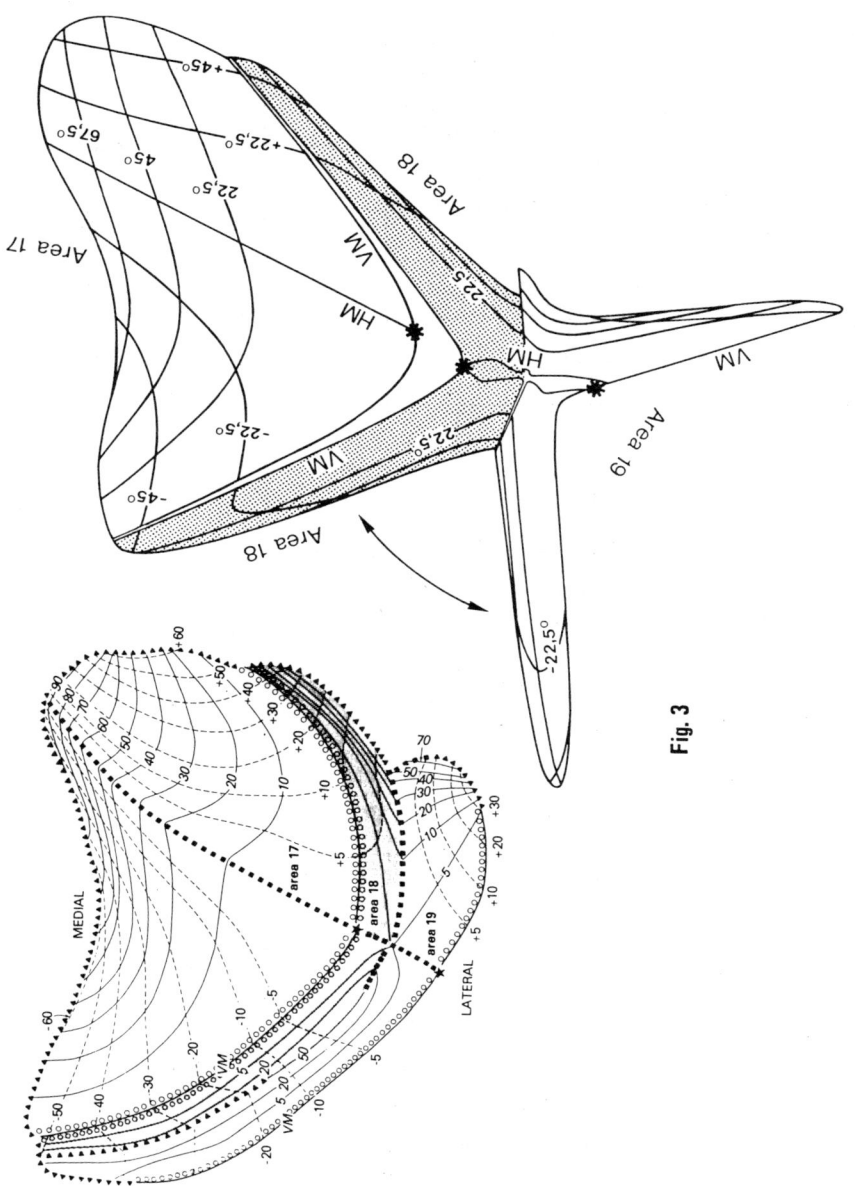

Fig. 3

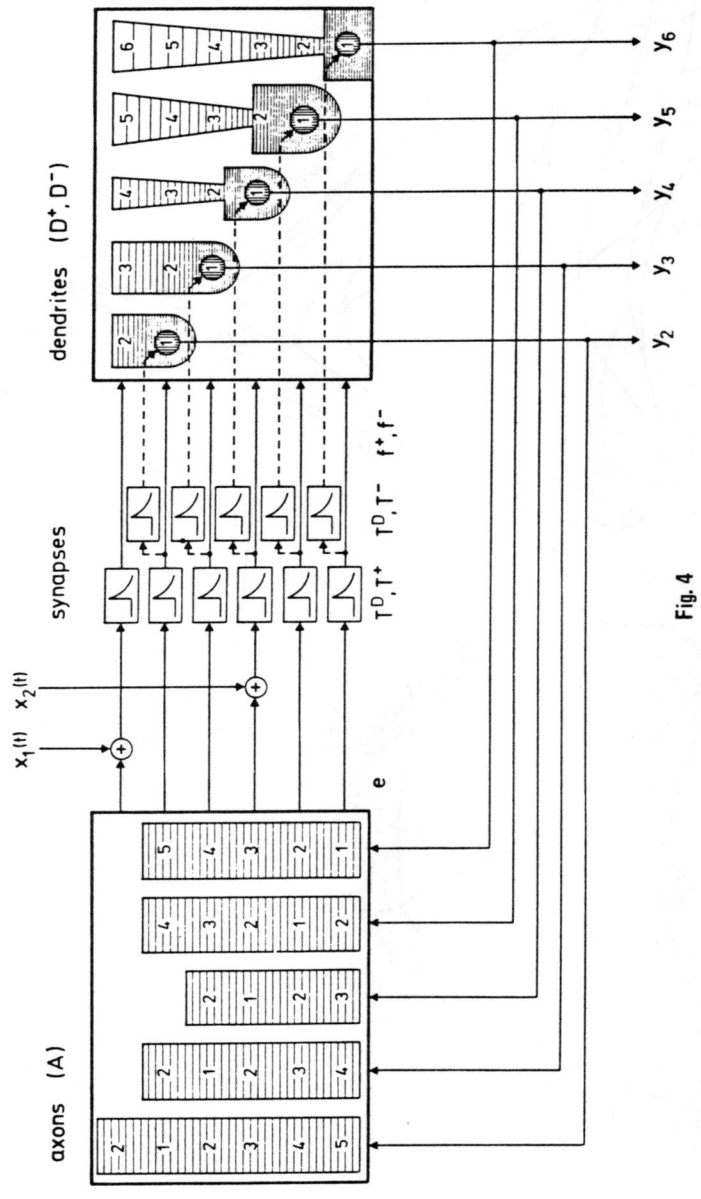

Fig. 4

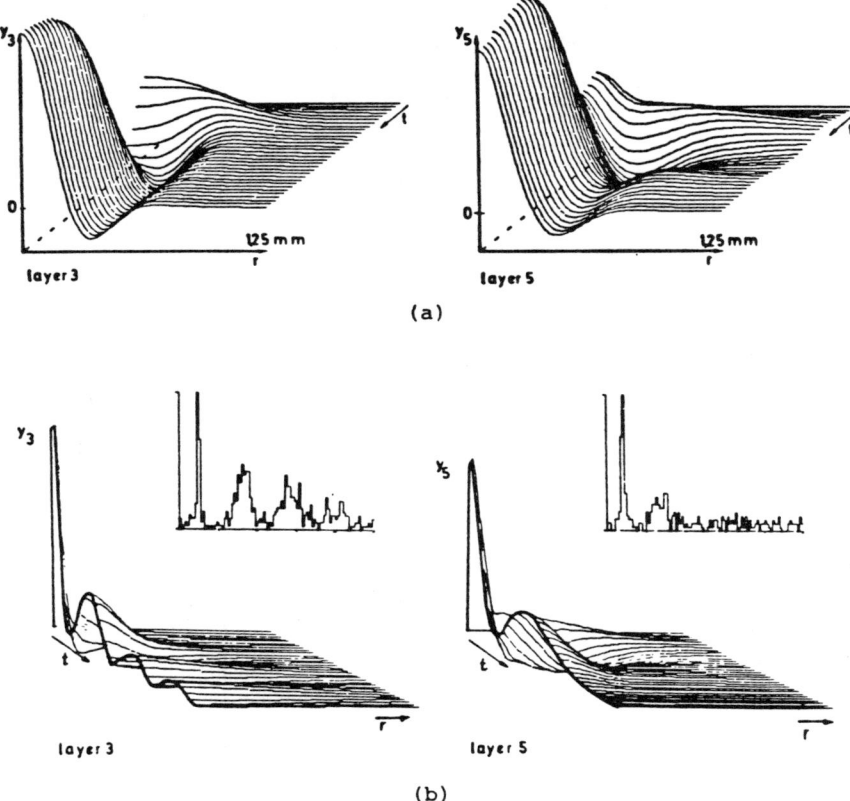

Fig. 5

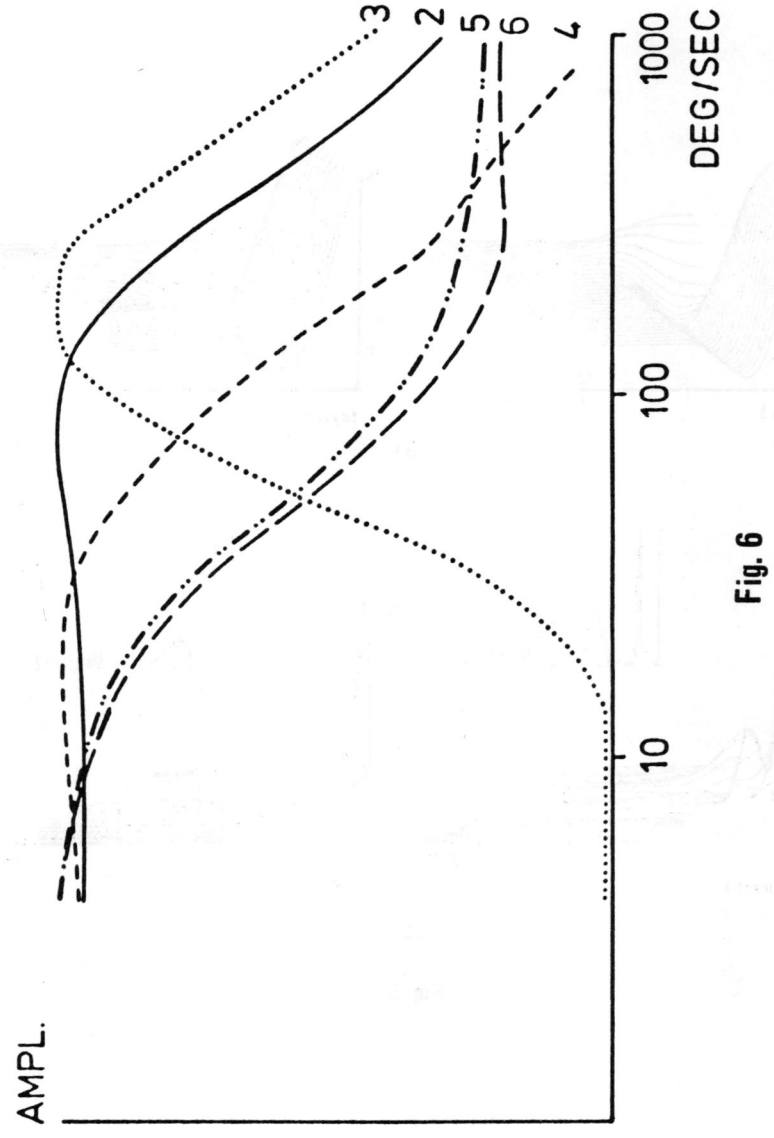

Fig. 6

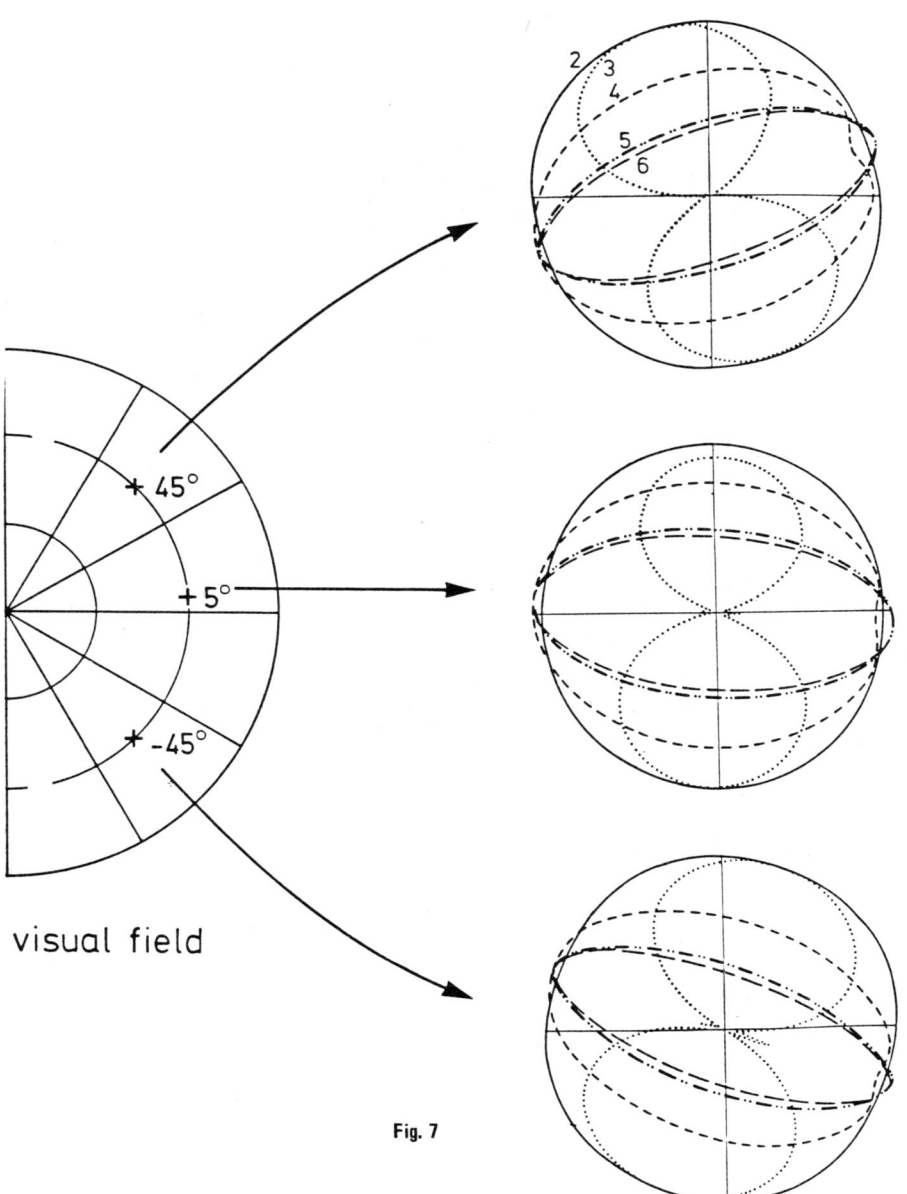

Fig. 7

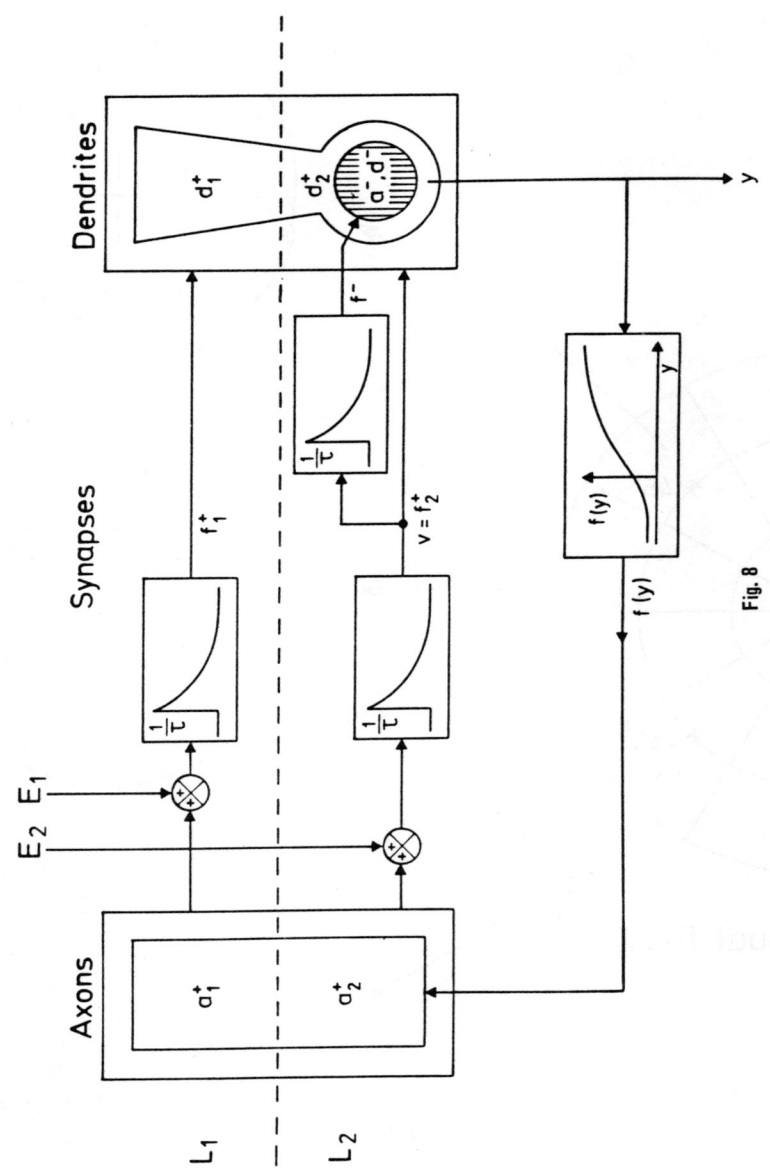

Fig. 8

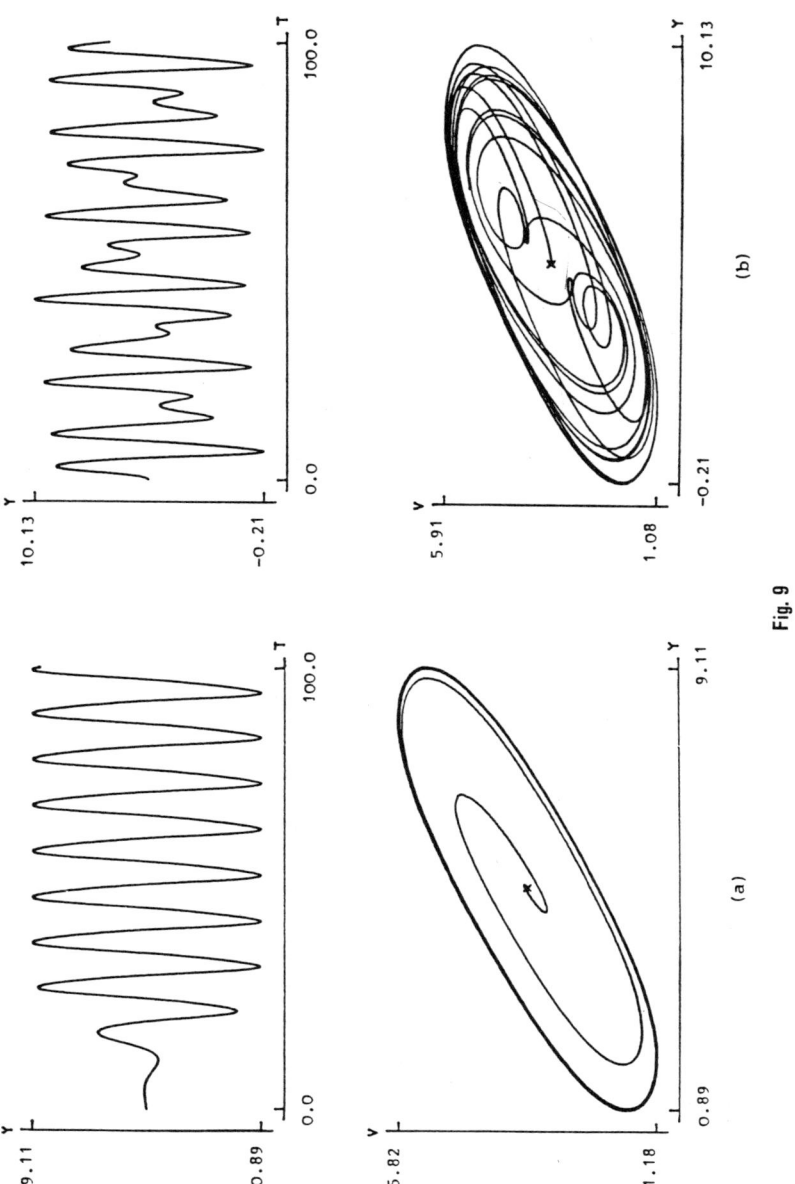

Fig. 9

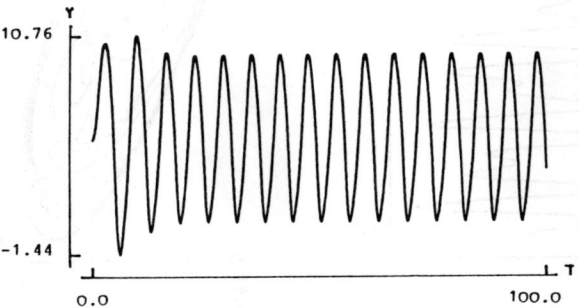

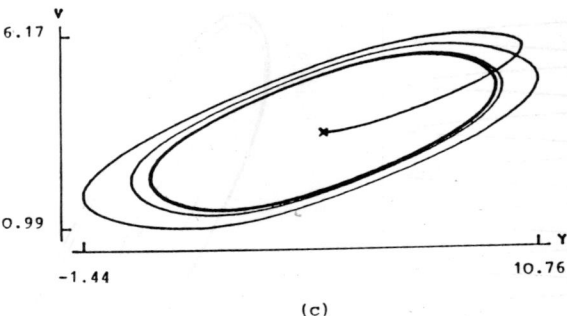

(c)

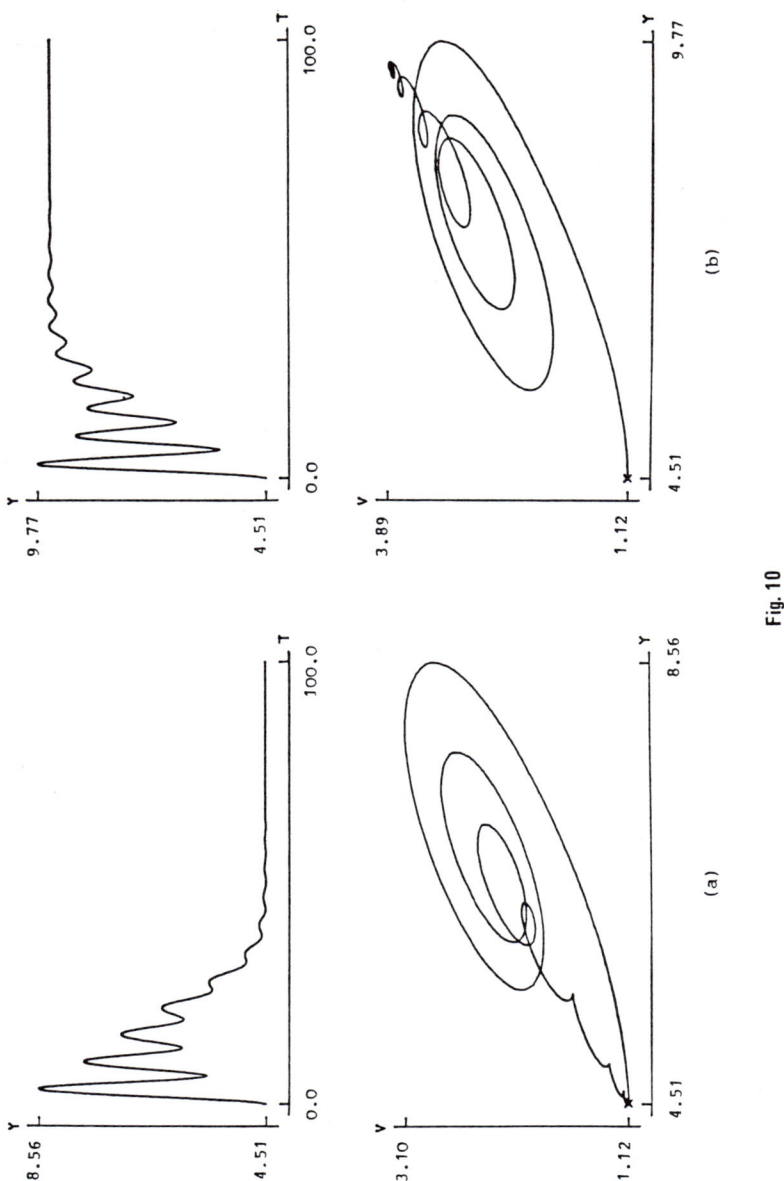

Fig. 10